Novel, Integrated and Revolutionary Well Test Interpretation Analysis

Authored by Freddy Humberto Escobar Macualo

Published in London, United Kingdom

IntechOpen

Supporting open minds since 2005

Novel, Integrated and Revolutionary Well Test Interpretation Analysis
http://dx.doi.org/10.5772/intechopen.81031
Authored by Freddy Humberto Escobar Macualo

Contributors
Freddy Humberto Escobar

First published in London, United Kingdom, 2019 by IntechOpen
IntechOpen is the global imprint of INTECHOPEN LIMITED, registered in England and Wales, registration number: 11086078, The Shard, 25th floor, 32 London Bridge Street
London, SE19SG - United Kingdom
Printed in Croatia

British Library Cataloguing-in-Publication Data
A catalogue record for this book is available from the British Library

Additional hard copies can be obtained from orders@intechopen.com

Novel, Integrated and Revolutionary Well Test Interpretation Analysis
Authored by Freddy Humberto Escobar Macualo
p. cm.
Print ISBN 978-1-78984-850-2
Online ISBN 978-1-78984-851-9

For EU product safety concerns:
IN TECH d.o.o., Prolaz Marije Krucifikse Kozulić 3, 51000 Rijeka, Croatia, info@intechopen.com or visit our website at intechopen.com.

Meet the author

Freddy Humberto Escobar finished his BSc studies at Universidad de America (Bogota-Colombia, 1989), and his MSc degree (1995) and a PhD (2002) at the University of Oklahoma. All his degrees are in Petroleum Engineering.
He is a professor at the Petroleum Engineering Department at Universidad Surcolombiana, USCO (Neiva-Colombia), where he has taught all the reservoir engineering area courses. His expertise focuses on well test analysis. He has also been a visiting professor at the University of Oklahoma. Dr. Escobar has written more than 150 technical papers, five book chapters, and two technological products. He has been Vice President for Research and Social Development, interim president, interim engineering college dean, and department director at USCO. He has also been a consultant at several oil and service companies.

Novel, Integrated and Revolutionary Well Test Interpretation and Analysis

Freddy Humberto Escobar Macualo

Acknowledgements

I am thankful to God and the Virgin Mary for all the blessing I have received during my whole life.

I also thank and dedicate this book to my wife Matilde and my children Jennifer Andrea, Freddy Alonso and Mary Gabriela, Universidad Surcolombiana (USCO), and Dr. Djebbar Tiab, creator of *TDS* Technique which forms the main topic of this book.

I truly thank Yuly Andrea Hernández, who has been one of the top most brilliant 12 students I have ever had, for taking part of her valuable time to review this book and write its foreword.

Contents

Foreword

It was in 2002, the first opportunity I had about being in touch with pressure well test interpretations. At that time, I could not be more inspired knowing the person from whose those words were coming from. That was exactly when I took my first well testing class with Doctor Escobar who just came from finishing his doctoral studies. Absolutely, he put the first stone of motivation and enthusiasm to end up working close to him applying Tiab's direct synthesis (*TDS*) technique to reservoir with channels and long structures. My passion for the subject even increased when publishing my first article from our research work in which we discovered a new flow regime: parabolic flow, and later, we saw the importance of geometrical skin factor which, so far, in spite of a long transition time, has not been yet incorporated into commercial software.

Since that time and for about 15 years, as part of my role as a reservoir engineer, definitely I could not be luckier, not only for sharing academic and professional discussion but also having the chance to be influenced by his creativity and deep thinking in different facets of his life. Dr. Escobar has been impacting the oil industry along his extensive research work. Without counting the numerous situations that as a professor who sees students as human beings with feelings that can go through multiple conditions of difficulties, can attest to the positive impact that as a person he has had in our lives. It is why to write short words about him is such a privilege to me.

As a reservoir engineer, I understand our challenge to describe, size, and develop hydrocarbon deposits in an efficient way, but oil and gas remains trapped in areas where we are restricted to have a direct recognition by our senses. We cannot see them. Neither, we cannot touch them nor we cannot design them. It requires the symbiosis between geologist and engineers to use their technical knowledge mixed with a great portion of imagination, creativity, and innovation to create models that help us to decrypt the way it will flow. This is the moment where Dr. Escobar attributes and his research work in well testing analysis is reflected.

Well testing analysis is an invaluable and low-cost tool in reservoir characterization that helps us to decode our reservoirs. From where we can obtain relevant information to model and understand them. Questions like: how many wells have to be drilled? How much hydrocarbon will they produce? What is the optimum strategy to obtain the best recovery factor with the highest interest return rate? Well, transient rate and pressure test analyses can be a contrivance for solving these issues. And, much better when there exist techniques such as Tiab's direct synthesis (*TDS*), which allows us to make the interpretations in a direct and simpler way, by using main features of the different flow regimes responses in combination with simple equations to determine the parameters of interest. Escobar presents in this book practical applications of this modern and revolutionary technique, helping all types of petroleum engineers to understand the concept and its benefits. I wish there were more examples, but for space-saving reasons, only the most practical ones are presented.

From this methodology I highlight the uncomplicated way to establish flow regimes types, diverse options to confirm the results, the simplicity and common sense of its calculations, and no need to use trial-and-error procedures, aspects that are not found in any other present methodology in the literature; however, the author in the book presents a comparison with other methodologies in an astonishing way.

The work of Dr. Escobar represents an important contribution to the hydrocarbon industry in the field of reservoir characterization, where his research extending the scope of the *TDS* technique plays an important role. I have also had the opportunity to apply his results and interpretation procedures in different types of reservoir, attesting to its usefulness and the multiple advances over the last years where this technique can be applied to diverse reservoir characteristics and flow regimes.

Enjoy it!!!

Yuly Andrea Hernández
Hocol S.A., Colombia[1]

[1] Yuly Andrea Hernández is a young Petroleum Engineer currently working for Hocol S.A. since July 2011. In 2004, she obtained a BSc degree diploma with honors in Petroleum Engineering under the author's supervision and received a MBA degree in 2014. She worked first for Hocol S.A. from August 2004 to September 2008 and went to Columbus Energy Sucursal Colombia from October 2008 to June 2011. Then, she moved back to Hocol. Yuly Andrea has been a very active engineer and she has gained experience in all the subjects of reservoir engineering. She is very familiar with the current and sophisticated software used in the oil industry. She is a devoted user of the application of *TDS* Technique to her engineering work and actually has a couple of publications on this subject.

Novel, Integrated and Revolutionary Well Test Interpretation and Analysis

Freddy Humberto Escobar Macualo

Abstract

Well test interpretation is an important tool for reservoir characterization. There exist four methods to achieve this goal, which are as follows: type-curve matching, conventional straight-line method, non-linear regression analysis, and *TDS* technique. The first method is basically a trial-and-error procedure; a deviation of a millimeter involves differences up to 200 psi and the difficulty of having so many matching charts. The second one, although very important, requires a plot for every flow regime, and there is no way for verification of the calculated parameters, and the third one has a problem of diversity of solutions but is the most used by engineers since it is automatically made by a computer program. This book focuses on the fourth method that uses a single plot of the pressure and pressure derivative plot for identifying different lines and feature for parameter estimation. It can be used alone and is applied practically to all the existing flow regime cases. In several cases, the same parameter can be estimated from different sources making a good way for verification. Combination of this method along with the second and third is recommended and widely used by the author.

TDS technique is quite versatile. The user finds the different flow regimes and, then, draws a line through it. From an arbitrary point on each flow regime, a given parameter can be calculated. Besides, the intersection point between the extrapolated flow regimes, although do not have a physical meaning, is excellent to find another reservoir parameter or the verification of others. For instance, well-drainage area can be readily estimated from the intersect point formed between the radial flow regime and the late pseudosteady state period. Every time someone starts working using *TDS* technique, he or she never stops. The reader is invited to give it a try.

Keywords: *TDS* technique, permeability, well-drainage area, flow regimes, intersection points, transient pressure analysis, conventional analysis

Author details

Freddy Humberto Escobar Macualo
Universidad Surcolombiana, Colombia

*Address all correspondence to: fescobar@usco.edu.co

Introduction

Well testing is a valuable and economical formation evaluation tool used in the hydrocarbon industry. It has been supported by mathematical modeling, computing, and the precision of measurement devices. The data acquired during a well test are used for reservoir characterization and description. However, the biggest drawback is that the system dealt with is neither designed nor seen by well test interpreters, and the only way to make contact with the reservoir is through the well by making indirect measurements.

Four methods are used for well test interpretation: (1) The oldest one is the conventional straight-line method which consists of plotting pressure or the reciprocal rate—if dealing with transient rate analysis—in the y-axis against a function of time in the x-axis. This time function depends upon the governing equation for a given flow. For instance, radial flow uses the logarithm of time and linear flow uses the square root of time. The slope and intercept of such plot are used to find reservoir parameters. The main disadvantage of this method is the lack of confirmation and the difficulty to define a given flow regime. The method is widely used nowadays. (2) Type-curve matching uses predefined dimensionless pressure and dimensionless time curves (some also use dimensionless pressure derivative), which are used as master guides to be matched with well pressure data to obtain a reference point for reservoir parameter determination. This method is basically a trial-and-error procedure which becomes into its biggest disadvantage. The method is practically unused. (3) Simulation of reservoir conditions and automatic adjustment to well test data by non-linear regression analysis is the method widely used by petroleum engineers. This method is also being widely disused since engineers trust the whole task to the computer. They even perform inverse modeling trying to fit the data to any reservoir model without taking care of the actual conditions. However, the biggest weakness of this method lies on the none uniqueness of the solution. Depending on the input starting values, the results may be different. (4) The newest method known as Tiab's direct synthesis (*TDS*) [1, 2] is the most powerful and practical one as will be demonstrated throughout the book. It employs characteristic points and features found on the pressure and pressure derivative versus time log-log plot to be used into direct analytic equations for reservoir parameters' calculation. It is even used, without using the original name, by all the commercial software. One of them calls it "Specialized lines." Because of its practicality, accuracy and application is the main object of this book. Conventional analysis method will be also included for comparison purposes.

The *TDS* technique can be easily implemented for all kinds of conventional or unconventional systems. It can be easily applied on cases for which the other methods fail or are difficult to be applied. It is strongly based on the pressure derivative curve. The method works by sector or regions found on the test. This means once a given flow regime is identified, a straight line is drawn throughout it, and then, any arbitrary point on this line and the intersection with other lines as well are used into the appropriate equations for the calculation of reservoir parameters.

The book contains the application and detailed examples of the *TDS* technique to the most common or fundamental reservoir/fluid scenarios. It is divided into seven chapters that are recommended to be read in the other they appear, especially for academic purposes in senior undergraduate level or master degree level. Chapter 1 contains the governing equation and the superposition principle. Chapter 2 is the longest one since it includes drawdown for infinite and finite cases, elongated system, multi-rate testing, and spherical/hemispherical flow. All the interpretation methods are studied in this chapter which covers about 45% of the book. Chapter 3 deals with pressure buildup testing and average reservoir pressure determination. Distance to barriers and interference testing are, respectively, treated in Chapters 4 and 5. Since the author is convinced that all reservoirs are naturally fractured, Chapter 6 covers this part which is also extended in hydraulically fractured wells in Chapter 7. In this last chapter, the most common flow regime shown in fractured wells: bilinear, linear, and elliptical are discussed with detailed for parameter characterization. The idea is to present a book on *TDS* technique as practical and short as possible; then, horizontal well testing is excluded here because of its complexity and extension, but the most outstanding and practical publications are named here.

My book entitled "*Recent Advances in Practical Applied Well Test Analysis*," published in 2015, was written for people having some familiarity with the *TDS* technique, so that, it can be read in any order. This is not the case of the present textbook. It is recommended to be read in order from Chapter 1 and take especial care in Chapter 2 since many equations and concepts will be applied in the remaining chapters. *TDS* technique applies indifferently to both pressure drawdown and pressure buildup tests.

Finally, this book is an upgraded and updated version of a former one published in Spanish. Most of the type curves have been removed since they have never been used by the author on actual well test interpretations. However, the first motivation to publish this book is the author's belief that *TDS* technique is the panacea for well test interpretation. *TDS* technique is such an easy and practical methodology that his creator, Dr. Djebbar Tiab, when day said to me "I still don't believe *TDS* works!" But, it really does. Well, once things have been created, they look easy.

Chapter 1

Fundamentals

1.1. Basic concepts

Pressure test fundamentals come from the application of Newton's law, especially the third one: Principle of action-reaction, since it comes from a perturbation on a well, as illustrated in **Figure 1.1**.

A well can be produced under any of two given scenarios: (a) by keeping a constant flow rate and recording the well-flowing pressure or (b) by keeping a constant well-flowing pressure and measuring the flow rate. The first case is known as pressure transient analysis, PTA, and the second one is better known as rate transient analysis, RTA, which both are commonly run in very low permeable formations such as shales.

Basically, the objectives of the analysis of the pressure tests are:

- Reservoir evaluation and description: well delivery, properties, reservoir size, permeability by thickness (useful for spacing and stimulation), initial pressure (energy and forecast), and determination of aquifer existence.

- Reservoir management.

There are several types of tests with their particular applications. DST and pressure buildup tests are mainly used in primary production and exploration. Multiple tests are most often used during secondary recovery projects, and multilayer and vertical permeability tests are used in producing/injectors wells. Drawdown, interference, and pulse tests are used at all stages of production. Multi-rate, injection, interference, and pulse tests are used in primary and secondary stages [3–7].

Pressure test analysis has a variety of applications over the life of a reservoir. DST and pressure buildup tests run in single wells are mainly used during primary production and exploration, while multiple tests are used more often during secondary recovery projects. Multilayer and vertical permeability tests are also run in producing/injectors wells. Drawdown, buildup, interference, and pulse tests are used at all stages of production. Multi-rate, injection, interference, and pulse testing are used in the primary and secondary stages. Petroleum engineers should take into

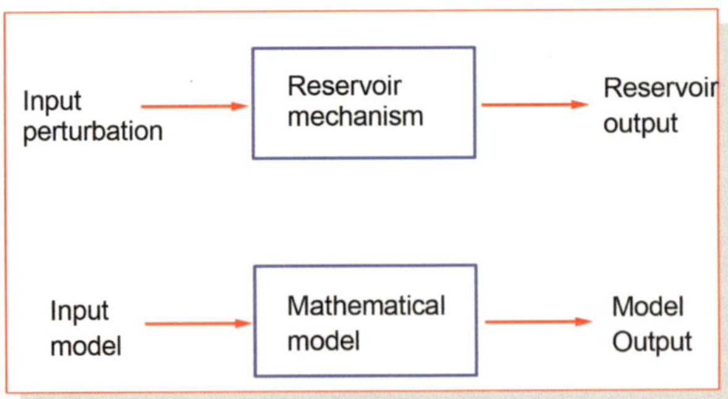

Figure 1.1.
Diagram of the mathematical representation of a pressure test.

account the state of the art of interpreting pressure tests, data acquisition tools, interpretation methods, and other factors that affect the quality of the results obtained from pressure test analysis.

Once the data have been obtained from the well and reviewed, the pressure test analysis comprises two steps: (1) To establish the reservoir model and the identification of the different flow regimes encountered during the test and (2) the parameter estimation. To achieve this goal, several plots are employed; among them, we have log-log plot of pressure and pressure derivative versus testing time (diagnostic tool), semilog graph of pressure versus time, Cartesian graph of the same parameters, etc. Pressure derivative will be dealt later in this chapter.

The interpretation of pressure tests is the primary method for determining average permeability, skin factor, average reservoir pressure, fracture length and fracture conductivity, and reservoir heterogeneity. In addition, it is the only fastest and cheapest method to estimate time-dependent variables such as skin factor and permeability in stress-sensitive reservoirs.

In general, pressure test analysis is an excellent tool to describe and define the model of a reservoir. Flow regimes are a direct function of the characteristics of the well/reservoir system, that is, a simple fracture that intercepts the well can be identified by detection of a linear flow. However, whenever there is linear flow, it does not necessarily imply the presence of a fracture. The infinite-acing behavior occurs after the end of wellbore storage and before the influence of the limits of the deposit. Since the boundaries do not affect the data during this period, the pressure behavior is identical to the behavior of an infinite reservoir. The radial flow can be recognized by an apparent stabilization of the value of the derivative.

1.2. Type of well tests

Well tests can be classified in several ways depending upon the view point. Some classifications consider whether or not the well produces or is shut-in. Other engineers focus on the number of flow rates. The two main pressure tests are (a) pressure drawdown and (b) buildup. While the first one involves only one flow rate, the second one involves two flow rates, one of which is zero. Then, a pressure buildup test can be considered as a multi-rate test.

1.2.1 Pressure tests run in producer wells

Drawdown pressure test (see **Figure 1.2**): It is also referred as a flow test. After the well has been shut-in for a long enough time to achieve stabilization, the well is placed in production, at a constant rate, while recording the bottom pressure against time. Its main disadvantage is that it is difficult to maintain the constant flow rate.

Pressure buildup test (see **Figure 1.2**): In this test, the well is shut-in while recording the static bottom-hole pressure as a function of time. This test allows obtaining the average pressure of the reservoir. Although since 2010, average reservoir pressures can be determined from drawdown tests. Its main disadvantage is economic since the shut-in entails the loss of production.

1.2.2 Pressure tests run in injector wells

Injection test (see **Figure 1.3**): Since it considers fluid flow, it is a test similar to the pressure drawdown test, but instead of producing fluids, fluids, usually water, are injected.

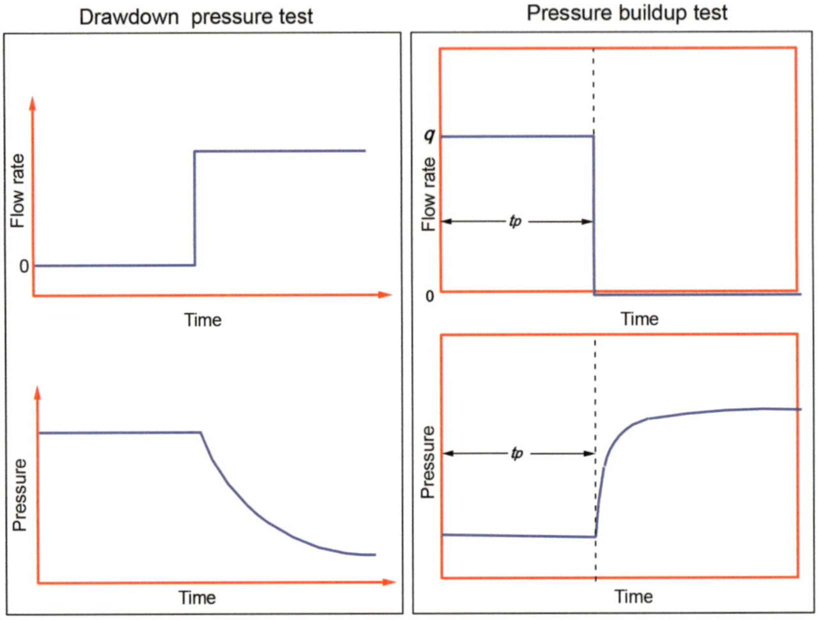

Figure 1.2.
Schematic representation of pressure drawdown and pressure buildup tests.

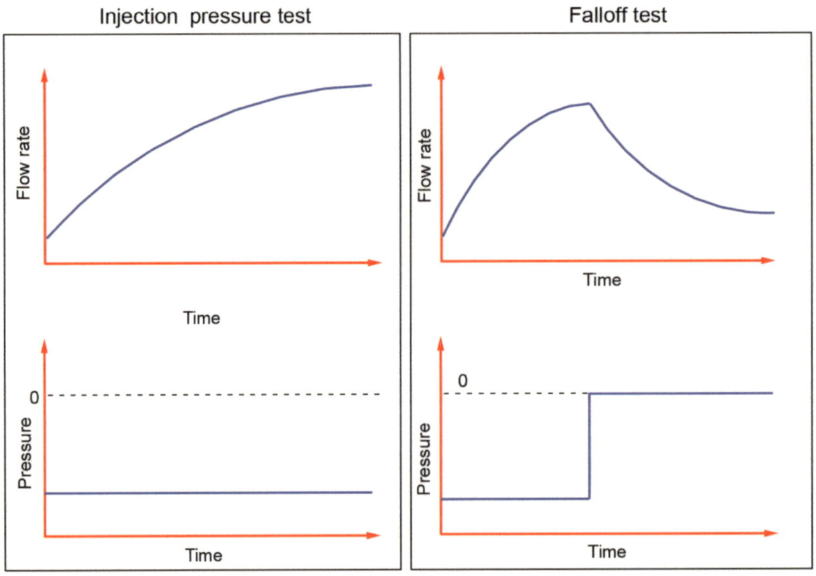

Figure 1.3.
Injection pressure test (left) and falloff test (right).

Falloff test (see **Figure 1.3**): This test considers a pressure drawdown immediately after the injection period finishes. Since the well is shut-in, falloff tests are identical to pressure buildup tests.

1.2.3 Other tests

Interference and/or multiple tests: They involve more than one well and its purpose is to define connectivity and find directional permeabilities. A well perturbation is observed in another well.

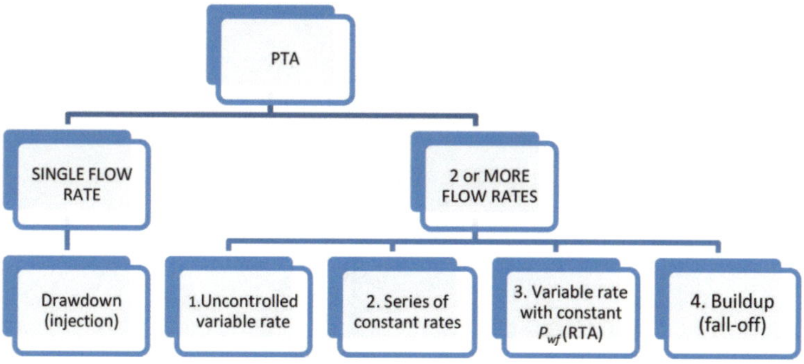

Figure 1.4.
Well test classification based on the number of flow rates.

Drill stem test (DST): This test is used during or immediately after well drilling and consists of short and continuous shut-off or flow tests. Its purpose is to establish the potential of the well, although the estimated skin factor is not very representative because well cleaning can occur during the first productive stage of the well (**Figure 1.4**).

Short tests: There are some very short tests mainly run in offshore wells. They are not treated in this book. Some of them are slug tests, general close chamber tests (CCTs), surge tests, shoot and pool tests, FasTest, and impulse tests.

As stated before, in a pressure drawdown test, the well is set to a constant flow rate. This condition is, sometimes, difficult to be fulfilled; then, multi-rate tests have to be employed. According to [8], multi-rate tests fit into four categories: (a) uncontrolled variable rate [9, 10], series of constant rates [11, 12], pressure buildup testing, and constant bottom-hole pressure with a continuous changing flow rate [13]. This last technique has been recently named as rate transient analysis (RTA) which is included in PTA, but its study is not treated in this book.

1.3 Diffusivity equation

At the beginning of production, the pressure in the vicinity of the well falls abruptly and the fluids near the well expand and move toward the area of lower pressure. Such movement is retarded by friction against the walls of the well and the inertia and viscosity of the fluid itself. As the fluid moves, an imbalance of pressure is created, which induces the surrounding fluids to move toward the well. The process continues until the pressure drop created by the production dissipates throughout the reservoir. The physical process that takes place in the reservoir can be described by the diffusivity equation whose deduction is shown below [5]:

According to the volume element given in **Figure 1.5**,

$$\begin{bmatrix}\text{Mass entering} \\ \text{the element}\end{bmatrix} - \begin{bmatrix}\text{Mass coming out} \\ \text{from the element}\end{bmatrix} = \begin{bmatrix}\text{System} \\ \text{accumulation rate}\end{bmatrix} \tag{1.1}$$

The right-hand side part of Eq. (1.1) corresponds to the mass accumulated in the volume element. Darcy's law for radial flow:

$$q = -\frac{kA}{\mu}\frac{dP}{dr} \tag{1.2}$$

The cross-sectional area available for flow is provided by cylindrical geometry, $2\pi rh$. Additionally, flow rate must be multiplied by density, ρ, to obtain mass flow. With these premises, Eq. (1.2) becomes:

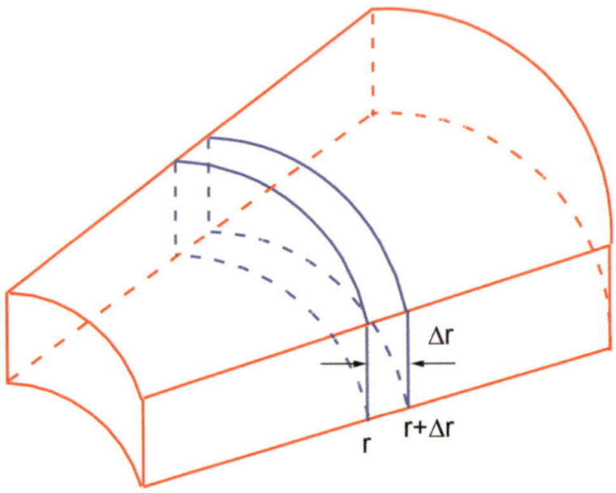

Figure 1.5.
Radial volume element.

$$q = -\frac{k}{\mu} 2\pi\, rh \frac{\partial P}{\partial r} \tag{1.3}$$

Replacing Eq. (1.3) into (1.1) yields:

$$-\frac{k\rho}{\mu}(2\pi\, rh)\frac{\partial P}{\partial r}\bigg|_r + \frac{k\rho}{\mu}(2\pi\, rh)\frac{\partial P}{\partial r}\bigg|_{r+dr} = \frac{\partial}{\partial t}([2\pi rhdr\phi]\rho) \tag{1.4}$$

If the control volume remains constant with time, then, Eq. (1.4) can be rearranged as:

$$-2\pi\, h\frac{k\rho}{\mu} r \frac{\partial P}{\partial r}\bigg|_r + 2\pi\, h\frac{k\rho}{\mu} r \frac{\partial P}{\partial r}\bigg|_{r+dr} = 2\pi rhdr\frac{\partial}{\partial t}(\phi\rho) \tag{1.5}$$

Rearranging further the above expression:

$$\frac{\frac{1}{r}\left[\frac{k\rho}{\mu} r \frac{\partial P}{\partial r}\bigg|_{r+dr} - \frac{k\rho}{\mu} r \frac{\partial P}{\partial r}\bigg|_r\right]}{dr} = \frac{\partial}{\partial t}(\phi\rho) \tag{1.6}$$

The left-hand side of Eq (1.6) corresponds to the definition of the derivative; then, it can be rewritten as:

$$\frac{1}{r}\frac{\partial}{\partial r}\left(\frac{k\rho}{\mu} r \frac{\partial P}{\partial r}\right) = \frac{\partial}{\partial t}(\phi\rho) \tag{1.7}$$

The definition of compressibility has been widely used;

$$c = -\frac{1}{V}\frac{\partial V}{\partial P} = \frac{1}{\rho}\frac{\partial\rho}{\partial P} \tag{1.8}$$

By the same token, the pore volume compressibility is given by:

$$c_f = \frac{1}{\phi}\frac{\partial\phi}{\partial P} \tag{1.9}$$

The integration of Eq. (1.8) will lead to obtain:

$$\rho = \rho_o e^{c(P-P_o)} \tag{1.10}$$

The right-hand side part of Eq. (1.7) can be expanded as:

$$\frac{\partial}{\partial t}(\phi\rho) = \phi\frac{\partial}{\partial t}\rho + \rho\frac{\partial}{\partial t}\phi = \phi\frac{\partial\rho}{\partial t} + \rho\frac{\partial\phi}{\partial P}\frac{\partial P}{\partial\rho}\frac{\partial\rho}{\partial t} \tag{1.11}$$

Using the definitions given by Eqs. (1.9) and (1.10) into Eq. (1.11) leads to:

$$\frac{\partial}{\partial t}(\phi\rho) = \phi\frac{\partial\rho}{\partial t} + \frac{\rho\,\phi\,c_f}{c\rho}\frac{\partial\rho}{\partial t} = \phi\frac{\partial\rho}{\partial t}\left[1+\frac{c_f}{c}\right] = \frac{\phi}{c}\left[c_f + c\right]\frac{\partial\rho}{\partial t} \tag{1.12}$$

Considering that the total compressibility, c_t, is the result of the fluid compressibility, c, plus the pore volume compressibility, c_f, it yields:

$$\frac{1}{r}\frac{\partial}{\partial r}\left(\frac{k\rho}{\mu}r\frac{\partial P}{\partial r}\right) = \frac{\phi c_t}{c}\frac{\partial\rho}{\partial t} \tag{1.13}$$

The gradient term can be expanded as:

$$\frac{\partial P}{\partial r} = \frac{\partial P}{\partial\rho}\frac{\partial\rho}{\partial r} = \frac{1}{c\rho}\frac{\partial\rho}{\partial r} \tag{1.14}$$

Combination of Eqs. (1.14) and (1.13) results in:

$$\frac{1}{r}\frac{\partial}{\partial r}\left(\frac{kr}{\mu c}\frac{\partial\rho}{\partial r}\right) = \frac{\phi}{c}c_t\frac{\partial\rho}{\partial t} \tag{1.15}$$

Taking derivative to Eq. (1.10) with respect to both time and radial distance and replacing these results into Eq. (1.15) yield:

$$\frac{1}{r}\frac{\partial}{\partial r}\left(\frac{kr}{\mu c}\rho_o e^{c(P-P_o)}c\frac{\partial P}{\partial r}\right) = \frac{\phi}{c}c_t\rho_o e^{c(P-P_o)}c\frac{\partial P}{\partial t} \tag{1.16}$$

After simplification and considering permeability and viscosity to be constant, we obtain:

$$\frac{1}{r}\frac{k}{\mu}\frac{\partial}{\partial r}\left(r\frac{\partial P}{\partial r}\right) = \phi\,c_t\frac{\partial P}{\partial t} \tag{1.17}$$

The hydraulic diffusivity constant is well known as

$$\frac{1}{\eta} = \frac{\phi\mu c_t}{k} \tag{1.18}$$

Then, the final form of the diffusivity equation in oilfield units is obtained by combination of Eqs. (1.17) and (1.18):

$$\frac{1}{r}\frac{\partial}{\partial r}\left(r\frac{\partial P}{\partial r}\right) = \frac{\phi\mu c_t}{0.0002637k}\frac{\partial P}{\partial t} = \frac{1}{\eta}\frac{\partial P}{\partial t} \tag{1.19}$$

In expanded form:

$$\frac{\partial^2 P}{\partial r^2} + \frac{1}{r}\frac{\partial P}{\partial r} = \frac{1}{0.0002637\eta}\frac{\partial P}{\partial t} \tag{1.20}$$

The final form of the diffusivity equation strongly depends upon the flow geometry. For cylindrical, [11, 14], spherical [14], and elliptical coordinates [15], the diffusivity equation is given, respectively,

$$\frac{\partial^2 P}{\partial r^2}+\frac{1}{r}\frac{\partial P}{\partial r}+\frac{k_\theta}{k_r}\frac{1}{r^2}\frac{\partial^2 P}{\partial \theta^2}+\frac{k_z}{k_r}\frac{\partial^2 P}{\partial z^2}=\frac{\phi\mu c_t}{k_r}\frac{\partial P}{\partial t} \tag{1.21}$$

$$\frac{1}{r}\left[\frac{\partial}{\partial r}\left(r^2\frac{\partial P}{\partial r}\right)+\frac{1}{\sin\theta}\frac{\partial}{\partial\theta}\left(\sin\theta\frac{\partial P}{\partial\theta}\right)+\frac{1}{\sin^2\theta}\frac{\partial^2 P}{\partial\phi^2}\right]=\frac{\phi c\mu}{k}\frac{\partial P}{\partial t} \tag{1.22}$$

$$\frac{\partial^2 P}{\partial\xi^2}+\frac{\partial^2 P}{\partial\eta^2}=\frac{1}{2}a^2(\cosh 2\xi-\cos 2\eta)\frac{\phi c\mu}{k}\frac{\partial P}{\partial t} \tag{1.23}$$

Here, ξ is a space coordinate and represents a family of confocal ellipses. The focal length of these ellipses is $2a$. The space coordinate, η, represents a family of confocal hyperbolas that represent the streamlines for elliptical flow. These two coordinates are normal to each other.

1.4. Limitations of the diffusivity equation

a. Isotropic, horizontal, homogeneous porous medium, permeability, and constant porosity

b. A single fluid saturates the porous medium

c. Constant viscosity, incompressible, or slightly compressible fluid

d. The well completely penetrates the formation. Negligible gravitational forces

The density of the fluid is governed by an equation of state (EOS). For the case of slightly compressible fluid, Eq. (1.8) is used as the EOS.

1.5. Multiphase flow

Similar to the analysis of gas well tests as will be seen later, multiphase tests can be interpreted using the method of pressure approximation (Perrine method), [6, 7, 16], which is based on phase mobility:

$$\lambda_t=\frac{k_o}{\mu_o}+\frac{k_g}{\mu_g}+\frac{k_w}{\mu_w}=\frac{k_{ro}}{\mu_o}+\frac{k_{rg}}{\mu_g}+\frac{k_{rw}}{\mu_w} \tag{1.24}$$

The total compressibility is defined by [17, 18]:

$$c_t=c_oS_o+c_gS_g+c_wS_w+c_f+\frac{S_oB_g}{5.615B_o}\frac{\partial R_s}{\partial P}+\frac{S_wB_g}{5.615B_w}\frac{\partial R_{sw}}{\partial P} \tag{1.25}$$

For practical purposes, Eq. (1.25) can be expressed as:

$$c_t\approx c_oS_o+c_gS_g+c_wS_w+c_f \tag{1.26}$$

As commented before Eq. (1.19) is limited to a single fluid. However, it can be extended to multiphase flow using the concept expressed by Eq. (1.24):

$$\frac{1}{r}\frac{\partial}{\partial r}\left(r\frac{\partial P}{\partial r}\right)=\frac{\phi c_t}{0.0002637}\frac{1}{\lambda_t}\frac{\partial P}{\partial t} \tag{1.27}$$

Perrine method assumes negligible pressure and saturation gradients. Martin [19] showed that (a) the method loses accuracy as the gas saturation increases, (b) the estimation of the mobility is good, and (c) the mobility calculations are sensitive to the saturation gradients. Better estimates are obtained when the saturation distribution is uniform and (d) underestimates the effective permeability of the phase and overestimates the damage factor.

1.6. Gas flow

It is well known that gas compressibility, gas viscosity, and gas density are highly dependent pressure parameters; then, the liquid diffusivity equation may fail to observe pressure gas behavior. Therefore, there exist three forms for a better linearization of the diffusivity equation to better represent gas flow: (a) the pseudopressure approximation [20], (b) the P2 approximation, and (c) linear approximation. The first one is valid for any pressure range; the second one is valid for reservoir pressures between 2000 and 4000 psia, and the third one is for pressures above 4000 psia [20].

Starting from the equation of continuity and the equation of Darcy:

$$\frac{1}{r}\frac{\partial}{\partial r}(r\rho u_r)=-\frac{\partial}{\partial t}(\phi\rho) \tag{1.28}$$

$$u_r=-\frac{k}{\mu}\frac{\partial P}{\partial r} \tag{1.29}$$

The state equation for slightly compressible liquids does not model gas flow; therefore, the law of real gases is used [21, 22]:

$$\rho=-\frac{PM}{zRT} \tag{1.30}$$

Combining the above three equations:

$$\frac{1}{r}\frac{\partial}{\partial r}\left(r\frac{kPM}{\mu zRT}\frac{\partial p}{\partial t}\right)=\frac{\partial}{\partial t}\left(\phi\frac{PM}{zRT}\right) \tag{1.31}$$

Since M, R, and T are constants and assuming that the permeability is constant, the above equation reduces to:

$$\frac{1}{r}\frac{\partial}{\partial r}\left(r\frac{P}{\mu z}\frac{\partial P}{\partial r}\right)=\frac{1}{k}\frac{\partial}{\partial t}\left(\phi\frac{P}{z}\right) \tag{1.32}$$

Applying the differentiation chain rule to the right-hand side part of Eq. (1.32) leads to:

$$\frac{1}{r}\frac{\partial}{\partial r}\left(r\frac{P}{\mu z}\frac{\partial P}{\partial r}\right)=\frac{1}{k}\left[\frac{P}{z}\frac{\partial\phi}{\partial t}+\frac{\partial\phi}{\partial t}\left(\frac{P}{z}\right)\right] \tag{1.33}$$

Expanding and rearranging,

$$\frac{1}{r}\frac{\partial}{\partial r}\left(r\frac{P}{\mu z}\frac{\partial P}{\partial r}\right) = \frac{P\phi}{zk}\frac{\partial P}{\partial t}\left[\frac{1}{\phi}\frac{\partial\phi}{\partial P} + \frac{z}{P}\frac{\partial}{\partial P}\left(\frac{P}{z}\right)\right] \tag{1.34}$$

Using the definition of compressibility for gas flow:

$$c_g = \frac{1}{\rho}\frac{\partial\rho}{\partial P} = \frac{zRT}{PM}\frac{\partial}{\partial P}\left(\frac{PM}{zRT}\right) = \frac{z}{P}\frac{\partial}{\partial P}\left(\frac{P}{z}\right) \tag{1.35}$$

Using Eqs. (1.9) and (1.35) into Eq. (1.34),

$$\frac{1}{r}\frac{\partial}{\partial r}\left(r\frac{P}{\mu z}\frac{\partial P}{\partial r}\right) = \frac{P\phi}{zk}\frac{\partial P}{\partial t}\left(c_f + c_g\right) \tag{1.36}$$

If $c_t = c_g + c_f$ then,

$$\frac{1}{r}\frac{\partial}{\partial r}\left(r\frac{P}{\mu z}\frac{\partial P}{\partial r}\right) = \frac{P\phi c_t}{zk}\frac{\partial P}{\partial t} \tag{1.37}$$

The above is a nonlinear partial differential equation and cannot be solved directly. In general, three limiting assumptions are considered for its solution, namely: (a) $P/\mu z$ is constant; (b) μc_t is constant; and (c) the pseudopressure transformation, [20], for an actual gas.

1.6.1 The equation of diffusivity in terms of pressure

Assuming the term $P/\mu z$ remains constant with respect to the pressure, Eq. (1.17) is obtained.

1.6.2 The equation of diffusivity in terms of pressure squared

Eq. (1.37) can be written in terms of squared pressure, P^2, starting from the fact that, [3–7, 9, 17, 21, 22]:

$$P\frac{\partial P}{\partial r} = \frac{1}{2}\frac{\partial P^2}{\partial r} \tag{1.38}$$

$$P\frac{\partial P}{\partial t} = \frac{1}{2}\frac{\partial P^2}{\partial t} \tag{1.39}$$

$$\frac{1}{r}\frac{\partial}{\partial r}\left(\frac{r}{\mu z}\frac{\partial P^2}{\partial r}\right) = \frac{\phi c_t}{kz}\frac{\partial P^2}{\partial t} \tag{1.40}$$

Assuming the term μz remains constant with respect to the pressure, and of course, the radius, then the above equation can be written as:

$$\frac{1}{r}\frac{\partial}{\partial r}\left(r\frac{\partial P^2}{\partial r}\right) = \frac{\phi\mu c_t}{k}\frac{\partial P^2}{\partial t} \tag{1.41}$$

This expression is similar to Eq. (1.37), but the dependent variable is P^2. Therefore, its solution is similar to Eq. (1.17), except that it is given in terms of P^2. This equation also requires that μc_t remain constant.

1.6.3 Gas diffusivity equation in terms of pseudopressure, *m(P)*

The diffusivity equation in terms of P^2 can be applied at low pressures, and Eq. (1.17) can be applied at high pressures without incurring errors. Therefore, a solution is required that applies to all ranges. Ref. [20] introduced a more rigorous linearization method called pseudopressure that allows the general diffusivity equation to be solved without limiting assumptions that restrict certain properties of gases to remain constant with pressure [3–7, 9, 17, 20–22]:

$$m(P) = 2\int_{P_0}^{P} \frac{P}{\mu z} dP \tag{1.42}$$

Taking the derivative with respect to both time and radius and replacing the respective results in Eq. (1.37), we obtain:

$$\frac{1}{r}\frac{\partial}{\partial r}\left[r\frac{P}{\mu z}\left(\frac{\mu z}{2P}\frac{\partial m(P)}{\partial r}\right)\right] = \frac{P\phi c_t}{zk}\left(\frac{\mu z}{2P}\frac{\partial m(P)}{\partial t}\right) \tag{1.43}$$

After simplification,

$$\frac{1}{r}\frac{\partial}{\partial r}\left(r\frac{\partial m(P)}{\partial r}\right) = \frac{\phi \mu c_t}{k}\frac{\partial m(P)}{\partial t} \tag{1.44}$$

Expanding the above equation and expressing it in oilfield units:

$$\frac{\partial^2 m(P)}{\partial r^2} + \frac{1}{r}\frac{\partial m(P)}{\partial r} = \frac{\phi \mu_{gi} c_t}{0.0002637 k_{gi}}\frac{\partial m(P)}{\partial t} \tag{1.45}$$

The solution to the above expression is similar to the solution of Eq. (1.17), except that it is now given in terms of $m(P)$ which can be determined by numerical integration if the PVT properties are known at each pressure level.

For a more effective linearization of Eq. (1.45), [23] introduced pseudotime, t_a, since the product $\mu_g c_t$ in Eq. (1.45) is not constant:

$$t_a = 2\int_0^t \frac{d\varsigma}{\mu c_t} \tag{1.46}$$

With this criterion, the diffusivity equation for gases is:

$$\frac{1}{r}\frac{\partial}{\partial r}\left(r\frac{\partial m(P)}{\partial r}\right) = \frac{2\phi (c_f + c_g)}{k\, c_g}\frac{\partial m(P)}{\partial t_a} \tag{1.47}$$

The incomplete linearization of the above expression leads to somewhat longer semilog slopes compared to those obtained for liquids. Sometimes it is recommended to use normalized variables in order to retain the units of time and pressure, [6]. The normalized pseudovariables are:

$$m(P)_n = P_i + \frac{\mu_i}{\rho_i}\int_{P_0}^{P} \frac{\rho(\varsigma)}{\mu(\varsigma)} d\varsigma \tag{1.48}$$

$$t_{an} = \mu_i c_{ti} + \int_0^t \frac{d\varsigma}{\mu(\varsigma)Z(\varsigma)} \tag{1.49}$$

1.7. Solution to the diffusivity equation

The line-source solution: The line-source solution assumes that the wellbore radius approaches zero. Furthermore, the solution considers a reservoir of infinite extent and the well produces as a constant flow rate. Ref. [4] presents the solution of the source line using the Boltzmann transform, the Laplace transform, and Bessel functions. The following is the combinations of independent variables method, which is based on the dimensional analysis of Buckingham's theorem [24]. This takes a function $f = f(x, y, z, t)$, it must be transformed into a group or function containing fewer variables, $f = f(s_1, s_2 ...)$. A group of variables whose general form is proposed as [24]:

$$s = ar^b t^c \tag{1.50}$$

The diffusivity equation is:

$$\frac{1}{r}\frac{\partial}{\partial r}\left(r\frac{\partial f}{\partial r}\right) = \frac{\partial f}{\partial t} \tag{1.51}$$

where f is a dimensionless term given by:

$$f = \frac{P - P_{wf}}{P_i - P_{wf}} \tag{1.52}$$

Eq. (1.51) is subjected to the following initial and boundary conditions:

$$f = 0, \ 0 \le r \le \infty, \ t = 0 \tag{1.53}$$

$$r\frac{\partial f}{\partial r} = 1, \ r = 0, \ t>0 \tag{1.54}$$

$$f = 0, \ r \to \infty, \ t>0 \tag{1.55}$$

Multiplying the Eq. (1.51) by $\partial s/\partial s$:

$$\frac{1}{r}\frac{\partial s}{\partial s}\frac{\partial}{\partial r}\left(r\frac{\partial s}{\partial s}\frac{\partial f}{\partial r}\right) = \frac{\partial s}{\partial s}\frac{\partial f}{\partial t} \tag{1.56}$$

Exchanging terms:

$$\frac{1}{r}\frac{\partial s}{\partial r}\frac{\partial}{\partial s}\left(r\frac{\partial s}{\partial r}\frac{\partial f}{\partial s}\right) = \frac{\partial s}{\partial t}\frac{\partial f}{\partial s} \tag{1.57}$$

The new derivatives are obtained from Eq. (1.50):

$$\frac{\partial s}{\partial r} = abr^{b-1}t^c \tag{1.58}$$

$$\frac{\partial s}{\partial t} = acr^b t^{c-1} \tag{1.59}$$

Replacing the above derivatives into Eq. (1.56) and rearranging:

$$\frac{1}{r}a^2b^2\frac{r^b}{r}t^{2c}\frac{\partial}{\partial s}\left(r\cdot\frac{r^b}{r}\frac{\partial f}{\partial s}\right) = acr^bt^{c-1}\frac{\partial f}{\partial s} \tag{1.60}$$

Solving from r^b from Eq. (1.50) and replacing this result into Eq. (1.6). After rearranging, it yields:

$$\frac{\partial}{\partial s}\left(s\frac{\partial f}{\partial s}\right) = \frac{c}{b^2}\left[r^2t^{-1}\right]s\frac{\partial f}{\partial s} \tag{1.61}$$

Comparing the term enclosed in square brackets with Eq. (1.50) shows that $b = 2$, $c = -1$, then

$$s = \frac{ar^2}{t} \tag{1.62}$$

From Eq. (1.61) follows $r^2t^{-1} = s/a$, then

$$\frac{\partial}{\partial s}\left(s\frac{\partial f}{\partial s}\right) = \left[\frac{c}{b^2a}\right]s\frac{\partial f}{\partial s} \tag{1.63}$$

The term enclosed in square brackets is a constant that is assumed equal to 1 for convenience. Since $c/(b^2a) = 1$, then $a = -1/4$. Therefore, the above expression leads to:

$$\frac{\partial}{\partial s}\left(s\frac{\partial f}{\partial s}\right) = s\frac{\partial f}{\partial s} \tag{1.64}$$

Writing as an ordinary differential equation:

$$\frac{d}{ds}\left(s\frac{df}{ds}\right) = s\frac{df}{ds} \tag{1.65}$$

The differential equation is now ordinary, and only two conditions are required to solve it. Applying a similar mathematical treatment to both the initial and boundary conditions to convert them into function of s. Regarding Eq. (1.62) and referring to the initial condition, Eq. (1.53), when the time is set to zero; then, then s function tends to infinite:

$$\text{at } t = 0,\ f = 0 \text{ when } s \to \infty \tag{1.66}$$

Darcy's law is used to convert the internal boundary condition. Eq. (1.54) multiplied by $\partial s/\partial s$ gives:

$$r\frac{\partial f}{\partial s}\frac{\partial s}{\partial r} = 1 \tag{1.67}$$

Replacing Eqs. (1.57) in the above equation; then, replacing Eq. (1.62) into the result, and after simplification, we obtain

$$\frac{\partial f}{\partial s}ab\frac{s}{at^c}t^c = 1 \tag{1.68}$$

Since $b = 2$, then,

$$s\frac{\partial f}{\partial s} = \frac{1}{2} \tag{1.69}$$

For the external boundary condition, Eq. (1.55), consider the case of Eq. (1.62) when $r \to \infty$ then:

$$s = \frac{ar^2}{t} \to \infty; f = 0, \; s \to \infty \tag{1.70}$$

Then, the new differential equation, Eq. (1.65) is subject to new conditions given by Eqs. (1.66), (1.69), and (1.70). Define now,

$$g = s\frac{df}{ds} \tag{1.71}$$

Applying this definition into the ordinary differential expression given by Eq. (1.65), it results:

$$\frac{d}{ds}g = g \tag{1.72}$$

Integration of the above expression leads to:

$$\ln g = s + c_1 \tag{1.73}$$

Rearranging the result and comparing to Eq. (1.71) and applying the boundary condition given by Eq. (1.69):

$$g = c_1 e^s = s\frac{df}{ds} = \frac{1}{2} \tag{1.74}$$

Solving for *df* and integrating,

$$\int df = c_1 \int \frac{e^s}{s} ds \tag{1.75}$$

Eq. (1.75) cannot be analytically integrated (solved by power series). Simplifying the solution:

$$f = c_1 \int \frac{e^s}{s} ds + c_2 \tag{1.76}$$

When $s = 0$, $e^s = 0$, then $c_1 = ½$ and Eq. (1.76) becomes:

$$f = \frac{1}{2} \int_0^s \frac{e^s}{s} ds + c_2 \tag{1.77}$$

Applying the external boundary condition, Eq. (1.69), when $s \to \infty, f = 0$, therefore, Eq. (1.77) leads,

$$c_2 = -\frac{1}{2} \int_0^{-\infty} \frac{e^s}{s} ds \tag{1.78}$$

Replacing c_1 and c_2 into Eq. (1.76) yields:

$$f = \frac{1}{2}\int_0^s \frac{e^s}{s} ds - \frac{1}{2}\int_0^{-\infty} \frac{e^s}{s} ds \tag{1.79}$$

This can be further simplified to:

$$f = -\frac{1}{2}\int_s^{\infty} \frac{e^{-s}}{s} ds \tag{1.80}$$

The integral given in Eq. (1.80) is well known as the exponential integral, $Ei(-s)$. If the f variable is changed by pressure terms:

$$P(r,t) = -\frac{1}{2}Ei\left(\frac{-r^2}{4t}\right) \tag{1.81}$$

In dimensionless form,

$$P_D(r_D, t_D) = -\frac{1}{2}Ei\left(-\frac{r_D^2}{4t_D}\right) = -\frac{1}{2}E_i(-x) \tag{1.82}$$

The above equation is a very good approximation of the analytical solution when it is satisfied (Mueller and Witherspoon [2, 9, 18, 19, 25, 26]) that $r_D \geq 20$ or $t_D/{r_D}^2 \geq 0.5$, see **Figure 1.6**. If $t_D/{r_D}^2 \geq 5$, an error is less than 2%, and if $t_D/{r_D}^2 \geq 25$, the error is less than 5%. **Figure 1.7** is represented by the following adjustment which has a correlation coefficient, R^2 of 0.999998. This plot can be easily rebuilt using the algorithm provided in **Figure 1.8**. The fitted equation was achieved with the data generated from simulation.

$$P_D = 10^{\frac{-0.2820668952451542+0.4472760048082251x+0.2581584173632316x^2+0.04998332927590892x^3}{1+1.047015081287319x+0.3493329681392351x^2+0.02955955788180784x^3-0.0001636047294307 38x^4}} \tag{1.83}$$

being $x = \log(t_D/{r_D}^2) > -1.13$.

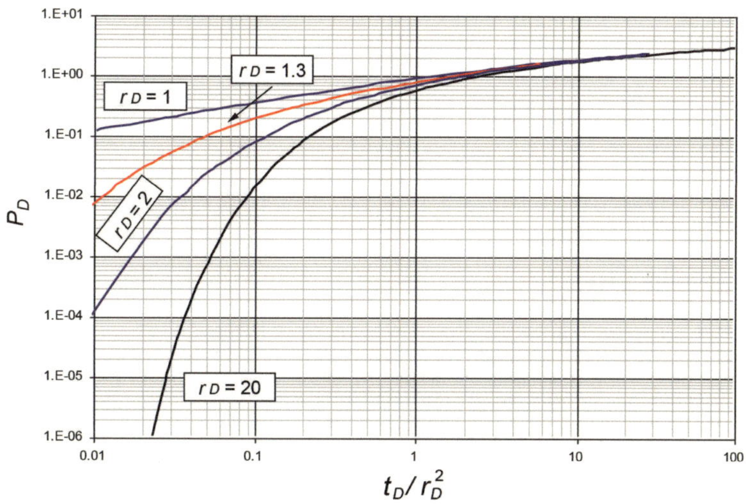

Figure 1.6.
Dimensionless pressure for different values of the dimensionless radius, taken from [9, 25].

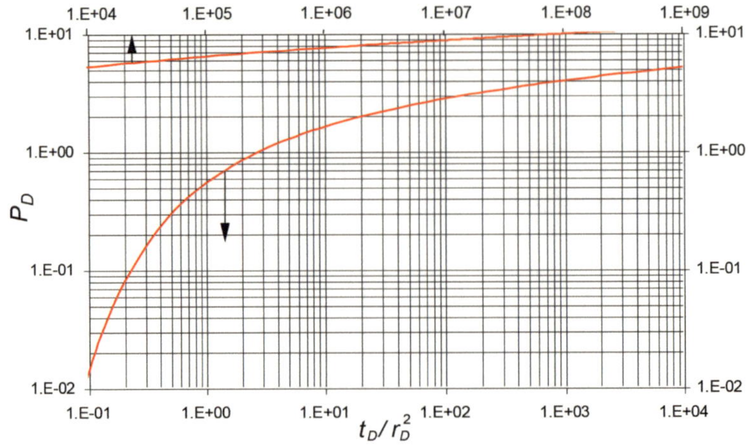

Figure 1.7.
Dimensionless well pressure behavior for a well without skin and storage effects in an infinite reservoir, taken from [9, 25].

The exponential function can be evaluated by the following formula, [27], for $x \leq 25$:

$$Ei(x) = 0.57721557 + \ln\, x - x + \frac{x^2}{2 \cdot 2!} - \frac{x^3}{3 \cdot 3!} + \frac{x^4}{4 \cdot 4!} \dots \tag{1.84}$$

Figure 1.8 shows a listing of a program code in Basic, which can be easily added as a function in Microsoft Excel to calculate the exponential function. **Figure 1.9** and **Table 1.1, 1.2, 1.3,** and **1.4** present solutions of the exponential function.

```
Function ei(x)
Dim dexp As Double, ARG As Double, dlog As Double
Dim Res1 As Double, Res2 As Double, Res3 As Double, Res4 As Double, Res5 As Double, Res As Double
        If x = 0 Then Exit Function
        dexp = Exp(-x)
        dlog = Log(x)
        If x > 60 Then
          ei = 0#
          Exit Function
        End If
        If x > 4# Then
            ARG = 4# / x
            Res = 0.011723273 + ARG * (-0.0049362007 + ARG * (0.00094427614))
            Res = -0.022951979 + ARG * (0.020412099 + ARG * (-0.017555779 + ARG * Res))
            Res = (0.24999999 + ARG * (-0.062498588 + ARG * (0.031208561 + ARG * Res)))
            Res = dexp * ARG * Res
          ei = Abs(Res)
        Else
        If x < 0 Then
          ei = 0#
          Exit Function
        End If
        If x = 0 Then
          ei = 1E+75
          Exit Function
        End If
            Res1 = -1.6826592E-10 + x * (1.5798675E-11 + x * (-1.0317602E-12))
            Res2 = 0.00000030726221 + x * (-0.00000002763583 + x * (2.1915699E-09 + x * Res1))
            Res4 = -0.00023148392 + x * (0.00002833759 + x * (-0.000003099604 + x * Res2))
            Res3 = -0.010416662 + x * (0.0016666906 + x * Res4)
            Res5 = x * (-0.25 + x * (0.05555552 + x * Res3))
            Res = -dlog - 0.57721566 + x * (1# + Res5)
            ei = Abs(Res)
        End If
End Function
```

Figure 1.8.
BASIC code function to calculate Ei *function, taken from [29].*

DOI: http://dx.doi.org/10.5772/intechopen.81078

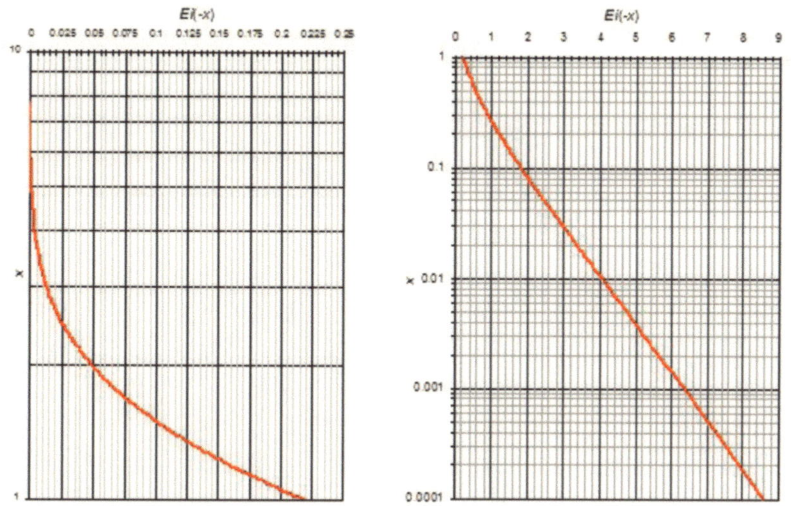

Figure 1.9.
Values of the exponential integral for 1 ≤ x ≤ 10 (left) and 0.0001 ≤ x ≤ 1 (right).

1.8. Dimensionless quantities

Dimensional parameters do not provide a physical view of the parameter being measured but rather a general or universal description of these parameters. For example, a real time of 24 hours corresponds to a dimensionless time of approximately 300 hours in very low permeability formations or more than 10^7 in very permeable formations [3, 9, 21, 25, 28].

A set number of *Ei* values for $0.0001 \le x \le 25$ with the aid of the algorithm given in **Figure 1.8**. Then, a fitting of these data was performed to obtain the polynomials given by Eqs. (1.85) and (1.90). The first one has a R^2 of 1, and the second one has a R^2 of 0.999999999 which implies accuracy up to the fifth digit can be obtained.

$$Ei(-x) = a + bx + cx^{2.5} + d\ln x + e \cdot \exp(-x); \quad x \le 1 \tag{1.85}$$

$$\ln\, Ei(-x) = \frac{a + cx + ex^2}{1 + bx + dx^2 + fx^3}; \quad x > 1 \tag{1.86}$$

Adapted from [29] and generated with the *Ei* function code given in **Figure 1.8**.
Define dimensionless radius, dimensionless time, and dimensionless pressure as:

$$r_D = r/r_w \tag{1.87}$$

$$t_D = \frac{t}{t_o} \tag{1.88}$$

$$P_D = \frac{kh(P_i - P)}{141.2q\mu B} \tag{1.89}$$

Adapted from [29] and generated with the *Ei* function code given in **Figure 1.8**.

For pressure drawdown tests, $\Delta P = P_i - P_{wf}$. For pressure buildup tests, $\Delta P = P_{ws} - P_{wf}\,(\Delta t = 0)$.

This means that the steady-state physical pressure drop for radial flow is equal to the dimensionless pressure multiplied by a scalable factor, which in this case depends on the flow and the properties of the reservoir, [3–7, 9, 21, 26, 30].

a	*b*	*c*	*d*	*e*	*f*
−0.0906765673563653	0.5133959845491270	−0.0243644307428167	−0.0000014346860800	−0.4865489789766050	–
0.7480202919199570	1.3629598993866700	−0.5960091961168400	0.0275653486990893	−0.7768782064908800	−0.0010740336145794

Table 1.1.
Constants for Eqs. (1.85) and (1.86).

x	0	1	2	3	4	5	6	7	8	9
0.000		8.63322	7.94018	7.53481	7.24723	7.02419	6.84197	6.68791	6.55448	6.43680
0.001	6.33154	6.23633	6.14942	6.06948	5.99547	5.92657	5.86214	5.80161	5.74455	5.69058
0.002	5.63939	5.59070	5.54428	5.49993	5.45747	5.41675	5.37763	5.33999	5.30372	5.26873
0.003	5.23493	5.20224	5.17059	5.13991	5.11016	5.08127	5.05320	5.02590	4.99934	4.97346
0.004	4.94824	4.92365	4.89965	4.87622	4.85333	4.83096	4.80908	4.78767	4.76672	4.74620
0.005	4.72610	4.70639	4.68707	4.66813	4.64953	4.63128	4.61337	4.59577	4.57847	4.56148
0.006	4.54477	4.52834	4.51218	4.49628	4.48063	4.46523	4.45006	4.43512	4.42041	4.40591
0.007	4.39162	4.37753	4.36365	4.34995	4.33645	4.32312	4.30998	4.29700	4.28420	4.27156
0.008	4.25908	4.24676	4.23459	4.22257	4.21069	4.19896	4.18736	4.17590	4.16457	4.15337
0.009	4.14229	4.13134	4.12052	4.10980	4.09921	4.08873	4.07835	4.06809	4.05793	4.04788
0.01	4.03793	3.94361	3.85760	3.77855	3.70543	3.63743	3.57389	3.51425	3.45809	3.40501
0.02	3.35471	3.30691	3.26138	3.21791	3.17634	3.13651	3.09828	3.06152	3.02614	2.99203
0.03	2.95912	2.92731	2.89655	2.86676	2.83789	2.80989	2.78270	2.75628	2.73060	2.70560
0.04	2.68126	2.65755	2.63443	2.61188	2.58987	2.56838	2.54737	2.52685	2.50677	2.48713
0.05	2.46790	2.44907	2.43063	2.41255	2.39484	2.37746	2.36041	2.34369	2.32727	2.31114
0.06	2.29531	2.27975	2.26446	2.24943	2.23465	2.22011	2.20581	2.19174	2.17789	2.16426
0.07	2.15084	2.13762	2.12460	2.11177	2.09913	2.08667	2.07439	2.06228	2.05034	2.03856
0.08	2.02694	2.01548	2.00417	1.99301	1.98199	1.97112	1.96038	1.94978	1.93930	1.92896
0.09	1.91874	1.90865	1.89868	1.88882	1.87908	1.86945	1.85994	1.85053	1.84122	1.83202
0.10	1.82292	1.81393	1.80502	1.79622	1.78751	1.77889	1.77036	1.76192	1.75356	1.74529
0.11	1.73711	1.72900	1.72098	1.71304	1.70517	1.69738	1.68967	1.68203	1.67446	1.66697
0.12	1.65954	1.65219	1.64490	1.63767	1.63052	1.62343	1.61640	1.60943	1.60253	1.59568

x	0	1	2	3	4	5	6	7	8	9
0.13	1.58890	1.58217	1.57551	1.56890	1.56234	1.55584	1.54940	1.54301	1.53667	1.53038
0.14	1.52415	1.51796	1.51183	1.50574	1.49970	1.49371	1.48777	1.48188	1.47603	1.47022
0.15	1.46446	1.45875	1.45307	1.44744	1.44186	1.43631	1.43080	1.42534	1.41992	1.41453
0.16	1.40919	1.40388	1.39861	1.39338	1.38819	1.38303	1.37791	1.37282	1.36778	1.36276
0.17	1.35778	1.35284	1.34792	1.34304	1.33820	1.33339	1.32860	1.32386	1.31914	1.31445
0.18	1.30980	1.30517	1.30058	1.29601	1.29147	1.28697	1.28249	1.27804	1.27362	1.26922
0.19	1.26486	1.26052	1.25621	1.25192	1.24766	1.24343	1.23922	1.23504	1.23089	1.22676
0.2	1.22265	1.21857	1.21451	1.21048	1.20647	1.20248	1.19852	1.19458	1.19067	1.18677

Table 1.2.
Values of the exponential integral for $0.0001 \le x \le 0.209$.

DOI: http://dx.doi.org/10.5772/intechopen.81078

x	0	1	2	3	4	5	6	7	8	9
4	37.7940000	37.7927530	33.4888052	29.6876209	26.3291192	23.3601005	20.7340078	18.4100584	16.3524950	14.5299393
5	11.4839049	11.4829557	10.2130008	9.0862158	8.0860830	7.1980442	6.4092603	5.7084015	5.0854647	4.5316127
6	3.6017735	3.6008245	3.2108703	2.8637634	2.5547143	2.2794796	2.0342987	1.8158374	1.6211385	1.4475779
7	1.1557663	1.1548173	1.0317127	0.9218812	0.8238725	0.7363972	0.6583089	0.5885877	0.5263261	0.4707165
8	0.3776052	0.3766562	0.3369951	0.3015486	0.2698641	0.2415382	0.2162112	0.1935625	0.1733060	0.1551866
9	0.1254226	0.1244735	0.1114954	0.0998807	0.0894849	0.0801790	0.0718477	0.0643883	0.0577086	0.0517267
10	0.0425187	0.0415697	0.0372704	0.0334186	0.0299673	0.0268747	0.0241031	0.0216191	0.0193925	0.0173966
11	0.0149520	0.0140030	0.0125645	0.0112746	0.0101178	0.0090804	0.0081498	0.0073151	0.0065663	0.0058946
12	0.0057001	0.0047511	0.0042658	0.0038303	0.0034395	0.0030888	0.0027739	0.0024913	0.0022377	0.0020099
13	0.0025709	0.0016219	0.0014570	0.0013090	0.0011761	0.0010567	0.0009495	0.0008532	0.0007667	0.0006890
14	0.0015056	0.0005566	0.0005002	0.0004496	0.0004042	0.0003633	0.0003266	0.0002936	0.0002640	0.0002373
15	0.0011409	0.00019186	0.00017251	0.00015513	0.00013950	0.00012545	0.00011282	0.00010146	9.1257E−05	8.2079E−05
16	0.0010155	6.6405E−09	5.9732E−09	5.3732E−09	4.8336E−09	4.3483E−09	3.9119E−09	3.5194E−09	3.1664E−09	2.8489E−09
17	0.0009725	2.3064E−09	2.0754E−09	1.8675E−09	1.6805E−09	1.5123E−09	1.3609E−09	1.2248E−09	1.1022E−09	9.9202E−10
18	0.0009563	8.0361E−10	7.2331E−10	6.5105E−10	5.8603E−10	5.2752E−10	4.7486E−10	4.2747E−10	3.8482E−10	3.4643E−10
19	0.0009511	2.8078E−10	2.5279E−10	2.2760E−10	2.0492E−10	1.8451E−10	1.6613E−10	1.4959E−10	1.3470E−10	1.2129E−10
20	0.0009526	9.8355E−11	8.8572E−11	7.9764E−11	7.1833E−11	6.4692E−11	5.8263E−11	5.2473E−11	4.7260E−11	4.2566E−11
21	0.0009248	3.4532E−11	3.1104E−11	2.8017E−11	2.5237E−11	2.2733E−11	2.0478E−11	1.8447E−11	1.6617E−11	1.4970E−11
22	0.0009183	1.2149E−11	1.0945E−11	9.8610E−12	8.8842E−12	8.0043E−12	7.2117E−12	6.4976E−12	5.8544E−12	5.2750E−12
23	0.0009464	4.2827E−12	3.8590E−12	3.4773E−12	3.1334E−12	2.8236E−12	2.5444E−12	2.2929E−12	2.0663E−2	1.8621E−12
24	0.0009316	1.5123E−12	1.3629E−12	1.2283E−12	1.1070E−12	9.9772E−13	8.9922E−13	8.1046E−13	7.3048E−13	6.5839E−13
25	0.0000779	5.3489E−13	4.8213E−13	4.3458E−13	3.9172E−13	3.5310E−13	3.1829E−13	2.8692E−13	2.5864E−13	2.3315E−13

Table 1.3.
Values of the exponential integral, $\mathrm{Ei}(-x) \times 10^{-4}$*, for* $4 \le x \le 25.9$.

x	0	1	2	3	4	5	6	7	8	9
0.20	1.222651	1.182902	1.145380	1.109883	1.076236	1.044283	1.013889	0.984933	0.957308	0.930918
0.30	0.905677	0.881506	0.858335	0.836101	0.814746	0.794216	0.774462	0.755442	0.737112	0.719437
0.40	0.702380	0.685910	0.669997	0.654614	0.639733	0.625331	0.611387	0.597878	0.584784	0.572089
0.50	0.559774	0.547822	0.536220	0.524952	0.514004	0.503364	0.493020	0.482960	0.473174	0.463650
0.60	0.454380	0.445353	0.436562	0.427997	0.419652	0.411517	0.403586	0.395853	0.388309	0.380950
0.70	0.373769	0.366760	0.359918	0.353237	0.346713	0.340341	0.334115	0.328032	0.322088	0.316277
0.80	0.310597	0.305043	0.299611	0.294299	0.289103	0.284019	0.279045	0.274177	0.269413	0.264750
0.90	0.260184	0.255714	0.251337	0.247050	0.242851	0.238738	0.234708	0.230760	0.226891	0.223100
1.00	0.2193840	0.2157417	0.2121712	0.2086707	0.2052384	0.2018729	0.1985724	0.1953355	0.1921606	0.1890462
1.10	0.1859910	0.1829936	0.1800526	0.1771667	0.1743347	0.1715554	0.1688276	0.1661501	0.1635218	0.1609417
1.20	0.1584085	0.1559214	0.1534793	0.1510813	0.1487263	0.1464135	0.1441419	0.1419107	0.1397191	0.1375661
1.30	0.1354511	0.1333731	0.1313314	0.1293253	0.1273541	0.1254169	0.1235132	0.1216423	0.1198034	0.1179960
1.40	0.1162194	0.1144730	0.1127562	0.1110684	0.1094090	0.1077775	0.1061734	0.1045960	0.1030450	0.1015197
1.50	0.1000197	0.0985445	0.0970936	0.0956665	0.0942629	0.0928822	0.0915241	0.0901880	0.0888737	0.0875806
1.60	0.0863084	0.0850568	0.0838252	0.0826134	0.0814211	0.0802477	0.0790931	0.0779568	0.0768385	0.0757379
1.70	0.0746547	0.0735886	0.0725392	0.0715063	0.0704896	0.0694888	0.0685035	0.0675336	0.0665788	0.0656387
1.80	0.0647132	0.0638020	0.0629048	0.0620214	0.0611516	0.0602951	0.0594516	0.0586211	0.0578032	0.0569977
1.90	0.0562045	0.0554232	0.0546538	0.0538960	0.0531496	0.0524145	0.0516904	0.0509771	0.0502745	0.0495824
2.00	0.0489006	0.0482290	0.0475673	0.0469155	0.0462733	0.0456407	0.0450173	0.0444032	0.0437981	0.0432019
2.10	0.0426144	0.0420356	0.0414652	0.0409032	0.0403493	0.0398036	0.0392657	0.0387357	0.0382133	0.0376986
2.20	0.0371912	0.0366912	0.0361984	0.0357127	0.0352340	0.0347622	0.0342971	0.0338387	0.0333868	0.0329414
2.30	0.0325024	0.0320696	0.0316429	0.0312223	0.0308077	0.0303990	0.0299961	0.0295988	0.0292072	0.0288210

x	0	1	2	3	4	5	6	7	8	9
2.40	0.0284404	0.0280650	0.0276950	0.0273301	0.0269704	0.0266157	0.0262659	0.0259210	0.0255810	0.0252457
2.50	0.0249150	0.0245890	0.0242674	0.0239504	0.0236377	0.0233294	0.0230253	0.0227254	0.0224296	0.0221380
2.60	0.0218503	0.0215666	0.0212868	0.0210109	0.0207387	0.0204702	0.0202054	0.0199443	0.0196867	0.0194326
2.70	0.0191820	0.0189348	0.0186909	0.0184504	0.0182131	0.0179790	0.0177481	0.0175204	0.0172957	0.0170740
2.80	0.0168554	0.0166397	0.0164269	0.0162169	0.0160098	0.0158055	0.0156039	0.0154050	0.0152087	0.0150151
2.90	0.0148241	0.0146356	0.0144497	0.0142662	0.0140852	0.0139066	0.0137303	0.0135564	0.0133849	0.0132155
3.00	0.0130485	0.0128836	0.0127209	0.0125604	0.0124020	0.0122457	0.0120915	0.0119392	0.0117890	0.0116408
3.10	0.0114945	0.0113502	0.0112077	0.0110671	0.0109283	0.0107914	0.0106562	0.0105229	0.0103912	0.0102613
3.20	0.0101331	0.0100065	0.0098816	0.0097584	0.0096367	0.0095166	0.0093981	0.0092811	0.0091656	0.0090516
3.30	0.0089391	0.0088281	0.0087185	0.0086103	0.0085035	0.0083981	0.0082940	0.0081913	0.0080899	0.0079899
3.40	0.0078911	0.0077935	0.0076973	0.0076022	0.0075084	0.0074158	0.0073244	0.0072341	0.0071450	0.0070571
3.50	0.0069702	0.0068845	0.0067999	0.0067163	0.0066338	0.0065524	0.0064720	0.0063926	0.0063143	0.0062369
3.60	0.0061605	0.0060851	0.0060106	0.0059371	0.0058645	0.0057929	0.0057221	0.0056523	0.0055833	0.0055152
3.70	0.0054479	0.0053815	0.0053160	0.0052512	0.0051873	0.0051242	0.0050619	0.0050003	0.0049396	0.0048796
3.80	0.0048203	0.0047618	0.0047041	0.0046470	0.0045907	0.0045351	0.0044802	0.0044259	0.0043724	0.0043195
3.90	0.0042672	0.0042157	0.0041647	0.0041144	0.0040648	0.0040157	0.0039673	0.0039194	0.0038722	0.0038255
4.00	0.0037794	0.0037339	0.0036890	0.0036446	0.0036008	0.0035575	0.0035148	0.0034725	0.0034308	0.0033896

Table 1.4.
Values of the exponential integral for $0.1 \le x \le 4.09$.

The same concept applies to transient flow and to more complex situations, but in this case, the dimensionless pressure is different. For example, for transient flow, the dimensionless pressure is always a function of dimensionless time.

Taking derivative to Eqs. (1.87) and (1.88),

$$\partial r = r_w \partial r_D \tag{1.90}$$

$$\partial t = t_o \partial t_D \tag{1.91}$$

Replacing the above derivatives into Eq. (1.20),
Adapted from [5] and generated with the *Ei* function code given in **Figure 1.8.**

$$\frac{\partial^2 P}{\partial r_D^2} + \frac{1}{r_D} \frac{\partial P}{\partial r_D} = \frac{\phi \mu c_t r_w^2}{k t_o} \frac{\partial P}{\partial t_D} \tag{1.92}$$

Definition of t_o requires assuming $\frac{\phi \mu c_t r_w^2}{k t_o} = 1$, [24], then;

$$t_o = \frac{\phi \mu c_t r_w^2}{k} \tag{1.93}$$

Replacing this definition into Eq. (1.88) and solving for the dimensionless time (oilfield units),

$$t_D = \frac{0.0002637 k t}{\phi \mu c_t r_w^2} \tag{1.94}$$

Replacing Eq. (1.93) in Eq. (1.92) leads, after simplification, to:

$$\frac{\partial^2 P}{\partial r_D^2} + \frac{1}{r_D} \frac{\partial P}{\partial r_D} = \frac{\partial P}{\partial t_D} \tag{1.95}$$

The dimensionless pressure is also affected by the system geometry, other well systems, storage coefficient, anisotropic characteristics of the reservoir, fractures, radial discontinuities, double porosity, among others. In general, the pressure at any point in a single well system that produces the constant rate, q, is given by [25]:

$$[P_i - P(r,t)] = \frac{qB\mu}{kh} P_D(t_D, r_D, C_D, \text{geometry}, \ldots.) \tag{1.96}$$

Taking twice derivative to Eq. (1.87), excluding the conversion factor, will provide:

$$\partial P_D = -\frac{kh}{qB\mu} \partial P \tag{1.97}$$

$$\partial^2 P_D = -\frac{kh}{qB\mu} \partial^2 P \tag{1.98}$$

Replacing Eqs. (1.97) and (1.98) in Eq. (1.95) and simplifying leads to:

$$\frac{\partial^2 P_D}{\partial r_D^2} + \frac{1}{r_D} \frac{\partial P_D}{\partial r_D} = \frac{1}{r_D} \frac{\partial}{\partial r_D} \left(r_D \frac{\partial P_D}{\partial r_D} \right) = \frac{\partial P_D}{\partial t_D} \tag{1.99}$$

If the characteristic length is the area, instead of wellbore radius, Eq. (1.92) can be expressed as:

$$t_{DA} = \frac{0.0002637kt}{\phi\,\mu\,c_t\,A} = t_D\left(\frac{r_w^2}{A}\right) \tag{1.100}$$

Example 1.1

A square shaped reservoir produces 300 BPD through a well located in the center of one of its quadrants. See **Figure 1.10**. Estimate the pressure in the well after 1 month of production. Other relevant data:

P_i = 3225 psia, h = 42 ft
k_o = 1 darcy, ϕ = 25%
μ_o = 25 cp, c_t = 6.1 × 10^{-6}/psia
B_o = 1.32 bbl/BF, r_w = 6 in
A = 150 Acres, q = 300 BPD

Solution

Assuming the system behaves infinitely, it means, during 1 month of production the transient wave has not yet reached the reservoir boundaries, the problem can be solved by estimating the *Ei* function. Replacing Eqs. (1.82) and (1.92) into the argument of Eq. (1.82), it results:

$$x = -\frac{r_D^2}{4t_D} = -\frac{948\phi\mu c_t r^2}{kt} \tag{1.101}$$

Using Eq. (1.101) with the above given reservoir and well data:

$$x = -\frac{948(0.25)(25)(6.1\times10^{-6})(0.5^2)}{(1000)(720)} = 1.25\times10^{-8}$$

This x value allows finding $Ei(-x)$ = 17.6163 using the function provided in **Figure 1.8**. From the application of Eq. (82), P_D = 8.808. This dimensionless pressure is meaningless for practical purposes. Converting to oilfield units by means of Eq. (1.87), the well-flowing pressure value after 1 month of production is given as:

$$8.808 = \frac{(1000)(42)}{(141.2)(300)(1.32)(25)}(3225 - P_{wf})$$

P_{wf} = 2931.84 psia.

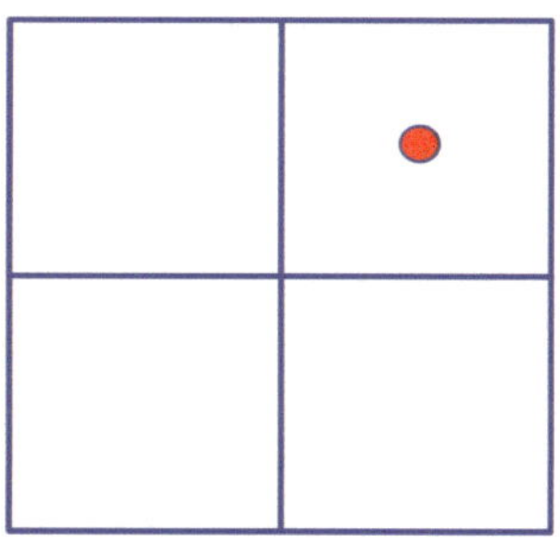

Figure 1.10.
Geometry of the reservoir for example 1.1.

How it can be now if the example was correctly done? A good approximation consists of considering a small pressure drop; let us say ± 0.002 psia (smallest value that can be read from current pressure recorders) at the closest reservoir boundary. Use Eq. (1.87) to convert from psia to dimensionless pressure:

$$P_D = \frac{(1000)(42)}{(141.2)(300)(1.32)(25)}(0.002) = 6.0091 \times 10^{-5}$$

Eq. (1.82) allows finding $Ei(-x) = 0.00012$. This value can be used to determine an x value from **Table 1.2**. However, a trial-and-error procedure with the function given in **Figure 1.8** was performed to find an x value of 6.97. Then, the time at which this value takes place at the nearest reservoir boundary is found from Eq. (1.101). The nearest boundary is obtained from one-fourth of the reservoir size area (3.7 Ac or 1663500 ft^2). Then, for a square geometry system (the system may also be approached to a circle):

$$L = \sqrt{1663500} = 1278.09 \text{ ft}$$

The radial distance from the well to the nearest boundary corresponds to one half of the square side, the r = 639.04 ft. Solving for time from Eq. (1.101);

$$t = \frac{948\phi\mu c_t r^2}{kx} = \frac{948(0.25)(25)(6.1 \times 10^{-6})(639.04^2)}{(1000)(6.97)} = 2.118 \text{ h}$$

This means that after 2 h and 7 min of flow, the wave has reached the nearest reservoir boundary; therefore, the infinite-acting period no longer exists for this reservoir, then, a pseudosteady-state solution ought to be applied (**Figures 1.11–1.14**). To do so, Eq. (1.98) is employed for the whole reservoir area:

$$t_{DA} = \frac{(0.0002637)(1000)(720)}{(0.25)(25)(6.1 \times 10^{-6})(6534000)} = 0.76$$

With this t_{DA} value of 0.76, the normal procedure is to estimate the dimensionless pressure for a given reservoir-well position configuration, which can be found in Figures C.13 through C.16 in [25] for which data were originally presented in [31]. These plots provide the pressure behavior for a well inside a

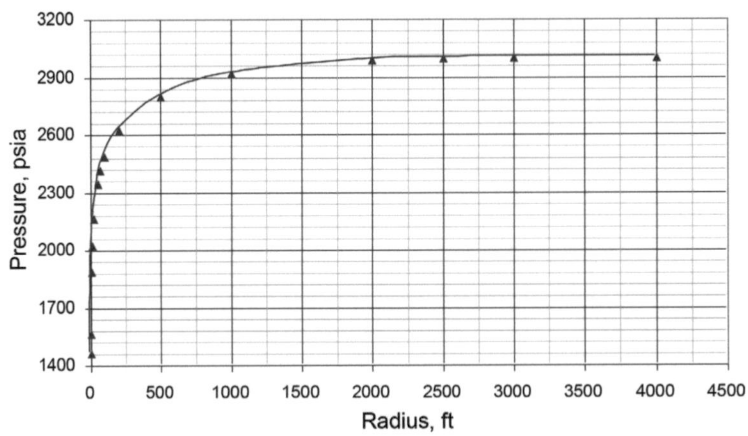

Figure 1.11.
Pressure versus distance plot for example 1.2.

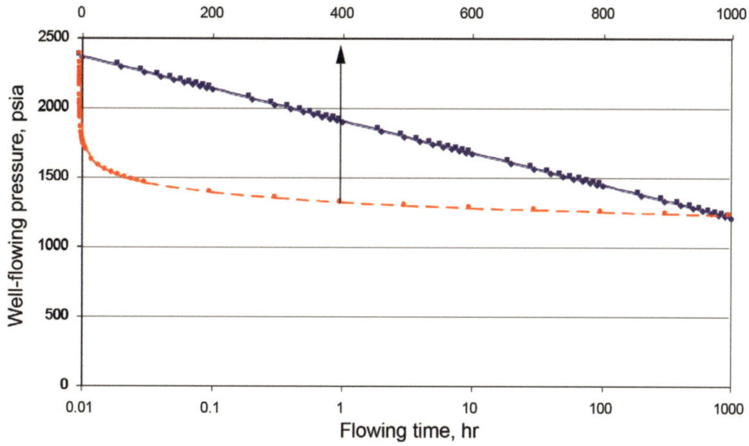

Figure 1.12.
Pressure versus time plot for example 1.3.

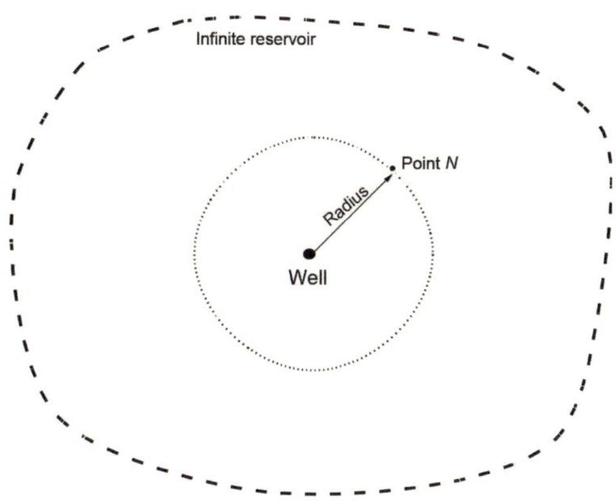

Figure 1.13.
Pressure distribution in the reservoir.

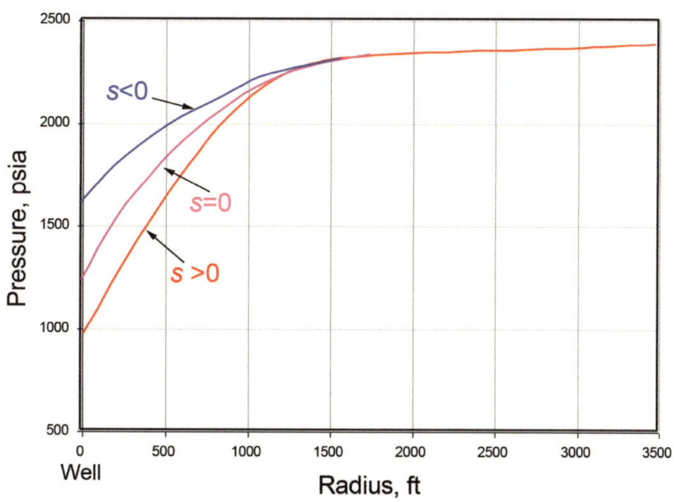

Figure 1.14.
Skin factor influence.

rectangular/square no-flow system, without storage wellbore and skin factor; $A^{0.5}/r_w$ = 2000 can also be found in [3, 9, 26]. This procedure is avoided in this textbook. Instead new set of data was generated and adjusted to the following polynomial fitting in which constants are reported in **Table 1.5**:

$$P_D = a + b^* t_{DA} + c^* t_{DA}^2 + d^* t_{DA}^{0.5} \ln t_{DA} + \frac{e}{t_{DA}^{0.5}} \tag{1.102}$$

Using Eq. (1.102) will result:

$$P_D = 4.4765 + 9.3437(12) - 0.2798(12^2) - 2.7516\sqrt{12}\ln(12) - \frac{0.016098}{\sqrt{12}}$$

P_D = 12.05597.

The well-flowing pressure is estimated with Eq. (1.87); thus,

$$12.056 = \frac{(1000)(42)}{(141.2)(300)(1.32)(25)}(P_i - P_{wf})$$

P_{wf} = 2823.75 psia.

1.9. Application of the diffusivity equation solution

A straight-line behavior can be observed in mostly the whole range on the right-hand plot of *Ei* versus *x* plot given in **Figure 1.9**. Then, it was concluded, [3–7, 9, 11, 19, 21, 26, 30], when $x < 0.0025$, the more complex mathematical representation of Eq. (1.82) can be replaced by a straight line function, given by:

$$E_i(-x) = \ln(1.781x) \tag{1.103}$$

this leads to,

$$E_i(-x) = \ln x + 0.5772 \tag{1.104}$$

Replacing this new definition into Eq. (1.82) will result in:

$$P_D = -\frac{1}{2}\left[\ln\left(\frac{r_D^2}{4t_D}\right) + 0.5772\right] \tag{1.105}$$

At the well $r_D = 1$, after rearranging,

$$P_D = \frac{1}{2}[\ln t_D + 0.80907] \tag{1.106}$$

The above indicates that the well pressure behavior obeys a semi-logarithmic behavior of pressure versus time.

Example 1.2

A well and infinite reservoir has the following characteristics:

q = 2000 STB/D,	μ = 0.72 cp,	$c_t = 1.5 \times 10^{-5}$ psia^{-1}
ϕ = 23%,	P_i = 3000 psia,	h = 150 ft
B = 1.475 bbl/STB,	k = 10 md,	r_w = 0.5 ft

DOI: http://dx.doi.org/10.5772/intechopen.81078

Reservoir configuration	a	b	c	d	e
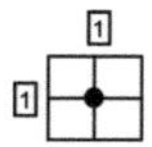	4.87003665	6.40209421	1.3534672	−2.055396	−0.0224277
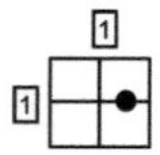	4.69798655	7.72167053	0.61940345	−2.3599896	−0.0196633
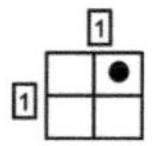	4.47647322	9.3437427	−0.2797873	−2.751624	−0.0160974
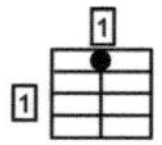	4.17068525	9.09489851	0.43617914	−3.5253409	−0.0124389
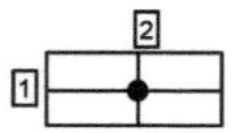	5.22696199	6.87113863	0.70426794	−1.6437779	−0.0351473
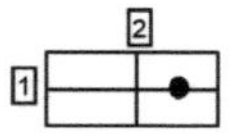	5.24504716	8.97155258	−0.6397168	−1.5198408	−0.0345266
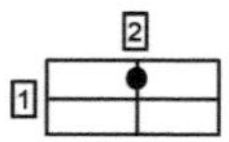	4.65131893	7.78361646	0.66507635	−2.6254729	−0.0221728
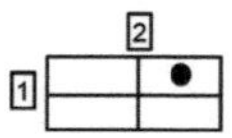	4.56520077	9.9133152	−0.5332832	−2.6610436	−0.0189231
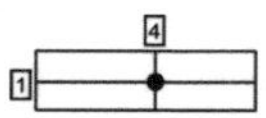	4.52748153	9.032441483	−0.09048377	−2.669107599	−0.017057535
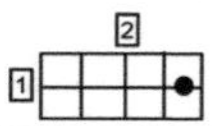	4.672817379	16.52873149	−3.575335215	−2.190064849	−0.018402371
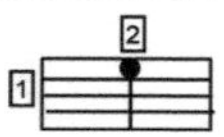	3.950602849	8.526796474	1.211519573	−4.232235498	−0.010580681

Reservoir configuration	*a*	*b*	*c*	*d*	*e*
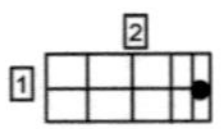	3.900271642	12.82720551	−1.31089423	−4.181344004	−0.009004582
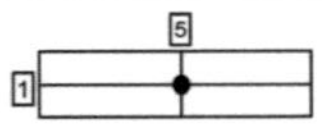	4.42301913	10.2404644	−0.7980823	−2.8540283	−0.0153571
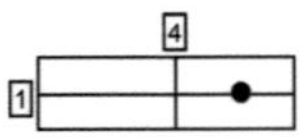	4.8024027	11.2579471	−1.2342193	−2.0818988	−0.0211339
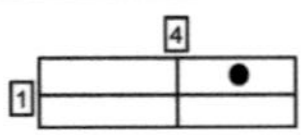	4.40840885	11.6323106	−0.8617361	−3.0003158	−0.0159063
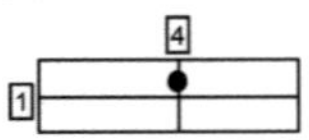	4.14494461	9.44257591	0.21833578	−3.5654996	−0.0120138

Table 1.5.
Constants for Eq. (1.102).

Estimate the well-flowing pressure at radii of 0.5, 1, 5, 10, 20, 50, 70, 100, 200, 500, 1000, 2000, 2500, 3000, and 4000 feet after 1 month of production. Plot the results.

Solution

For the wellbore radius, find x with Eq. (1.101);

$$x = \frac{948(0.23)(0.72)(1.5 \times 10^{-5})(0.5^2)}{(10)(720)} = 8.177 \times 10^{-8}$$

Using the function given in **Figure 1.9** or Eq. (1.103), a value of $Ei(-x)$ of 15.7421 is found. Then, Eq. (1.82) indicates that P_D = 7.871. Use of Eq. (1.87) allows estimating both pressure drop and well-flowing pressure:

$$\Delta P = P_i - P_{wf} = \frac{141.2 q\mu B}{kh} P_D = \frac{141.2(2000)(0.72)(1.475)}{(10)(150)} 7.871 = 1573.74 \text{ psia}$$

The remaining results are summarized in **Table 1.6** and plotted in **Figure 1.11**. From this, it can be inferred that the highest pressure drop takes place in the near-wellbore region which mathematically agrees with the continuity equation stating that when the area is reduced, the velocity has to be increased so the flow rate can be constant. The higher the fluid velocity, the higher the pressure drops.

Example 1.3

Re-work example 1.2 to estimate the sand-face pressure at time values starting from 0.01 to 1000 h. Show the results in both Cartesian and semilog plots. What does this suggest?

Solution

Find x with Eq. (1.101);

r, ft	*x*	*Ei*(−*x*)	*P*, psia	P_{wf}, psia
0.5	8.18E−08	15.7421	1537.74	1462.26
1	3.27E−07	14.3558	1435.15	1564.85
5	8.18E−06	11.137	1113.36	1886.64
10	3.27E−04	9.75	974.78	2025.22
20	1.31E−04	8.365	836.2	2163.8
50	8.18E−04	6.533	653.07	2346.93
70	1.60E−03	5.86	585.87	2414.13
100	3.27E−03	5.149	514.72	2485.28
200	1.31E−02	3.772	377.11	2622.89
500	8.17E−02	2.007	200.616	2799.384
1000	3.27E−01	0.8425	84.225	2915.775
2000	1.31E+00	0.1337	13.368	2986.632
2500	2.04E+00	0.046	4.6	2995.4
3000	2.94E+00	0.014	1.401	2998.599
4000	5.23E+00	0.0009	0.087	2999.913

Table 1.6.
Summarized results for example 1.2.

$$x = \frac{948(0.23)(0.72)(1.5\times10^{-5})(0.5^2)}{(10)(0.01)} = 0.000948$$

A value of $Ei(-x)$ of 6.385 is found with Eq. (1.103). Then, Eq. (1.82) gives a P_D value of 3.192 and Eq. (1.87) leads to calculate a well-flowing pressure of;

$$P_{wf} = P_i - \frac{141.2q\mu B}{kh}P_D = 3000 - \frac{141.2(2000)(0.72)(1.475)}{(10)(150)}3.192 = 2361.71 \text{ psia}$$

The remaining well-flowing pressure values against time are given in **Table 1.7** and plotted in **Figure 1.12**. The semilog behavior goes in the upper part of the plot (solid line), and the Cartesian plot corresponds to the lower dashed line. The semilog line behaves linearly while the Cartesian curve does not. This situation perfectly agrees with Eq. (1.106), which ensures that the behavior of pressure drop versus time obeys a semilog trend. In other word, in a transient radial system, pressure drops is a linear function of the logarithm of time.

t, h	x	Ei(−x)	P_D	P_{wf}, Psia	t, h	x	Ei(−x)	P_D	P_{wf}, psia
0.01	9.480E−04	6.385	3.192	2361.71	6	1.580E−06	12.781	6.390	1722.30
0.02	4.740E−04	7.078	3.539	2292.46	7	1.354E−06	12.935	6.468	1706.89
0.03	3.160E−04	7.483	3.741	2251.94	8	1.185E−06	13.069	6.534	1693.54
0.04	2.370E−04	7.770	3.885	2223.19	9	1.053E−06	13.186	6.593	1681.77
0.05	1.896E−04	7.994	3.997	2200.89	10	9.480E−07	13.292	6.646	1671.23
0.06	1.580E−04	8.176	4.088	2182.66	20	4.740E−07	13.985	6.992	1601.94
0.07	1.354E−04	8.330	4.165	2167.25	30	3.160E−07	14.390	7.195	1561.41
0.08	1.185E−04	8.464	4.232	2153.91	40	2.370E−07	14.678	7.339	1532.65
0.09	1.053E−04	8.581	4.291	2142.13	50	1.896E−07	14.901	7.451	1510.34
0.1	9.480E−05	8.687	4.343	2131.60	60	1.580E−07	15.083	7.542	1492.11
0.2	4.740E−05	9.380	4.690	2062.31	70	1.354E−07	15.238	7.619	1476.70
0.3	3.160E−05	9.785	4.893	2021.78	80	1.185E−07	15.371	7.686	1463.35
0.4	2.370E−05	10.073	5.036	1993.02	90	1.053E−07	15.489	7.744	1451.58
0.5	1.896E−05	10.296	5.148	1970.71	100	9.480E−08	15.594	7.797	1441.05
0.6	1.580E−05	10.478	5.239	1952.49	200	4.740E−08	16.287	8.144	1371.75
0.7	1.354E−05	10.632	5.316	1937.08	300	3.160E−08	16.693	8.346	1331.22
0.8	1.185E−05	10.766	5.383	1923.73	400	2.370E−08	16.981	8.490	1302.46
0.9	1.053E−05	10.884	5.442	1911.95	500	1.896E−08	17.204	8.602	1280.15
1	9.480E−06	10.989	5.495	1901.42	600	1.580E−08	17.386	8.693	1261.92
2	4.740E−06	11.682	5.841	1832.13	700	1.354E−08	17.540	8.770	1246.51
3	3.160E−06	12.088	6.044	1791.59	800	1.185E−08	17.674	8.837	1233.17
4	2.370E−06	12.375	6.188	1762.84	900	1.053E−08	17.792	8.896	1221.39
5	1.896E−06	12.599	6.299	1740.53	1000	9.480E−09	17.897	8.948	1210.86

Table 1.7.
Summarized results for example 1.3.

1.10. Pressure distribution and skin factor

Once the dimensionless parameters are plugged in Eq. (1.82), this yields:

$$P(r,t) = P_i - 70.6\frac{qB\mu}{kh}E_i\left\{-\frac{948\phi\mu c_t r^2}{kt}\right\} \tag{1.107}$$

At point N, **Figure 1.13**, the pressure can be calculated by Eq. (1.107). At the wellbore $r_D = r/r_w = 1$, then, $r = r_w$ and $P(r,t) = P_{wf}$. Note that application of the line-source solution requires the reservoir to possess an infinite extent, [3, 9, 18, 21, 25, 26].

There are several ways to quantify damage or stimulation in an operating well (producer or injector). These conditions are schematically represented in **Figure 1.14**. The most popular method is to represent a well condition by a steady-state pressure drop occurring at the wellbore, in addition to the transient pressure drop normally occurring in the reservoir. This additional pressure drop is called "skin pressure drop" and takes place in an infinitesimally thin zone: "damage zone," [4, 5, 9, 11, 19, 30]. It can be caused by several factors:

1. Invasion of drilling fluids
2. Partial well penetration
3. Partial completion
4. Blocking of perforations
5. Organic/inorganic precipitation
6. Inadequate drilling density or limited drilling
7. Bacterial growth
8. Dispersion of clays
9. Presence of cake and cement
10. Presence of high gas saturation around the well

Skin factor is a dimensionless parameter; then, it has to be added to the dimensionless pressure in Eq. (1.87), so that:

$$P_i - P_{wf} = 141.2\frac{q\mu B}{kh}(P_D + s) \tag{1.108}$$

From the above expression can be easily obtained:

$$P_i - P_{wf} = 141.2\frac{q\mu B}{kh}P_D + 141.2\frac{q\mu B}{kh}s \tag{1.109}$$

Therefore, the skin factor pressure drop is given by:

$$\Delta P_s = 141.2\frac{q\mu B}{kh}s \tag{1.110}$$

Assuming steady state near the wellbore and the damage area has a finite radius, r_s, with an altered permeability, k_s, the pressure drop due to the damage is expressed as the pressure difference between the virgin zone and the altered zone, that is to say:

$$\Delta P_s = 141.2\frac{q\mu B}{k_s h}\ln\frac{r_s}{r_w} - 141.2\frac{q\mu B}{kh}\ln\frac{r_s}{r_w} \tag{1.111}$$

Rearranging;

$$\Delta P_s = 141.2\frac{q\mu B}{kh}\left(\frac{k}{k_s} - 1\right)\ln\frac{r_s}{r_w} \tag{1.112}$$

Comparing Eqs (1.112) and (1.107), the following can be concluded:

$$s = \left(\frac{k}{k_s} - 1\right)\ln\frac{r_s}{r_w} \tag{1.113}$$

r_s and k_s are not easy to be obtained.

Equation (1.82) and (1.106) can be respectively written as:

$$P_D + s = -\frac{1}{2}E_i(-x) \tag{1.114}$$

$$P_D + s = \frac{1}{2}[\ln t_D + 0.80907] \tag{1.115}$$

Replacing the dimensionless quantities given by Eqs. (1.87) and (1.95) in Eq. (1.115) will result:

$$P_i = P_{wf} + \frac{70.6q\mu B}{kh}\left[\ln\left(\frac{0.0002637kt}{\phi\mu c_t r_w^2}\right) + 0.80908 + 2s\right] \tag{1.116}$$

Taking natural logarithm to 0.0002637 and adding its result to 0.80908 results in:

$$P_i = P_{wf} + \frac{70.6q\mu B}{kh}\left[-7.4316 + \ln\left(\frac{kt}{\phi\mu c_t r_w^2}\right) + 2s\right] \tag{1.117}$$

Multiplying and dividing by the natural logarithm of 10 and solving for the well-flowing pressure:

$$P_{wf} = P_i - \frac{162.6q\mu B}{kh}\left[\log\left(\frac{kt}{\phi\mu c_t r_w^2}\right) - 3.2275 + 0.8686s\right] \tag{1.118}$$

Thus, a straight line is expected to develop from a semilog plot of pressure against the time, as seen on the upper curve of **Figure 1.12**.

1.11. Finite reservoirs

In closed systems, the radial flow is followed by a transition period. This in turn is followed by the pseudosteady, semi-stable, or quasi-stable state, which is a transient flow regime where the pressures change over time, dP/dt, is constant at all points of the reservoir:

$$\frac{dP}{dt} = \frac{-q}{cV_p} \tag{1.119}$$

Eq. (1.99) is now subjected to the following initial and boundary conditions:

$$P_D(r_D, t_D = 0) = 0 \tag{1.120}$$

$$\left(\frac{\partial P_D}{\partial r_D}\right)_{r_{eD}} = 0 \tag{1.121}$$

$$\left(\frac{\partial P_D}{\partial r_D}\right)_{r_D=1} = -1 \tag{1.122}$$

Which solution is [9, 30]:

$$P_D(r_D, t_D) = \frac{2}{(r_{eD}^2 - 1)}\left(\frac{r_D^2}{4} + t_D\right) - \frac{r_{eD}^2 \ln r_D}{(r^2{}_{eD} - 1)} - \frac{(3r_{eD}^4 - 4r_{eD}^4 \ln r_{eD} - 2r_{eD}^2 - 1)}{4(r_{eD}^2 - 1)^2}$$
$$+\pi \sum_{n=1}^{\infty} \left\{ \frac{e^{-a_n^2 t_D} J_1^2(a_n r_{eD})[J_1(a_n)Y_0(a_n r_D) - Y_1(a_n)(J_0)(a_n r_D)]}{a_n[J_1^2(a_n r_{eD}) - J_1^2(a_n)]} \right\} \tag{1.123}$$

The pseudosteady-state period takes place at late times ($t > 948\phi\mu c_t r_e^{\,2}/k$), so that as time tends to infinity, summation tends to zero, then:

$$P_D(r_D, t_D) = \frac{2}{(r_{eD}^2 - 1)}\left(\frac{r_D^2}{4} + t_D\right) - \frac{r_{eD}^2 \ln\ r_D}{(r_{eD}^2 - 1)} - \frac{(3r_{eD}^4 - 4r_{eD}^4 \ln\ r_{eD} - 2r_{eD}^2 - 1)}{4(r_{eD}^2 - 1)^2} \tag{1.124}$$

At the well, $r_D = 1$ and as $r_{eD} >>>> 1$, the above expression is reduced to:

$$P_D(t_D) = \frac{2}{r_{eD}^2} + \frac{2t_D}{r_{eD}^2} - \frac{3}{4} + \ln\ r_{eD} + \frac{1}{2r_{eD}^2} - \frac{1}{4r_{eD}^4} \tag{1.125}$$

This can be approximated to:

$$P_D(t_D) \cong \frac{2t_D}{r_{eD}^2} + \ln\ r_{eD} - \frac{3}{4} \tag{1.126}$$

Invoking Eq. (1.98) for a circular reservoir area,

$$t_{DA} = t_D \frac{r_w^2}{\pi r_e^2} = \frac{t_D}{\pi r_{eD}^2} \tag{1.127}$$

It follows that;

$$\pi t_{DA} = \frac{t_D}{r_{eD}^2} \tag{1.128}$$

The final solution to the pseudosteady-state diffusivity equation is obtained from using the definition given by Eq. (1.128) in Eq. (1.129):

$$P_D(t_D) = 2\pi t_{DA} + \ln\ r_{eD} - \frac{3}{4} \tag{1.129}$$

The derivative with respect to time of the above equation in dimensional form allows obtaining the pore volume:

$$\frac{dP(r,t)}{dt} = -\frac{1.79qB}{h\phi c_t r_e^2} \tag{1.130}$$

An important feature of this period is that the rate of change of pressure with respect to time is a constant, that is, $dP_D/dt_{DA} = 2\pi$.

When the reservoir pressure does not change over time at any point, the flow is said to be stable. In other words, the right side of Eq. (1.99) is zero, [3]:

$$\frac{1}{r_D}\frac{\partial}{\partial r_D}\left(r_D\frac{\partial P_D}{\partial r_D}\right) = 0 \tag{1.131}$$

Similar to the pseudosteady-state case, steady state takes place at late times. Now, its initial, external, and internal boundary conditions are given by:

$$P_D(r_D, t_D = 0) = 0 \tag{1.132}$$

$$P_D(r_{De}, 0) = 0 \tag{1.133}$$

$$\left(\frac{\partial P_D}{\partial r_D}\right)_{r_D=1} = -1 \tag{1.134}$$

The solution to the steady-state diffusivity equation is [3]:

$$P_D(r_D, t_D) = \ln\ r_{eD} - 2\sum_{n=1}^{\infty}\left\{\frac{e^{-\beta_n^2 t_{Dn} J_0^2(\beta_n r_{eD})}}{\beta_n^2\left[J_1^2(\beta_n) - J_0^2(\beta_n r_{eD})\right]}\right\} \tag{1.135}$$

As time tends to infinity, the summation tends to infinity, then:

$$(P_D)_{ssr} = \ln r_{eD} = \ln\frac{r_e}{r_w} \tag{1.136}$$

In dimensional terms, the above expression is reduced to Darcy's equation. The dimensionless pressure function for linear flow is given by:

$$(P_D)_{ssL} = 2\pi\frac{Lh}{A} \tag{1.137}$$

Steady state can occur in reservoirs only when the reservoir is fully recharged by an aquifer or when injection and production are balanced. However, a reservoir with a very active aquifer will not always act under steady-state conditions. First, there has to be a period of unsteady state, which will be followed by the steady state once the pressure drop has reached the reservoir boundaries. Extraction of fluids from a pressurized reservoir with compressible fluids causes a pressure disturbance which travels throughout the reservoir. Although such disturbance is expected to travel at the speed of sound, it is rapidly attenuated so that for a given duration of production time, there is a distance, the drainage radius, beyond which no substantial changes in pressure will be observed. As more fluid is withdrawn (or injected), the disturbance moves further into the reservoir with continuous pressure decline at all points that have experienced pressure decline. Once a reservoir boundary is found, the pressure on the boundary continues to decline but at a faster rate than when the boundary was not detected. On the other hand, if the pressure transient reaches an open boundary (water influx), the pressure remains constant at some point; the pressure closest to the well will decline more slowly than if a closed boundary were found. Flow changes or the addition of new wells cause additional pressure drops that affect both the pressure decline and the pressure distribution. Each well will establish its own drainage area that supplies fluid. When a flow boundary is found, the pressure gradient—not the pressure level—tends to stabilize after sufficiently long production time. For the closed boundary case, the pressure reaches the pseudosteady state with a constant pressure gradient and general pressure drop everywhere, which is linear over time. For constant-pressure boundaries, steady state is obtained; both the pressure and its gradient remain constant over time.

1.12. The pressure derivative function

Pressure derivative has been one of the most valuable tools ever introduced to the pressure transient analysis field. In fact, [32] affirms that pressure derivative and deconvolution have been the best elements added for well test interpretation. However, here it is affirmed that besides these two "blessings," *TDS* technique, [1, 2], is the best and practical well test interpretation method in which application will be very devoted along this textbook. Actually, in the following chapters, *TDS* is extended for long, homogeneous reservoirs, [33], interference testing [34], drainage area determination in constant-pressure reservoirs, [35], and recent applications on fractured vertical wells, [36], among others. More complex scenarios, for instance finite-conductivity faults, [37], are treated extensively in [38].

Attempts to introduce the pressure derivative are not really new. Some of them try to even apply the derivative concept to material balance. Just to name a few of them, [39] in 1961, tried to approach the rate of pressure change with time for detection of reservoir boundaries. Later, in 1965, [40] presented drawdown curves of well pressure change with time for wells near intersecting faults (36 and 90°). These applications, however, use numerical estimations of the pressure rate change on the field data regardless of two aspects: (1) an understanding of the theoretical situation behind a given system and (2) noise in the pressure data.

Between 1975 and 1976, Tiab's contributions on the pressure derivative were remarkable. Actually, he is the father of the pressure derivative concept as used nowadays. Refs. [41, 42] include detailed derivation and application of the pressure derivative function. These results are further summarized on [41–45]. Ref. [46] applied Tiab's finding to provide a type-curve matching technique using the natural logarithm pressure derivative.

It was required to obtain the pressure derivative from a continuous function, instead of attempting to work on discrete data in order to understand the pressure derivative behavior in an infinite system. Then, Tiab decided to apply the Leibnitz's rule of derivation of an integral to the *Ei* function.

$$\frac{\partial}{\partial x}\int_{f(x)}^{h(x)} g(u)du\left\{g[h(x)]\frac{\partial[h(x)]}{\partial x}-g[f(x)]\frac{\partial[f(x)]}{\partial x}\right\} \quad (1.138)$$

Applying Leibnitz's rule to the *Ei* function in Eq. (1.81) to differentiate with respect to t_D (see Appendix B in [42]),

$$\frac{\Delta}{\Delta t_D}\left[E_i\left(\frac{-r_D^2}{4t_D}\right)\right] = -\int_{\frac{r_D^2}{4t_D}}^{\infty}\frac{e^{-u}}{u}\Delta u = -\frac{e^{-u}}{u}\frac{\Delta u}{\Delta t_D}\Bigg|_{\frac{r_D^2}{4t_D}}^{\infty} \tag{1.139}$$

Taking the derivative $\Delta u/\Delta t_D$ and replacing u by $r_D^2/4t_D$,

$$\frac{\Delta}{\Delta t_D}\left[E_i\left(-\frac{r_D^2}{4t_D}\right)\right] = \frac{e^{-\left(r_D^2/4t_D\right)}}{r_D^2/4t_D}\left(-\frac{r_D^2}{4t_D^2}\right) \tag{1.140}$$

After simplification,

$$\frac{\Delta}{\Delta t_D}\left[E_i\left(-\frac{r_D^2}{4t_D}\right)\right] = -\frac{1}{t_D}e^{-\left(r_D^2/4t_D\right)} \tag{1.141}$$

From inspection of Eq. (1.81) results:

$$\frac{\partial P_D}{\partial t_D} = -\frac{1}{2}\frac{1}{t_D}e^{-\frac{r_D^2}{4t_D}} \tag{1.142}$$

In oilfield units,

$$\frac{\Delta P_{wf}}{\Delta t} = \frac{70.6q\mu B}{kht}e^{\left(-\frac{948\phi\mu c_t}{kt}\right)} \tag{1.143}$$

At the well, $r_D = 1$, then, Eq. (1.142) becomes:

$$P_D' = \frac{1}{2t_D}e^{-\frac{1}{4t_D}} \tag{1.144}$$

For $t_D > 250$, $e^{-1/4tD} = 1$; then, Eq. (1.144) reduces to

$$P_D' = \frac{1}{2t_D} \tag{1.145}$$

The derivative of equation (1.145) is better known as the Cartesian derivative. The natural logarithmic derivative is obtained from:

$$t_D{}^*P_D' = t_D\frac{\partial P_D}{\partial t_D} = t_D\frac{\partial P_D}{(\partial \ln t_D)/t_D} = \frac{\partial P_D}{\partial \ln t_D} \tag{1.146}$$

Later on, [46] use the natural logarithmic derivative to develop a type-curve matching technique.

Appendix C in [42] also provides the derivation of the second pressure derivative:

$$\partial P_D'' = P_D'\frac{1}{t_D}\left(\frac{r_D^2}{4t_D} - 1\right) \tag{1.147}$$

Conversion of Eq. (1.145) to natural logarithmic derivative requires multiplying both sides of it by t_D; then, it results:

$$t_D{}^*P_D' = \frac{1}{2} \tag{1.148}$$

Eq. (1.148) suggests that a log-log plot of dimensionless pressure derivative against dimensionless time provides a straight line with zero slope and intercept of ½. Taking logarithm to both sides of Eq. (1.145) leads to:

$$\log P_D{}' = -\log t_D - 0.301 \tag{1.149}$$

$$P_{wf}{}' = \frac{\partial P_{wf}}{\partial t} = \frac{1}{t}\left(\frac{70.6q\mu B}{kh}\right) \tag{1.150}$$

The above expression corresponds of a straight line with negative unit slope. In dimensional form:

Taking logarithm to both sides of the above expression:

$$\log P_{wf}{}' = -\log t + \log\left(\frac{70.6q\mu B}{kh}\right) \tag{1.151}$$

As shown in **Figure 1.15**, Eq. (1.151) corresponds to a straight line with negative unit slope and intercept of:

$$P'_{1hr} = \frac{70.6q\mu B}{kh} \tag{1.152}$$

Eq. (1.152) is applied to find permeability from the intersect plot of the Cartesian pressure derivative versus time plot. This type of plot is also useful to detect the presence of a linear boundary (fault) since the negative unit slope line displaces when the fault is felt as depicted in **Figure 1.16**.

The noise that occurs in a pressure test is due to such factors as (1) turbulence, (2) tool movements, (3) temperature variations, (4) opening and closing wells in the field, and (5) gravitational effects of the sun and moon on the tides (near the great lakes the noise is about 0.15 psia and offshore up to 1 psia).

The estimation of the pressure derivative with respect to time to actual data, of course, must be performed numerically since data recorded from wells are always discrete. During the derivative calculation, the noise is increased by the rate of change that the derivative imposes, so it is necessary to soften the derivative or to use smoothing techniques. The low resolution of the tool and the log-log paper also increase or exaggerate the noise. Therefore, calculating the derivative of pressure

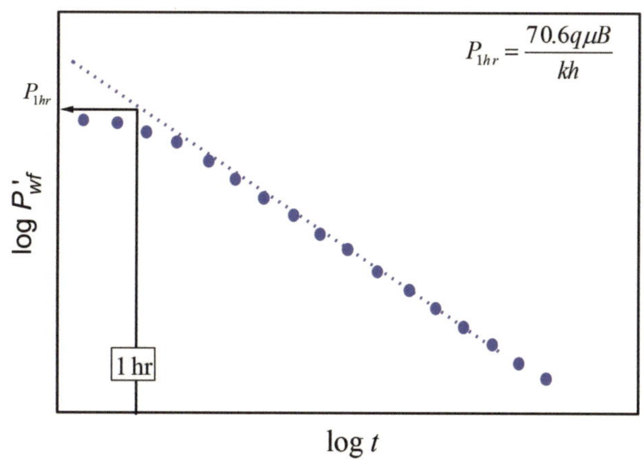

Figure 1.15.
Log-log plot of P_{wf}' *against* t.

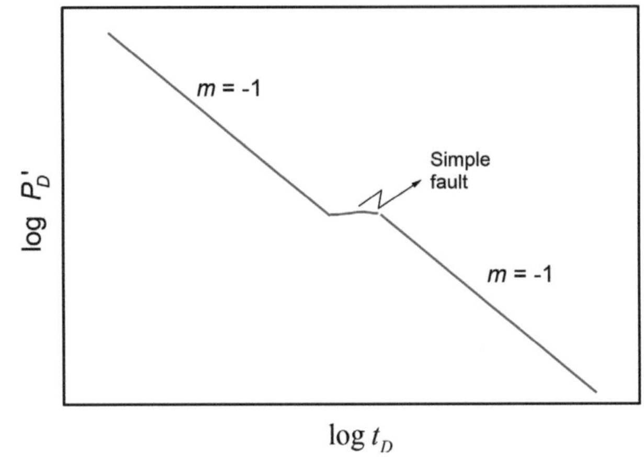

Figure 1.16.
Fault identification by means of a log-log plot of P_D' *vs.* t_D.

requires some care because the process of data differentiation can amplify any noise that may be present. Numerical differentiation using adjacent points will produce a very noisy derivative, [8, 47, 48].

Ref. [8] conducted a comparative study of several algorithms for estimation of the pressure derivative. They obtained synthetic pressure derivatives for seven different reservoir and well configuration scenarios and, then, estimated the pressure derivative using several comparative methods. They found that the Spline algorithm (not presented here) is the best procedure to derive pressure versus time data since it produces minimal average errors. It is the only algorithm of polynomial character that to be continuous can be smoothed during any derivation process and the form of the curve obtained is in agreement with the worked model. The Horne and Bourdet algorithms when the smoothing window is of either 0.2 or 0.4 are good options for derivation processes. Ref. [8] also found the best procedure for data analysis of pressure against time is to differentiate and then smooth the data.

By itself, the central finite difference formula fails to provide good derivative computation. Instead, some modifications are introduced by [18, 20, 46], respectively:

Horne equation [32]:

$$t\left(\frac{\partial P}{\partial t}\right)_i = t\left(\frac{\partial P}{\partial \ln t}\right)_i = \left\{\begin{array}{l} \dfrac{\ln(t_i/t_{i-k})\Delta P_{i+j}}{\ln(t_{i+j}/t_i)\ln(t_{i+j}/t_{i-k})} + \dfrac{\ln(t_{i+j}t_{i-k}/t_i^2)\Delta P_i}{\ln(t_{i+1}/t_i)\ln(t_i/t_{i-1})} - \\ \dfrac{\ln(t_{i+j}/t_i)\Delta P_{i-1}}{\ln(t_i/t_{i-k})\ln(t_{i+j}/t_{i-k})} \end{array}\right\} \tag{1.153}$$

$\ln t_{i+j} - \ln t_i \geq 0.2$ and $\ln t_i - \ln t_{i-k} \geq 0.2$

When the data are distributed in a geometrical progression (with the time difference from one point to the next much larger as the test passes), then the noise in the derivative can be reduced using a numerical differentiation with respect to the logarithm of time. The best method to reduce noise is to use data that is separated by at least 0.2 logarithmic cycles, rather than points that are immediately adjacent. This procedure is recognized as smoothing and is best explained in **Figure 1.17**.

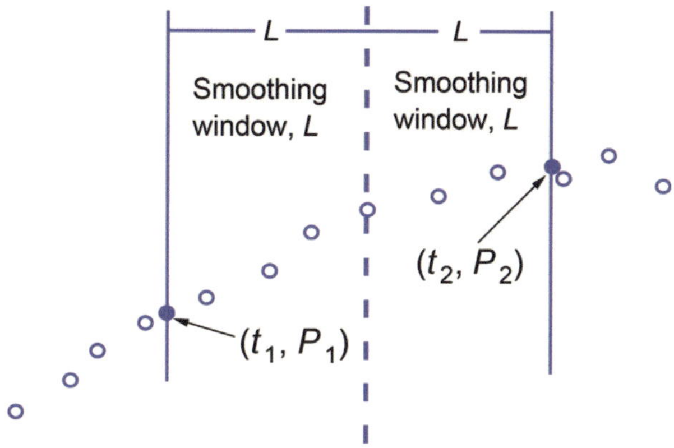

Figure 1.17.
Smoothing diagram.

Equation of Bourdet et al. [46]:

$$\left(\frac{dP}{dx}\right)_i = \frac{\frac{P_i - P_{i-1}}{X_i - X_{i-1}}(X_{i+1} - X_i) + \frac{P_{i+1} - P_i}{X_{i+1} - X_i}(X_i - X_{-1i})}{X_{i+1} - X_{i-1}} \tag{1.154}$$

Let X is the natural logarithm of the time function.

This differentiation algorithm reproduces the test type curve over the entire time interval. It uses a point before and a point after the point of interest, i, to calculate the corresponding derivative and places its weighted mean for the objective point. Smoothing can also be applied.

1.13. The principle of superposition

This principle is not new. It was first introduced to the petroleum literature by van Everdingen and Hurst in 1949, [49]. However, its application is too important and many field engineers fail or neglect to use it. Superposition is too useful for systems having one well producing at variable rate or the case when more than one well produces at different flow rates.

As quoted from [25], the superposition principle is defined by:

"*Adding solutions to the linear differential equation will result in a new solution of that differential equation but for different boundary conditions,*" which mathematically translates to:

$$\psi = \psi_1 f_1 + \psi_2 f_2 + \psi_3 f_3 \dots \tag{1.155}$$

where ψ is the general solution and $\psi_1 f_1$, $\psi_2 f_2$ and $\psi_3 f_3$... are the particular solutions.

1.13.1 Space superposition

If the wells produce at a constant flow rate, the pressure drop at point N, **Figure 1.18**, will be [3, 9, 19, 21, 25]:

$$\Delta P_N = \Delta P_{N,1} + \Delta P_{N,2} + \Delta P_{N,3} \tag{1.156}$$

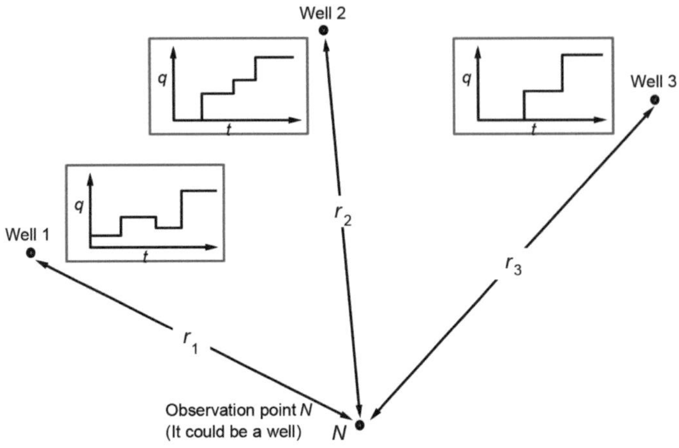

Figure 1.18.
Pressure at the point N.

If reservoir and fluid properties are considered constant, then, Eq. (1.87) can be applied to the above expression, so that:

$$\Delta P_N = \frac{141.2\mu}{kh}\left[(qB_o)_1 P_D(r_{D1}, t_D) + (qB_o)_2 P_D(r_{D2}, t_D) + (qB_o)_3 P_D(r_{D3}, t_D)\right] \tag{1.157}$$

The dimensionless radii are defined by:

$$r_{Dn} = \frac{r_n}{r_w};\ n = 1, 2, 3 \tag{1.158}$$

Extended to n number of wells:

$$\Delta P_N = \sum_{i=1}^{n} \frac{141.2q\mu B}{kh}[P_D(r_{Di}, t_D)] \tag{1.159}$$

If point N is an active well, its contribution to the total pressure drop plus the skin factor pressure drop, Eq. (1.108), must be included in Eq. (1.159), then,

$$\Delta P_N = \sum_{i=1}^{n} \frac{141.2q\mu B}{kh}[P_D(r_{DNi}, t_D)] + \frac{141.2q\mu B}{kh} s|_N \tag{1.160}$$

Notice that in Eqs. (1.159) and (1.160), changes of pressures or dimensionless pressures are added. If the point of interest is a well in operation, the damage factor should be added to the dimensionless pressure of that well only.

1.13.2 Time superposition

Sometimes there are changes in flow rate when a well produces as referred in **Figures 1.19** and **1.22**. Then, the superposition concept must be applied. To do this, [25], a single well is visualized as if there were two wells at the same point, one with a production rate of q_1 during a time period from t = 0 to t and another imaginary well with a production rate of $q_2 - q_1$ for a time frame between t_1 and $t - t_1$. The total rate after time t_1 is $q_1 + (q_2 - q_1) = q_2$. The change in well pressure due to the rate change [19, 25] is,

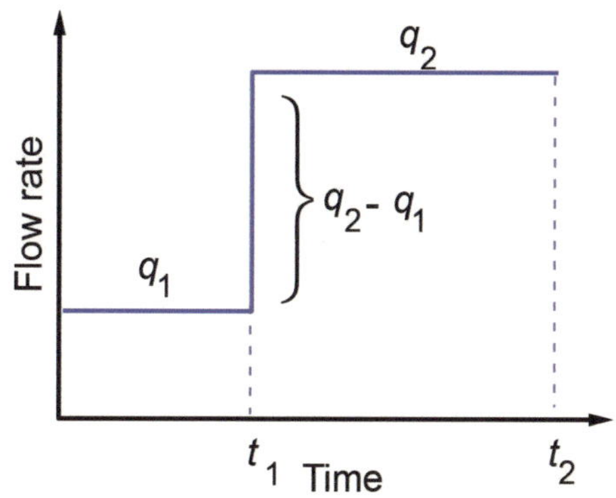

Figure 1.19.
Time superposition.

$$\Delta P = \frac{141.2\mu B}{kh}\left[q_1 P_D(r_D, t_{D1}) + (q_2 - q_1)P_D(r_D, t_{D2} + s)\right] \tag{1.161}$$

where $t_{D2} = (t-t_1)_D$. If there are more variations in flow rate,

$$\Delta P = \frac{141.2\mu}{kh}\sum_{i=1}^{n}\left[(qB)_i - (qB)_{i-1}\right]\left\{P_D(r_D, (t - t_i)_D + s\right\} \tag{1.162}$$

Example 1.4

This example is taken [25]. The below data and the schematic given in **Figure 1.20** correspond to two wells in production:

k = 76 md, ϕ = 20 %, B = 1.08 bbl/STB
P_i = 2200 psia, μ = 1 cp, $c_t = 10 \times 10^{-6}$/psia
h = 20 ft

Calculate the pressure in (a) well 1 after 7 h of production and (b) in well 2 after 11 h of production. Assume infinite behavior.

Solution

Part (a):

$\Delta P_{(7\ \mathrm{hr})} = \Delta P$ caused by production from well 1 to well 1 + ΔP caused by production from well 2 to well 1. Mathematically,

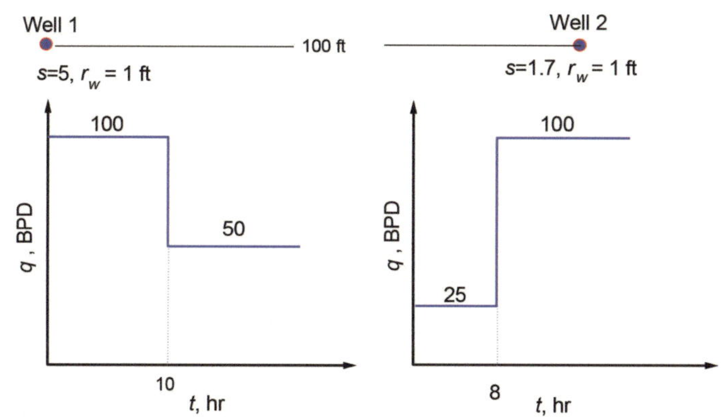

Figure 1.20.
Flow rate changes for example 1.4.

$$\Delta P_{7\text{hr @ well1}} = \frac{141.2\mu q_1 B}{kh}[P_D(r_{D1}, t_D) + s] + \frac{141.2\mu q_2 B}{kh}(P_D(r_{D2}, t_D))$$

Using Eq. (1.101) for the well,

$$x = -\frac{948\phi\mu c_t r^2}{kt} = \frac{948(0.2)(1)(1 \times 10^{-5})1^2}{(76)(7)} = 3.56 \times 10^{-6}$$

Since $x <<<< 0.0025$, it implies the use of Eq. (1.82) with Eq. (1.103); then,

$$P_D(r_D, t_D) = -\frac{1}{2}\ln(1.781x) \tag{1.163}$$

$$P_D = \frac{1}{2}\left|\left[\ln\left(1.781^{*}3.56 \times 10^{-6}\right)\right]\right| = 5.98$$

In well 2, $x = 0.03564$ from Eq. (1.101). Interpolating this value in **Table 1.2**, $Ei(-x) = 2.7924$; then, $P_D \cong 1.4$. Estimating ΔP in well 1 will result:

$$\Delta P_{7hr, r_D=1} = \frac{141.2(100)(1.08)(1)}{(76)(20)}(5.98 + 5) + \frac{141.2(100)(1.08)(1)}{(76)(20)}(1.4) = 113.7$$

$P_{wf\ @\ well1} = 2200 - 113.7 = 2086.4$ psia (notice that skin factor was only applied to well 1)

Part (b);

At 11 h, it is desired to estimate the pressure in well 2. Two flow rates should be considered for in each well. Then, the use of Eq. (1.162) will provide:

$$\Delta P_{(11\text{ hr,well 2})} = \Delta P_{\text{well 1}\rightarrow\text{Well 2},t=11\text{ hr},q=100\text{ BPD},r_D=100} + \Delta P_{\text{well 1}\rightarrow\text{Well 2},t=(11-10)\text{ hr},q=(50\text{-}100)\text{ BPD},r_D=100}$$
$$\Delta P_{\text{well 1}\rightarrow\text{Well 1},t=11\text{ hr},q=25\text{ BPD},r_D=1,s_2} + \Delta P_{\text{well 1}\rightarrow\text{Well 1},t=(11-8)\text{ hr},q=(100\text{-}25)\text{ BPD},r_D=1,s_2}$$

Using Eq. (1.101), the four respective values of x are: $x = 0.02268$, 0.2494, 2.268×10^{-6}, and 8.316×10^{-6}. Estimation of Ei requires the use of **Table 1.2** for the first two values and use of Eq. (1.103) for the last two values. The four values of $Ei(-x)$ are: 0.0227, 0.811, 12.42, and 11.12. Therefore, the respective values of P_D are 1.605, 0.405, 6.209, and 5.56. The total pressure drop is found with Eq. (1.161) as follows:

$$\Delta P_{\text{well 2, 11 hr})} = \frac{141.2(1)(1.08)}{(76)(20)}\left\{\begin{array}{l}(100)(1.605) + (50 - 100)(0.405) + \\ (25)(6.209 + 1.7) + (100 - 25)(5.56 + 1.7)\end{array}\right\} = 87.75 \text{ psia}$$

$P_{wf\ @\ well2} = 2200 - 87.75 = 2112.25$ psia

1.13.3 Space superposition—method of images

The method of images applies to deal with either no-flow or constant-pressure boundaries. If a well operates at a constant flow rate at a distance, d, from an impermeable barrier (fault), the systems acts as if there were two wells separated $2d$ from each other [3, 25]. For no-flow boundaries, the image well corresponds to the same operating well. For constant-pressure boundary, the resulting image corresponds to an opposite operating well. In other words, if the well is a producer near a fault, the image well corresponds to an injector well. These two situations are sketched in **Figure 1.21**. For the no-flow boundary, upper system in **Figure 1.21**, the dimensionless pressure can be expressed as:

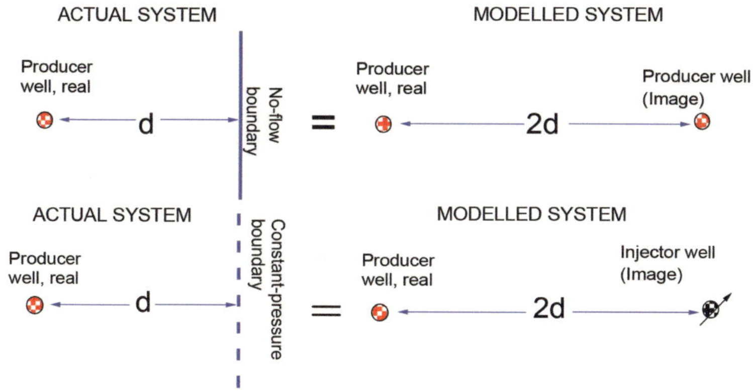

Figure 1.21.
Well near a linear barrier.

$$P_{D\text{ at real well}} = P_{D\text{ at real well, }r_D=1,s} + P_{D\text{ at image well}\rightarrow\text{real well, }r_D=2d/r_w} \tag{1.164}$$

For the constant-pressure boundary, lower part in **Figure 1.21**, the dimensionless pressure can be expressed as:

$$P_{D\text{ at real well}} = P_{D\text{ at real well, }r_D=1,s} - P_{D\text{ at image well}\rightarrow\text{real well, }r_D=2d/r_w} \tag{1.165}$$

The negative sign in Eq. (1.165) is because of dealing with an imaginary injector well.

For the case of two intersecting faults, the total number of wells depends on the value of the angle formed by the two faults, thus:

$$n_{\text{wells}} = \frac{360}{\theta} \tag{1.166}$$

The image method is limited to one well per quadrant. If this situation fails to be fulfilled, then, the method cannot be applied. In the system of **Figure 1.22**, an angle of 90° is formed from the intersecting faults. According to Eq. (1.166), n_{wells} = 360/90 = 4 wells, as shown there. The ratio of the distances from the well to each fault is given by:

$$y_D = b_y/b_x \tag{1.167}$$

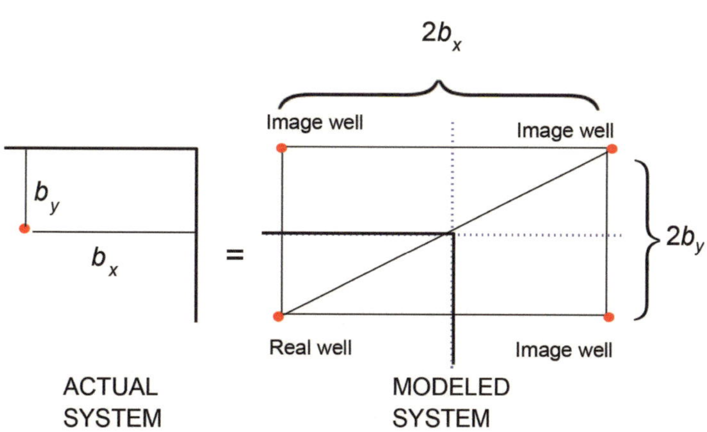

Figure 1.22.
Well between two intersecting faults.

The practical way to apply space superposition for generating the well system resulting from two intersecting faults consist of extending the length of the faults and setting as many divisions as suggested by Equation (1.166); that is, for example, 1.5, **Figure 1.23** left, six well spaces are obtained. Then, draw a circle with center at the fault intersection and radius at well position. This guarantees that the total length corresponds to the double length value from the well to the fault. Draw from the well a line to be perpendicular to the nearest fault and keep drawing the line until the circle line has been reached. See **Figure 1.24** left. Set the well. A sealing fault provides the same type of well as the source well, that is, a producing well generates another producing well to the other side of the fault. A constant-pressure boundary provides the opposite well type of the source well, that is, a producing well generates an injector well on the other side of the line. Draw a new line from the just drawn imaginary well normal to the fault and keep drawing the line until the line circle is reached. See **Figure 1.24** right. Repeat the procedure until the complete well set system has been drawn.

For more than six well spaces generated, that is angles greater than 60°, as the case of example 1.5, when a fault intersects a constant-pressure boundary injector and producer imaginary wells ought to be generated. What type of line should be drawn? A solid line representing a sealing fault, or a dash line, representing a constant-pressure boundary? The answer is any of both. The lines should be

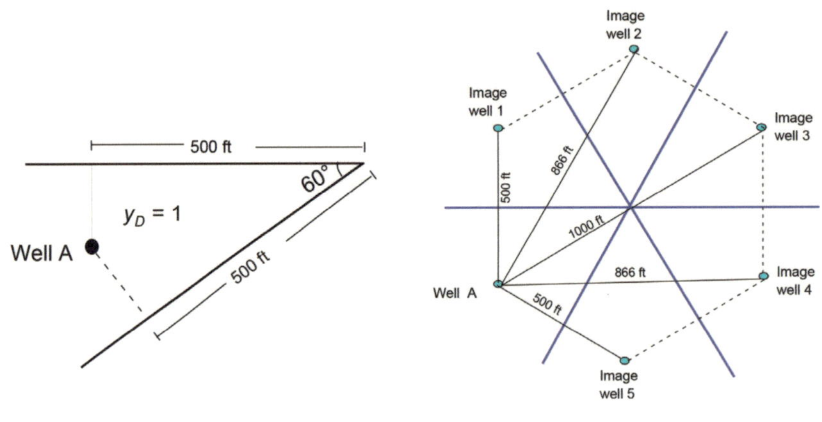

Figure 1.23.
Location of well A and resulting well number system for example 1.5.

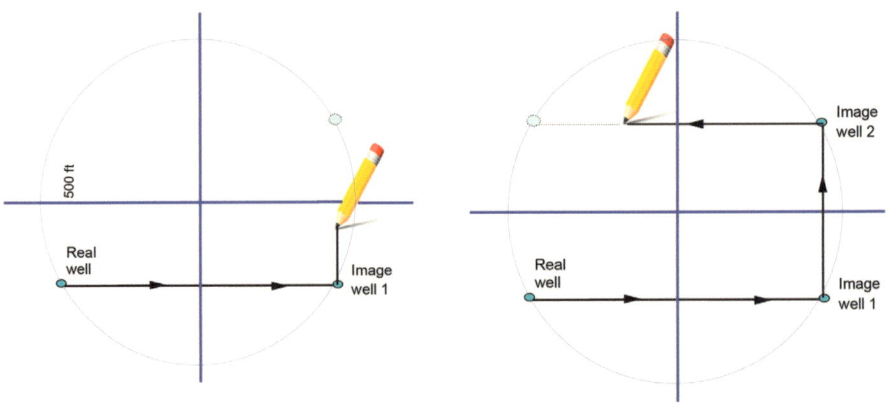

Figure 1.24.
Generating the well system for two intersecting faults.

drawn alternatively and as long as the system closes correctly, superposition works well.

Example 1.5

Well A in **Figure 1.23** has produced a constant rate of 380 BPD. It is desired (a) to estimate the well-flowing pressure after one week of production. The properties of the reservoir, well and fluid are given as follows:

P_i = 2500 psia,	B = 1.3 bbl/STB,	μ = 0.87 cp
h = 40 ft,	$c_t = 15\times10^{-6}$/psia,	ϕ = 18 %
r_w = 6 in,	k = 220 md, s = −5	

(b) What would be the well-flowing pressure after a week of production if the well were in an infinite reservoir?

Solution

Part (a)

The pressure drop in well A is affected by its own pressure drop and pressure drop caused by its well images. The distance from well A to its imaginary wells is shown in **Figure 1.23** (right-hand side). The total pressure drop for well A is:

$$\Delta P_A = \Delta P_{A,r=rw} + \Delta P_{image\ 1\to \text{well}\ A,r=500\ ft} + \Delta P_{image\ 2\to \text{well}\ A,r=866\ ft} + \Delta P_{image\ 3\to \text{well}\ A,r=1000\ ft} + \Delta P_{image\ 4\to \text{well}\ A,r=866\ ft} + \Delta P_{image\ 5\to \text{well}\ A,r=500\ ft}$$

By symmetry, the above expression becomes:

$$\Delta P_A = \Delta P_{A,r=rw} + 2\Delta P_{image\ 1\to \text{well}\ A,r=500\ ft} + 2\Delta P_{image\ 2\to \text{well}\ A,r=866\ ft} + \Delta P_{image\ 3\to \text{well}\ A,r=1000\ ft}$$

Using Eq. (1.101) for the well:

$$\frac{948(0.18)(0.87)(1.5\times10^{-5})(0.5)^2}{(220)(168)} = 1.5\times10^{-8}$$

Since $x <<<< 0.0025$, Eq. (1.163) applies:

$$P_D(r_D,t_D) = -\frac{1}{2}\left|\ln\left[1.781(1.5\times10^{-8})\right]\right| = 8.72$$

Estimation for the image wells are given below. In all cases, x > 0.0025, then, **Table 1.2** is used to find *Ei* and the resulting below divided by 2 for the estimation of P_D,

$$x_{\text{image well 1 or 5}} = \frac{948(0.18)(0.87)(1.5\times10^{-5})(500)^2}{(220)(168)} = 0.015,\ P_D = 1.816$$

$$x_{\text{image well 2 or 4}} = \frac{948(0.18)(0.87)(1.5\times10^{-5})(866^2)}{(220)(168)} = 0.0452,\ P_D = 1.282$$

$$x_{\text{image well 3}} = \frac{948(0.18)(0.87)(1.5\times10^{-5})(1000^2)}{(220)(168)} = 0.06,\ P_D = 1.145$$

Then, the pressure drop in A will be:

$$\Delta P_A = 141.2\frac{q\mu B}{kh}\left[(P_{DA,r=rw} + s) + 2P_{D\text{ image 1},r=500ft} + 2P_{D\text{ image 2},r=866ft} + P_{D\text{ image 3},r=1000ft}\right]$$

$$\Delta P_A = \frac{141.2(380)(0.87)(1.3)}{(220)(40)}[(8.72-5)+2(1.816)+2(1.282)+1.145] = 76.3\text{ psia}$$

$P_{wf\text{ @ well }A}$ = 2500 – 76.3 = 2423.7 psia
Part (b)
If the well were located inside an infinite reservoir, the pressure drop would not include imaginary wells, then:

$$\Delta P_A = 141.2\frac{q\mu B}{kh}[P_{D,r=rw} + s]$$

$$\Delta P_A = \frac{141.2(380)(0.87)(1.3)}{(220)(40)}[8.72-5] = 25.63\text{ psia}$$

The well-flowing pressure would be (2500 – 25.3) = 2474.4 psia. It was observed that the no-flow boundaries contribute with 66.4% of total pressure drop in well A.

Nomenclature

A	area, ft^2 or Ac
B_g	gas volume factor, ft^3/STB
B_o	oil volume factor, bbl/STB
B_w	oil volume factor, bbl/STB
b_x	distance from closer lateral boundary to well along the x-direction, ft
b_y	distance from closer lateral boundary to well along the y-direction, ft
c	compressibility, 1/psia
cf	pore volume compressibility, 1/psia
c_t	total or system compressibility, 1/psia
d	distance from a well to a fault, ft
f	a given function
h	formation thickness, ft
k	permeability, md
k_s	permeability in the damage zone, md
k_{rf}	phase relative permeability, f = oil, water or gas
L	reservoir length, ft
m	slope
$m(P)$	pseudopressure function, $psia^2$/cp
M	gas molecular weight, lb/lbmol
P	pressure
dP/dr	pressure gradient, psia/ft
P_D'	dimensionless pressure derivative
P_D''	dimensionless second pressure derivative
P_D	dimensionless pressure
P_i	initial reservoir pressure, psia
P_{wf}	well flowing pressure, psia
q	flow rate, bbl/D. For gas reservoirs the units are Mscf/D
R_s	gas dissolved in crude oil, SCF/STB
R_{sw}	gas dissolved in crude water, SCF/STB
r_D	dimensionless radius

r_{De}	dimensionless drainage radius = r_e/r_w
r	radial distance, radius, ft
r_e	drainage radius, ft
rs	radius of the damage zone, ft
r_w	well radius, ft
S_f	fluid saturation, f = oil, gas or water
s	skin factor
T	reservoir temperature, °R
t	time, h
t_a	pseudotime, psia h/cp
t_o	dummy time variable
u_r	radial flow velocity, ft/h
t_D	dimensionless time based on well radius
t_{DA}	dimensionless time based on reservoir area
$t_D{}^*P_D{}'$	logarithmic pressure derivative
V	volume, ft^3
z	vertical direction of the cylindrical coordinate, real gas constant

Greek

Δ	change, drop
Δt	shut-in time, h
ϕ	porosity, fraction. Spherical coordinate
λ	phase mobility, md/cp
η	hydraulic diffusivity constant, md-cp/psia
ρ	density, lbm/ft^3
θ	cylindrical coordinate
μ	viscosity, cp
ζ	time function

Suffices

1 hr	reading at time of 1 h
D	dimensionless
DA	dimensionless with respect to area
f	formation
g	gas
i	initial conditions
o	oil, based condition
w	well, water
p	pore

References

[1] Tiab D. Analysis of pressure and pressure derivative without type-curve matching: 1-skin and wellbore storage. Journal of Petroleum Science and Engineering; 1995;**12**:171-181

[2] Tiab D. Analysis of pressure and pressure derivatives without type-curve matching: I-skin and wellbore storage. Society of Petroleum Engineers. 1993. DOI: 10.2118/25426-MS

[3] Lee J, Rollins JB, Spivey JP. Pressure Transient Testing. Society of Petroleum Engineers. SPE Textbook. Vol. 9. Richardson, TX; 2003. 336 p

[4] Sabet M. Well Testing. Houston, TX: Gulf Publishing Co.; 1991. 460 p

[5] Slider HC. Wordlwide Practical Petroleum Reservoir Engineering Methods. Revised ed. Tulsa, Oklahoma, USA: PennWell Publishing Co.; 1983. 826 p

[6] Stanislav JF, Kabir CS. Pressure Transient Analysis. New Jersey: Prentice Hall; 1990. 320 p

[7] Strelsova TD. Well Testing in Heterogeneous Formations. New York: John Wiley and Sons; 1988. 377 p

[8] Escobar FH, Navarrete JM, Losada HD. Evaluation of pressure derivative algorithms for well-test analysis. Society of Petroleum Engineers. 2004. DOI: 10.2118/86936-MS

[9] Mathews CS, Russell DG. Pressure buildup and flow tests in wells. SPE Monograph Series. 1967;**1**:163

[10] Odeh AS, Jones LG. Pressure drawdown analysis, variable-rate case. Society of Petroleum Engineers. 1965: 960-964. DOI: 10.2118/1084-PA

[11] Russell DG. Determination of formation characteristics from two-rate flow tests. Society of Petroleum Engineers. 1963:1347-1355. DOI: 10.2118/645-PA

[12] Doyle RE, Sayegh EF. Real gas transient analysis of three-rate flow tests. Society of Petroleum Engineers. 1970:1347-1356. DOI: 10.2118/2607-PA

[13] Jacob CE, Lohman SW. Nonsteady flow to a well of constant drawdown in an extensive aquifer. Transactions, American Geophysical Union. 1952; **33**(4):559-569. https://doi.org/10.1029/TR033i004p00559

[14] Joseph JA. Unsteady-state cylindrical, spherical and linear flow in Porous Media [Ph.D. dissertation]. University of Missouri-Rolla; 1984

[15] Kucuk F, Brigham WE. Transient flow in elliptical systems. SPE Journal. 1979:401-410. DOI: 10.2118/7488-PA

[16] Dykstra H, Kazemi H, Raghavan R, Gulati MS. Pressure Transient Testing Methods, SPE Reprint Series No. 14. Dallas, Texas: Society of Petroleum Engineers; 1967. 264 p

[17] Perrine RL. Analysis of Pressure-buildup Curves. American Petroleum Institute, Drilling and Production Practices, API, Washington, D.C.; 1956. pp. 482-509. ID API-56-482

[18] Horne RN. Modern Well Test Analysis: A Computer-aided Approach. Palo Alto, CA: Petroway Inc.; 1990. 185 p

[19] Dake LP. The Practice of Reservoir Engineering (Revised edition). Amsterdam, Netherlands: Elsevier Developments in Petroleum Science (Second Printing); 2004. 544 p

[20] Al-Hussainy R, Ramey HJ Jr, Crawford PB. The flow of real gases through porous media. Journal of

Petroleum Technology. 1966;**237**: 624-636. DOI: 10.2118/1243-A-PA

[21] Lee J. Well Testing. SPE Textbook Series. Society of Petroleum Engineers, Richardson, TX. Vol. 1. 1982. 159 p

[22] Lee J, Wattenberger RA. Gas Reservoir Engineering. Society of Petroleum Engineers. SPE Textbook Series No. 05. Second Printing. Richardson, TX; 2002. 349 p

[23] Agarwal RG. Real gas pseudo-time—A new function for pressure buildup analysis of MHF gas wells. Society of Petroleum Engineers. 1979. DOI: 10.2118/8279-MS

[24] Civan F. PE-5143: Fluid Flow in Porous Media. Lecture Notes. Fall 1993. The University of Oklahoma, Norman, OK; 1993

[25] Earlougher RC Jr. Advances in Well Test Analysis. Monograph Series. Vol. 5, Dallas, TX: SPE; 1977

[26] Horner DR. Pressure buildup in wells. In: Proceedings of the Third World Petroleum Congress. Section II, Preprint 7. Leiden, Netherlands; 1951. pp. 25-43

[27] Nisle RG. How to use the exponential integral. Petroleum Engineering. 1956:171-173

[28] Tracy GW. Pressure Analysis Methods, SPE Reprint Series No. 9. Society of Petroleum Engineers of AIME, Richardson, TX. 1967

[29] Sprott JC. Numerical recipes routines and examples in BASIC. In: Companion Manual to Numerical Recipes Routines and Examples in BASIC. 3rd ed. Cambridge, UK: Cambridge University Press; 2007. 1256 p

[30] Rhagavan R. Well Test Analysis. New Jersey: Prentice Hall; 1993. 558 p

[31] Earlougher RC, Ramey HJ. Interference Analysis in Bounded Systems Technology. Montreal: Petroleum Society of Canada; 1973. DOI: 10.2118/73-04-04

[32] Gringarten AC. From straight lines to deconvolution: The evolution of the state of the art in well test analysis. SPE Reservoir Evaluation and Engineering. 2008:41-62. DOI: 10.2118/102079-PA

[33] Escobar FH, Hernández YA, Hernández CM. Pressure transient analysis for long homogeneous reservoirs using TDS technique. Journal of Petroleum Science and Engineering. 2007;**58**:68-82. DOI: 10.1016/j.petrol.2006.11.010

[34] Escobar FH, Cubillos J, Montealegre M. Estimation of horizontal reservoir anisotropy without type-curve matching. Journal of Petroleum Science and Engineering 2008;**60**:31-38. DOI: 10.1016/j.petrol.2007.05.003

[35] Escobar F-H, Hernández YA, Tiab D. Determination of reservoir drainage area for constant-pressure systems using well test data. CT&F—Ciencia, Tecnologia y Futuro. 2010;**4**:51-62

[36] Escobar FH, Ghisays-Ruiz A, Bonilla LF. New model for elliptical flow regime in hydraulically-fractured vertical wells in homogeneous and naturally-fractured systems. ARPN Journal of Engineering and Applied Sciences. 2014;**9**:1629-1636

[37] Escobar FH, Martínez JA, Montealegre M. Pressure transient analysis for a reservoir with a finite-conductivity fault. CT&F—Ciencia, Tecnologia y Futuro. 2013;**5**:5-18

[38] Escobar FH. Recent Advances in Practical Applied Well Test Analysis. New York: Nova Science Publishers, Inc.; 2015. 422 p

[39] Jones P. Reservoir limit test on gas wells. Journal of petroleum Technology. 1962:613-619. DOI: 10.2118/24-PA

[40] Van Poolen HK. Drawdown curves give angle between–intersecting faults. The Oil & Gas Journal. 1965;**20**:71-75

[41] Tiab D. A new approach to detect and locate multiple reservoir boundaries by transient well pressure data [M.S. thesis]. Socorro, New Mexico: New Mexico Institute of Mining and Technology; 1975. 274 p

[42] Tiab D. Analysis of multiple-sealing-fault systems and closed rectangular reservoirs by type curve matching [PhD dissertation]. The University of Oklahoma; 1976. 343 p

[43] Tiab D, Crichlow HB. Pressure analysis of multiple-sealing-fault systems and bounded reservoirs by type-curve matching. SPE Journal. 1979: 378-392. DOI: 10.2118/6755-PA

[44] Tiab D, Kumar A. Application of the P_D' function to interference analysis. Journal of Petroleum Technology. 1980: 1465-1470. DOI: 10.2118/6053-PA

[45] Tiab D, Kumar A. Detection and location of two parallel sealing faults around a well. Journal of Petroleum Technology. 1980:1701-1708. DOI: 10.2118/6056-PA

[46] Bourdet D, Ayoub JA, Pirard YM. Use of pressure derivative in well test interpretation. Society of Petroleum Engineers. 1989:293-302. DOI: 10.2118/12777-PA

[47] Escobar FH, Alzate HD, Moreno L. Effect of extending the radial superposition function to other flow regimes. Journal of Engineering and Applied Sciences. 2013;**8**(8):625-634

[48] Escobar FH, Alzate HD, Moreno-Collazos L. Effect of Extending the Radial Superposition Function to Other Flow Regimes. Society of Petroleum Engineers; 2014. DOI: 10.2118/169473-MS

[49] Van Everdingen AF, Hurst W. The application of the Laplace transformation to flow problems in reservoirs. Society of Petroleum Engineers. 1949;**186**:305-324. DOI: 10.2118/949305-G

Chapter 2

Pressure Drawdown Testing

As can be seen in **Figure 1.4**, well pressure test analysis (PTA) considers this as the most basic and simple test, which does not mean that it is not important. In these tests, bottom-hole well-flowing pressure, P_{wf}, is continuously recorded keeping the flow constant. These tests are also referred as flow tests. Similar to an injection test, these tests require either production/injection from/into the well.

These tests are performed with the objective of (a) obtaining pore volume of the reservoir and (b) determining heterogeneities (in the drainage area). In fact, what is obtained is (*a*) transmissibility and (*b*) porous volume by total compressibility. In fact, a recent study by Agarwal [1] allows using drawdown tests to estimate the average permeability in the well drainage area. To run a pressure decline test, the following steps are generally followed:

- The well is shut-in for a long enough time to achieve stabilization throughout the reservoir, if this is not achieved, multirate testing is probably required;

- The recording pressure tool is lowered to a level immediately above the perforations. This is to reduce Joule-Thompson effects. It is important to have at least two pressure sensors for data quality control purposes;

- The well opens in production at constant flow and in the meantime the well-flowing pressure is continuously recorded.

Ideally, the well is closed until the static reservoir pressure. The duration of a drawdown test may last for a few hours or several days, depending upon the test objectives and reservoir characteristics. There are extensive pressure drawdown tests or reservoir limit tests (RLT) that run to delimit the reservoir or estimate the well drainage volume. Other objectives are the determination of: well-drainage area permeability, skin factor, wellbore storage coefficient (WBS), porosity, reservoir geometry, and size of an adjacent aquifer.

2.1. Wellbore storage coefficient

It is the continuous flow of the formation to the well after the well has been shut-in for stabilization. It is also called after-flow, postproduction, postinjection, loading, or unloading (for flow tests). The flow occurs by the expansion of fluids in the wellbore. In pressure buildup tests, after-flow occurs. **Figure 2.1** illustrates the above [2].

Traditional pressure tests had to be long enough to cope with both wellbore storage and skin effects so that a straight line could be obtained indicating the radial flow behavior. Even this approach has disadvantages since more than one apparent line can appear and analysts have problems deciding which line to use. In addition, the scale of the graph may show certain pressure responses as straight lines when in fact they are curves. To overcome these issues, analysts developed the method the type-curve matching method.

There is flow in the wellbore face after shutting-in the well in surface. Wellbore storage affects the behavior of the pressure transient at early times. Mathematically,

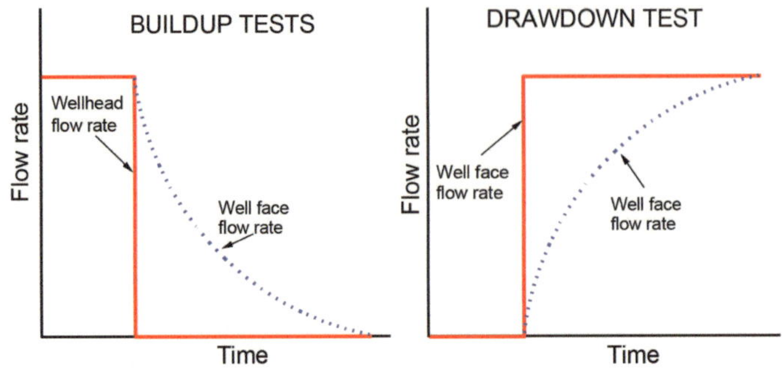

Figure 2.1.
Effects of wellbore storage on buildup and drawdown tests, taken from [2].

the storage coefficient is defined as the total volume of well fluids per unit change in bottom-hole pressure, or as the capacity of the well to discharge or load fluids per unit change in background pressure:

$$C = \frac{\Delta V}{\Delta P} \tag{2.1}$$

As commented by Earlougher [2], wellbore storage causes the flow rate at the face of the well to change more slowly than the surface flow rate. **Figure 2.2** schematizes the relation q_{sf}/q when the surface rate is changed from 0 to q, when $C = 0$, $q_{sf}/q = 1$, while for $C > 0$, the relation q_{sf}/q gradually changes from 0 to 1. The greater the value of C, the greater the transition is. As the storage effects become less severe, the formation begins to influence more and more the bottom-hole pressure until the infinite behavior is fully developed. Pressure data that are influenced by wellbore storage can be used for interpretation purposes since fluids unload or load has certain dependence on reservoir transmissibility; however, this analysis is risky and tedious. *TDS* technique, presented later in this chapter, can provide a better solution to this problem.

Typically, the flow rate is surface-controlled (unless there is a bottom shut-in tool), the fluids in the well do not allow an immediate transmission of the disturbance from the subsurface to the surface, resulting in uneven surface and wellbore face flow [2–7]. Wellbore storage can change during a pressure test in both injector and producer wells. Various circumstances cause changes in storage, such as phase redistribution and increase or decrease in storage associated with pressure

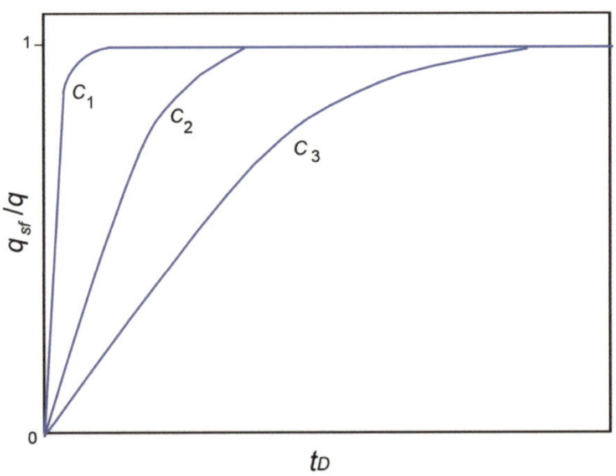

Figure 2.2.
Effect of storage on the flow rate at the face of the well, $C_3 > C_2 > C_1$, *taken from [2].*

tests in injector wells. In injector wells, once the well is closed, the surface pressure is high but could decrease to atmospheric pressure and go to vacuum if the static pressure is lower than the hydrostatic pressure. This causes an increase in storage (up to 100 times) of an incompressible system to one in a system where the liquid level drops [2]. The inverse situation occurs in injector wells with a high level of increase of liquid storage level and in producing wells with a high gas-oil ratio or by redissolution of the free gas. Both for increase or decrease of storage, the second storage coefficient determines the beginning of the semilogarithmic straight line.

When the relationship between ΔV and ΔP does not change during the test, the wellbore storage coefficient is constant and can be estimated from completion data [2–4].

$$C = \left(\frac{144}{\rho}\right) V_u \tag{2.2}$$

where V_u is the wellbore volume/unit length, bbl/ft, r is the density of the fluid in the wellbore, lbm/ft^3, and C is the wellbore storage coefficient, bbl/psia.

For injector wells or wells completely filled with fluids:

$$C = c_{wb} V_{wb} \tag{2.3}$$

where C_{wb} is the wellbore fluid compressibility = $1/P_{wb}$, V_{wb} is the total wellbore volume, and V_u can be estimated with internal casing, ID_{csg}, and external tubing, OD_{tbg}, diameters.

$$V_u = 0.0009714 (ID_{csg}^2 - OD_{tbg}^2) \tag{2.4}$$

When opening a well, see **Figure 2.3**, the oil production will be given by the fluid that is stored in the well, q_{sf} = 0. As time goes by, q_{sf} tends to q and storage is neglected and the amount of liquid in the wellbore will be constant. The net accumulation volume will be (assuming constant B) [3, 5]:

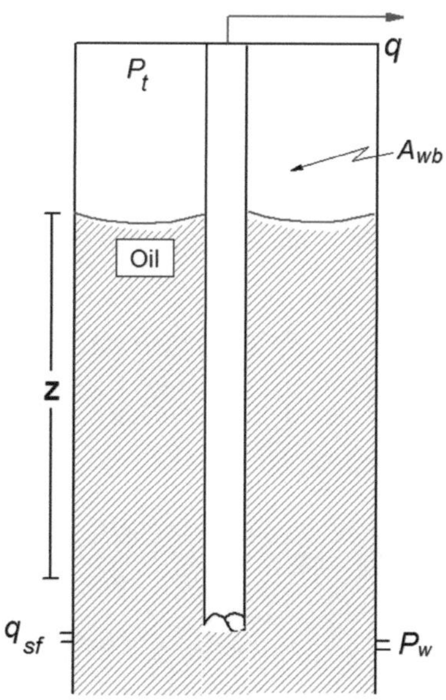

Figure 2.3.
Schematic representation of wellbore storage, taken from [3].

$$V_{wb} = A_{wb}(Z) \tag{2.5}$$

The flow rate is given by:

$$\frac{dV_{wb}}{dt} = A_{wb}\frac{dZ}{dt} \tag{2.6}$$

The rate of volume change depends upon the difference between the subsurface and surface rates:

$$\frac{dV_{wb}}{dt} = (q_{sf} - q)B = \frac{24}{5.615}A_{wb}\frac{dZ}{dt} \tag{2.7}$$

Since (assuming $g/g_c = 1$):

$$P_w - P_t = \frac{\rho Z}{144} \tag{2.8}$$

Taking the derivative to Eq. (2.8),

$$\frac{d}{dt}(P_w - P_t) = \frac{\rho}{144}\frac{dZ}{dt} \tag{2.9}$$

Combining Eqs. (2.7) and (2.9) will result:

$$(q_{sf} - q)B = \frac{24(144)}{5.615}\frac{A_{wb}}{\rho}\frac{d(P_w - P_t)}{dt} \tag{2.10}$$

Define

$$C = \frac{144}{5.615}\frac{A_{wb}}{\rho} \tag{2.11}$$

Assuming constant, P_t, replacing the definition given by Eq. (2.11) and solving for the wellbore face flow rate, q_{sf}, leads to:

$$q_{sf} = q + \frac{24C}{B}\frac{dP_w}{dt} \tag{2.12}$$

Taking derivative to Eqs. (1.89) and (1.94) with respect to time and taking the ratio of these will yield:

$$\frac{dP_w}{dt} = \left(-\frac{0.0373qB}{\phi\, h\, c_t\, r_w^2}\right)\frac{dP_D}{dt_D} \tag{2.13}$$

Combining Eqs. (2.12) and (2.13);

$$q_{sf} = q - \frac{0.894qC}{\phi\, c_t h r_w^2}\frac{dP_{wD}}{dt_D} \tag{2.14}$$

Defining the dimensionless wellbore storage coefficient;

$$C_D = \frac{0.894C}{\phi c_t h r_w^2} \tag{2.15}$$

Rewriting Eq. (2.14);

$$\frac{q_{sf}}{q} = 1 - C_D \frac{dP_{wD}}{dt_D} \tag{2.16}$$

The main advantage of using downhole shut-in devices is the minimization of wellbore storage effects and after-flow duration.

Rhagavan [5] presents the solution for the radial flow diffusivity equation considering wellbore storage and skin effects in both Laplace and real domains, respectively:

$$\overline{P}_D = \frac{K_0(\sqrt{u}) + s\sqrt{u}K_1(\sqrt{u})}{u\{\sqrt{u}K_1(\sqrt{u}) + C_D u[K_0(\sqrt{u}) + s\sqrt{u}K_1(\sqrt{u})]\}} \tag{2.17}$$

$$P_D = \frac{4}{\pi^2}\int_0^\infty \frac{1 - e^{-x^2 t_D}}{x^3\left\{[vC_D J_0(x) - f(x)J_1(x)]^2 + [xC_D Y_0(x) - f(x)Y_1(x)]^2\right\}} dv \tag{2.18}$$

where $f(x) = 1 - C_D(s)\, x^2$, and K_0, K_1, J_0, J_1, Y_0, and Y_1 are Bessel functions.

2.2. Well test interpretation methods

There exist four methods for well test interpretation as follows: (a) conventional straight-line, (b) type-curve matching, (c) regression analysis, and (d) modern method: *TDS* technique. Although they were named chronologically, from oldest to most recent, they will be presented in another way:

2.2.1 Regression analysis

This is the most widely used method. It consists of automatically matching the pressure versus time data to a given analytical solution (normally) of a specific reservoir model. The automatic procedure uses nonlinear regression analysis by taking the difference between a given matching point and the objective point from the analytical solution.

This method has been also widely misused. Engineers try to match the data with any reservoir model without considering the reservoir physics. The natural problem arid=sing with this method is the none-uniqueness of the solution. This means that for a given problem, the results are different if the starting simulation values change. This can be avoided if the starting values for the simulation values are obtained from other techniques, such as *TDS* technique or conventional analysis, and then, the range of variation for a given variable is reduced. This technique will not be longer discussed here since this book focused on analytical and handy interpretation techniques.

2.2.2 Type-curve matching

As seen before, this technique was the second one to appear. Actually, it came as a solution to the difficulty of identity flow regimes in conventional straight-line plots. However, as observed later, the technique is basically a trial-and-error procedure. This makes the technique tedious and risky to properly obtain reservoir parameters.

The oldest type-curve method was introduced by Ramey [2, 8, 9]. If $C_D = 0$ in Eq. (2.16), then, $q_{sf} = q$. Therefore;

$$1 - C_D \frac{dP_D}{dt_D} = 0 \tag{2.19}$$

By integration between 0 and a given P_{wD} and from dimensionless time zero to t_D, and taking logarithm to both terms, it yields:

$$\log P_D = \log t_D - \log C_D \tag{2.20}$$

Suffix w is used to emphasize that the pressure drop takes place at the wellbore bottom-hole. This will be dropped for practical purposes. It is clearly observed in Eq. (2.19) that the slope is one. Then in any opportunity that is plotted P_D vs. t_D and a straight line with a unitary slope is observed at early times, is a good indication that storage exists. Substituting the dimensionless quantities given by Eqs. (1.89), (1.94) and (2.15) in Eq. (2.20), we have:

$$C = \frac{qB}{24} \frac{t}{\Delta P} = \frac{qB}{24} \frac{t_N}{(P_i - P_{wf})_N} \tag{2.21}$$

Eq. (2.21) serves to determine the storage coefficient from data from a pressure decline test using a log-log plot of ΔP versus time. Any point N is taken from the unit-slope straight line portion. The value of C obtained using Eq. (2.21) must match the value obtained from Eq. (2.5). Otherwise, there may be an indication that the liquid level is going down or rising inside the well. The reasons most commonly attributed to this phenomenon are high gas-oil ratios, highly stimulated wells, exhaust gaskets or spaces in the well connections caused by formation collapse or poor cementation and wells used for viscous fluid injection. In conclusion, the properties of Ramey's type curves allow (a) a unitary slope to be identified which indicates wellbore storage and (b) the fading of wellbore storage effects.

It can also be seen that each curve deviates from the unitary slope and forms a transition period lasting approximately 1.5 logarithmic cycles. This applies only to constant wellbore storage, otherwise, refer to [10]. If every ½ cycle is equal to ($10^{0.5}$ = 3.1622), it means that three half cycles (3.16223 = 31.62) represent approximately a value of 30. That is to say that a line that deviates at 2 min requires 1 h forming the transient state or radial flow regime. In other words, the test is masked for 1 h by wellbore storage effects [2, 5, 11]. It is also observed that a group of curves that present damage are mixed at approximately a dimensionless time,

$$t_D \cong (60 + 3.5s)C_D \tag{2.22}$$

After which time, the test is free of wellbore storage effects [2, 5, 6]. Along with *TDS* technique [10, 12–73] which will be discussed later in this chapter, type-curve matching is the only manual procedure that can be applied in short tests where radial flow has not been developed (semilog line). However, type-curve matching is risky because it is a trial-and-error technique, but can provide approximate results even when conventional methods fail. One millimeter shifting can cause pressure differences of up to 200 psia. The procedure is as follows [2, 9]:

1. Prepare a plot of DP vs. t on logarithmic paper using the same scale as the master curve given in **Figure 2.4**. This is recognized as the field data plot, fdp.

2. Place the fdp on the master curve so that the axes are parallel.

3. Find the best match with one of the curves in **Figure 2.4**.

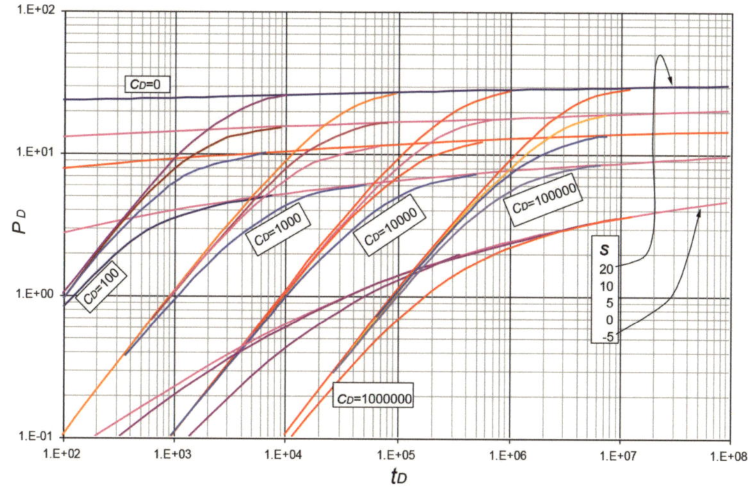

Figure 2.4.
Type curve of dimensionless pressure against dimensionless time for a well in an infinite reservoir (wellbore storage and skin), taken from [2, 9].

4. Choose a suitable match point and read the corresponding coordinates DPM, tM, PDM, tDM, and CDM. The two first parameters are read from the fdp. The remaining from the type-curve (**Figure 2.4**).

5. Estimate permeability, porosity, and wellbore storage coefficient, respectively:

$$k = 141.2\frac{q\mu B}{h}\left(\frac{P_{DM}}{\Delta P_M}\right) \tag{2.23}$$

$$\phi = \frac{0.0002637k}{\mu\, c_t\, r_w^2}\left(\frac{t_M}{t_{DM}}\right) \tag{2.24}$$

$$C = \frac{\phi c_t h\, r_w^2}{0.8936} C_{DM} \tag{2.25}$$

The results from the Ramey's type curve must be verified with some other type curve. For instance, Earlougher and Kersch [8], formulated another type curve, **Figure 2.5**, which result should agree with those using Ramey method. The procedure for this method [8] is outlined as follows:

1. Plot ΔP/t vs. t (fdp) on logarithmic paper using the same scale as the master curve given in **Figure 2.5**. Match the plotted curve, fdp, with the appropriate curve of **Figure 2.5**. Choose any convenient point and read from the master graph $(CDe2s)_M$, $(\Delta P/t\ 24C/qB)_M$ and $(kh/\mu\ t/C)_M$. Read from the fdp: $(\Delta P/t)_M$ and t_M.

Find wellbore storage coefficient, formation permeability, and skin factor using, respectively, the below expressions:

$$C = \frac{qB}{24}\left(\frac{\Delta P}{t}\frac{24C}{qB}\right)_M \Big/ \left(\frac{\Delta P}{t}\right)_M \tag{2.26}$$

$$k = \frac{\mu C}{h}\left(\frac{kh}{\mu}\frac{t}{c}\right)_M \Big/ t_M \tag{2.27}$$

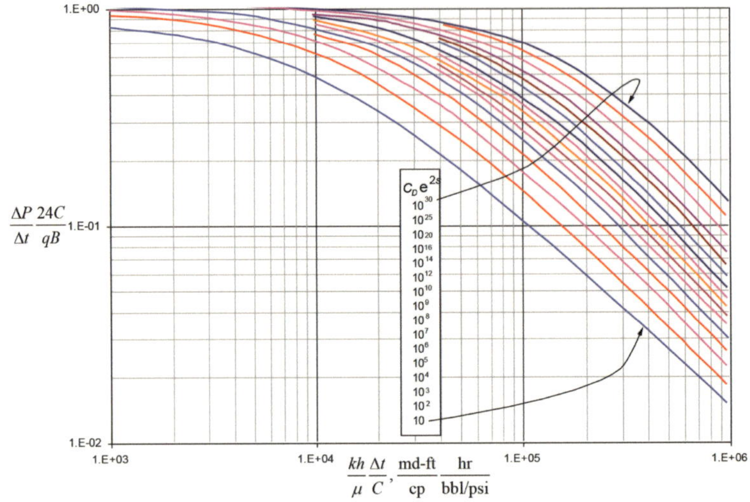

Figure 2.5.
Earlougher and Kersch type-curve for a well in infinite reservoir with wellbore storage and skin, taken from [2, 8].

$$s = \frac{1}{2}\ln\left[\frac{\phi\,\mu\,c_t\,h\,r_w^2}{0.89359C}(C_D e^{2s})_M\right] \tag{2.28}$$

Another important type curve that is supposed to provide a better match was presented by Bourdet et al. [73], **Figure 2.6**. This includes both pressure and pressure derivative curves. The variables to be matched are ΔP_M, $(t^*\Delta P')_M$, $(P_D)_M$, $[(t_D/C_D)P_D']_M$, t_M, $(t_D/C_D)_M$, and $(C_D e^{2s})_M$. The equations use after the matching are [73]:

$$k = \frac{141.2q\mu\,B}{h}\frac{P_{DM}}{\Delta P_M} \tag{2.29}$$

$$C = \left(0.000295\frac{kh}{\mu}\right)\frac{t_M}{(t_D/C_D)_M} \tag{2.30}$$

$$k = \frac{\frac{1412q\mu B}{h}\,[(t_D/C_D)P_D']_M}{(t*\Delta P')_M} \tag{2.31}$$

$$s = \frac{1}{2}\ln\frac{(C_D e^{2s})_M}{C} \tag{2.32}$$

2.2.3 Straight-line conventional analysis

The conventional method implies plotting either pressure or pressure drop against a given time function. The intercept and slope of such plot is used for reservoir and well parameters estimation. When the fluid initiates its path from the farthest reservoir point until the well head, several states and flow regimes are observed depending on the system geometry. For instance, if the reservoir has an elongated shape, probably linear flow will be observed. Linear flow obeys a pressure dependency on the square-root of time, or, if the fluid experiences radial flow regime, the relation between pressure and time observes a semilog behavior, or, either inside the well or the limitation of the reservoir boundaries imply a pseudosteady-state condition, then, pressure is a linear function of time.

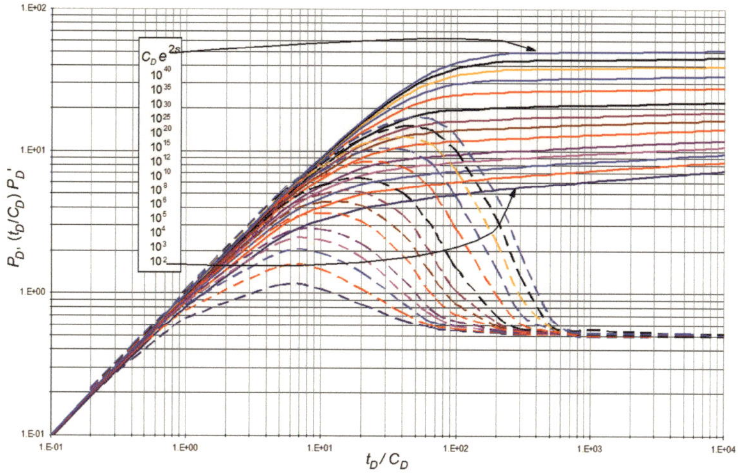

Figure 2.6.
Bourdet et al. [73] pressure and pressure-derivative versus time-type curve.

$$P\,\alpha\,f(t) \tag{2.33}$$

The time function depends on the system geometry and could be any of the kinds described by Eq. (2.34).

Normally, the pressure or pressure drop are plotted in Cartesian coordinates, except certain few cases as for the Muskat method, see Chapter 3, which requires a potential plot, meaning, logarithm scale of pressure drop in the y-axis and Cartesian scale for time in the x-direction.

2.2.3.1 Semilog analysis

It is commonly referred as the "semilog method" since the radial flow is the most important regime found on a pressure test. Then, a semilogarithm plot is customary used in well test analysis.

$$f(t)=\begin{cases}\log t\ -\text{Radial flow}\\ \log\dfrac{t_p+\Delta t}{\Delta t}\ -\text{Radial flow (Horner plot)}\\ t\ -\text{Pseudosteady state}\\ t^{0.135}-\text{Ellipsoidal flow}\\ t^{0.25}-\text{Bilinear flow}\\ t^{0.36}-\text{Birradial flow}\\ t^{0.5}-\text{Linear flow}\\ 1/t^{0.5}-\text{Spherical/Hemispherical flow}\\ (t_p+\Delta t)^{\xi}-\Delta t^{\xi};\ \xi=0.135-\textit{Ellipsoidal},\ 0.25,\ 0.36,\ 0.5\\ 1/\sqrt{t_p}+1/\sqrt{\Delta t}-1/\sqrt{t_p+\Delta t}-\text{Spherical/Hemispherical flow}\\ 1/\sqrt{\Delta t}-1/\sqrt{t_p+\Delta t}-\text{Spherical/Hemispherical flow}\end{cases} \tag{2.34}$$

Starting by including the skin factor in Eq. (1.106);

$$P_D=\frac{1}{2}[\ln t_D+0.80907]+2s \tag{2.35}$$

Replacing the dimensionless terms given by Eqs. (1.89) and (1.94) into Eq. (2.35) and dividing both terms by ln 10 will lead to:

$$\frac{kh(P_i - P_{wf})}{162.6q\mu B} = \left[\log\left(\frac{0.0002637kt}{\phi\mu c_t r_w^2}\right) + 0.3514 + 0.8686s\right] \tag{2.36}$$

Solving for the well-flowing pressure;

$$P_{wf} = P_i - \frac{162.6\,q\mu B}{kh}\left[\log\left(\frac{kt}{\phi\mu c_t r_w^2}\right) - 3.2275 + 0.8686s\right] \tag{2.37}$$

Eq. (2.37) suggests a straight-line behavior which is represented in the central region of **Figure 2.7**. The other two regions are affected by wellbore storage and skin effects, at early times and boundary effects at late times. Reservoir transmissivity, mobility, or permeability can be determined from the slope;

$$m = T = \frac{kh}{\mu} = \left|\frac{162.6\,qB}{m}\right| \tag{2.38}$$

The intercept of Eq. (2.34) is used for the determination of the mechanical skin factor. For practical purposes, the well-flowing pressure at time of 1 h, P_{1hr}, is read from the straight-line portion of the semilog behavior, normally extrapolated as sketched in **Figure 2.7**, so solving for skin factor, s, from Eq. (2.34) results:

$$s = 1.1513\left[\frac{P_{1hr} - P_i}{m} - \log\left(\frac{k}{\phi\mu c_t r_w^2}\right) + 3.23\right] \tag{2.39}$$

Since the slope possesses a negative signed, so does the $P_{1hr} - P_i$ term. Therefore, the first fractional in the above equation is always positive unless the well is highly stimulated.

According to Eq. (2.39), the contribution to the pressure drop caused by the mechanical skin factor is included to the last term: 0.8686s multiplied by the slope. Then:

$$\Delta P_s = |0.87(m)|s, \begin{cases} \text{if } s>0 \rightarrow \Delta P_s>0 \\ \text{if } s<0 \rightarrow \Delta P_s<0 \end{cases} \tag{2.40}$$

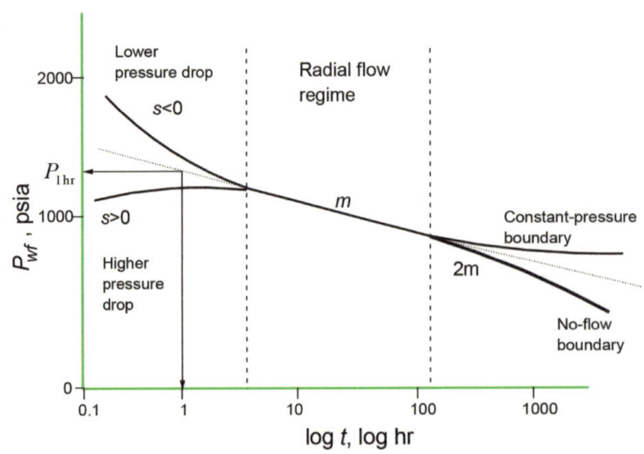

Figure 2.7.
Behavior of the well-flowing pressure observed in a semilog graph, taken from [68].

Eq. (2.40) is similar to Eq. (1.110) and works for either pressure drawdown or pressure buildup tests.

$$\Delta P_s = -0.87(m)\left\{\frac{k}{k_s} - 1\right\} \ln \frac{r_s}{r_w} \tag{2.41}$$

Eq. (1.110) is useful to find either skin factor, s, formation damaged permeability, k_s, or the damaged or affected skin zone radius, r_s. However, since the skin zone covers an infinitely thin area and the pressure wave travels at high speed, it is difficult to detect transmissivity changes, then, r_s and k_s are difficult to be measured.

Eqs. (1.110) and (2.37) imply the skin factor along flow rate just increases or decreases the well pressure drop. However, this occurs because the well radius behaves as if its radius was modified by the value of the skin factor. Brons and Miller [74] defined the apparent or effective wellbore radius, r_{wa}, to be used in Eqs. (1.89), (1.94), and (1.100)

$$r_{wa} = r_w e^{-s} \tag{2.42}$$

Example 2.1

A well with a radius of 0.25 ft was detected to have a skin factor of 2. A skin factor of −2 was obtained after a stimulation procedure. Find the apparent radii and the percentage of change in the radius due to the stimulation. What conclusion can be drawn?

Solution

Application of Eq. (2.42) for the damaged-well case gives:

$$r_{wa} = r_w e^{-s} = 0.25e^{-2} = 0.034 \text{ ft} = 0.406 \text{ in}$$

Application of Eq. (2.42) for the damaged-well case gives:

$$r_{wa} = r_w e^{-s} = 0.25e^{-(-2)} = 1.848 \text{ ft} = 22.17 \text{ in}$$

It can be observed that $1.847 \times 100/0.034 \cong 5460\%$, meaning that the stimulation helps the well to increase its radius 55 times. It can be concluded from the example that for positive skin factor values, the effective wellbore radius decreases ($r_{wa} < r_w$) and for negative skin factor values, the effective wellbore radius increases ($r_{wa} > r_w$).

The starting time of the semilog straight line defined by Ramey [9] in Eq. (2.22) allows determining mathematically where the radial flow starts, i.e., the moment wellbore storage effects no longer affect the test. Replacing into Eq. (2.22) the dimensionless parameters given by Eqs. (1.94) and (2.15) results [2]:

$$t_{SSL} = \frac{(200000 + 12000s)\mu C}{kh} \tag{2.43}$$

The application of Eq. (2.40) is twofolded. (1) It can be used for test design purposes. The duration of a pressure drawdown test should be last 10 times the value of t_{SSL}, so a significant portion of the radial flow regime can be observed and analyzed and (2) finding the semilog slope can be somehow confusing. Once the semilog line is drawn and permeability, skin factor, and wellbore storage are calculated, then, Eq. (2.40) can be used to find the starting point of the radial flow regime. Radial flow is correctly found if the t_{SSL} value agrees with the one chosen in the plot. This last situation is avoid if the pressure and pressure derivative plot is available since radial flow is observed once the pressure derivative curve gets flat as seen in **Figure 2.6**.

The declination stabilization time (time required to reach the boundaries and develop the pseudosteady-state period) during the test can be from the maximum time at which the maximum pressure drops (not shown here) take place. This is:

$$t_{\max} = \frac{948\phi\mu c_t r^2}{k} \tag{2.44}$$

From which;

$$t_{pss} = \frac{948\phi\mu c_t r_e^2}{k} \tag{2.45}$$

For square or circular geometries, $t_{DA} = 0.1$ from **Table 2.1**. Replacing this value in Eq. (1.100) and solving for time leads to:

$$t_{pss} = \frac{1190\phi\mu c_t r_e^2}{k} \tag{2.46}$$

from

$$r_{inv} = 0.0325\sqrt{\frac{kt_p}{\mu\phi c_t}} \tag{2.47}$$

For any producing time, t_p, the radius of investigation—not bigger than r_e—can be found.

The point reached by the disturbance does not imply fluid movement occurs there. The drainage radius is about 90% that value, then

$$r_d = 0.029\sqrt{\frac{kt_p}{\mu\phi c_t}} \tag{2.48}$$

Skin factor is a dimensionless quantity. This does not necessarily reflect the degree of either damage or stimulation of a well. Then, more practical measurement parameters ought to be used. One of this is the flow efficiency, *FE*, which implies what percentage of the total pressure drawdown is due to skin factor. The flow efficiency is defined as the ratio between the actual productivity index, *J*, and the ideal productivity index. The productivity index involves money since it is defined as the amount of pressure drop needed to produce a barrel of fluid per day. In other words, it is the energy required to produce one BPD. Mathematically;

$$J = \frac{q}{\overline{P} - P_{wf}} \tag{2.49}$$

$$J_{ideal} = \frac{q}{\overline{P} - P_{wf} - \Delta P_s} \tag{2.50}$$

$$FE = \frac{J}{J_{ideal}} = 1 - \frac{\Delta P_s}{\overline{P} - P_{wf}} \tag{2.51}$$

$FE < 1$ is an indication that well damage exists, otherwise there is stimulation. The productivity index can be increased by:

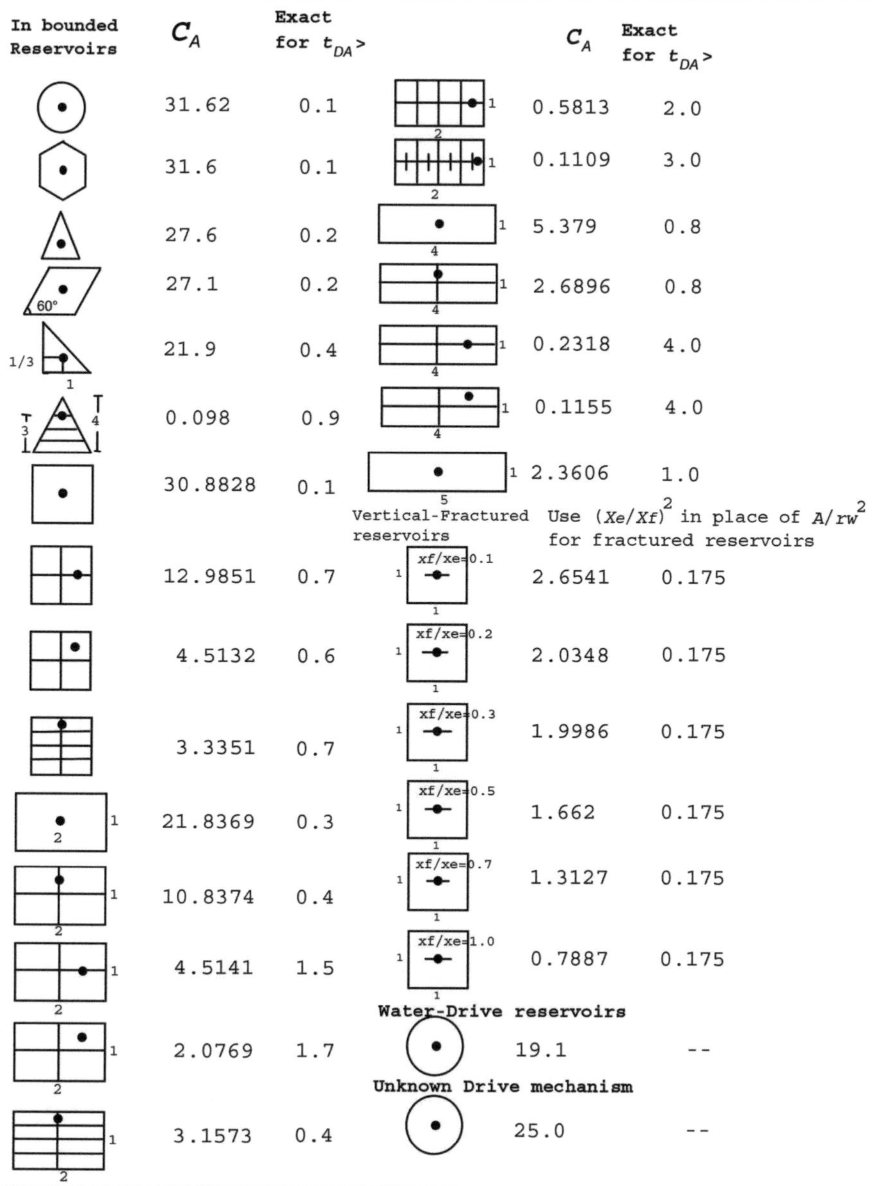

In bounded Reservoirs	C_A	Exact for t_{DA}>
	31.62	0.1
	31.6	0.1
	27.6	0.2
60°	27.1	0.2
1/3, 1	21.9	0.4
3, 4	0.098	0.9
	30.8828	0.1
	12.9851	0.7
	4.5132	0.6
	3.3351	0.7
2, 1	21.8369	0.3
2, 1	10.8374	0.4
2, 1	4.5141	1.5
2, 1	2.0769	1.7
2, 1	3.1573	0.4

	C_A	Exact for t_{DA}>
2, 1	0.5813	2.0
2, 1	0.1109	3.0
4, 1	5.379	0.8
4, 1	2.6896	0.8
4, 1	0.2318	4.0
4, 1	0.1155	4.0
5, 1	2.3606	1.0
Vertical-Fractured reservoirs	Use $(Xe/Xf)^2$ in place of A/rw^2 for fractured reservoirs	
xf/xe=0.1	2.6541	0.175
xf/xe=0.2	2.0348	0.175
xf/xe=0.3	1.9986	0.175
xf/xe=0.5	1.662	0.175
xf/xe=0.7	1.3127	0.175
xf/xe=1.0	0.7887	0.175
Water-Drive reservoirs	19.1	--
Unknown Drive mechanism	25.0	--

Table 2.1.
Shape factors for different drainage areas, taken from [8, 75].

- Increasing the permeability in the zone near the well—hydraulic fracturing;
- Reduce viscosity—steam injection, dissolvent, or in situ combustion;
- Damage removing—acidification;
- Increase well penetration;
- Reduce volumetric factor—choosing correct surface separators.

Other parameters to quantify well damage are [68]:

Damage ratio, *DR*

$$DR = 1/FE \tag{2.52}$$

Damage ratios less than the unity indicate stimulation.

Damage factor, *DF*

$$DR = 1 - FE \tag{2.53}$$

Negative values of damage factors indicate stimulation. The damage factor can also be estimated from [68]:

$$DF = \frac{s}{s + \ln\left(r_e/r_w\right)} = 1 - \frac{q(\text{actual})}{q(\text{ideal})} \tag{2.54}$$

Eq. (2.54) applies to circular-shaped reservoir.

Productivity ratio, *PR*

$$PR = \frac{q}{q_a} = \frac{q(\text{ideal})}{q(\text{actual})} = \frac{\ln\left(r_e/r_w\right)}{\ln\left(r_e/r_w\right) + s} \tag{2.55}$$

Annual loss income, *FD$L* (USD$)

$$FD\$L = 365q(OP)DF \tag{2.56}$$

where *OP* is oil price.

Example 2.2

What will be the annual loss of a well that produces 500 BFD, which has a damage factor of 8, drains an area of 120 acres and has a radius of 6 inches? Assume circular reservoir area and a price of oil crude of USD $ 55/barrel.

Solution

120 acres = 5,227,200 ft^2. If the area is circular, then: r = 1290 ft. Find the damage factor from Eq. (2.54);

$$DF = \frac{s}{s + \ln\left(r_e/r_w\right)} = \frac{8}{8 + \ln\left(1290/0.5\right)} = 0.5046$$

Find the yearly loss income using Eq. (2.56)

$$FD\$L = 365q(OP)DF = (365)(500)(55)(0.5046) = \text{USD\$ } 5064922$$

This indicates that the well requires immediate stimulation.

2.2.3.2 Reservoir limit test, RLT

It is a drawdown test run long enough to reach the reservoir boundaries. Normal pressure drawdown tests, during either radial flow or transient period test, are used to estimate formation permeability and artificial well conditions (C and s), while an RLT test—introduced by [76]—deals with boundaries and is employed to determine well drainage area or well drainage pore volume. In a Cartesian graph for a closed boundary system, **Figure 2.8**, three zones are distinguished [8, 68]: (i) skin and wellbore storage dominated zone, (ii) transient zone (radial flow), and (iii) pseudosteady-state zone. As indicated by Eq. (1.129), the pressure drop is a linear function of time. Eq. (1.129) is given for circular reservoir geometry. For any geometry, the late time pseudosteady-state solution involves the Dietz shape factor, [75], to extent the use of Eq. (1.129) for other reservoir geometries, as described in **Table 2.1**. Under this condition, Eq. (1.129) becomes [77]:

$$P_D = 2\pi t_{DA} + \frac{1}{2}\ln\left(\frac{A}{r_w^2}\right) + \ln\left(\frac{2.2458}{C_A}\right) \tag{2.57}$$

Replacing in the above expression the dimensionless quantities given by Eqs. (1.89) and (1.94), it results:

$$P_{wf} = -\left[\frac{0.23395qB}{\phi c_t Ah}\right]t + P_i - \frac{70.6q\mu B}{kh}\left[\ln\frac{A}{r_w^2} + \ln\left(\frac{2.2458}{C_A}\right) + 2s\right] \tag{2.58}$$

From the slope, m^*, and intercept, P_{INT}, of Eq. (2.58), the reservoir pore volume and Dietz shape factor [74] can be obtained from either:

$$V_p = -\frac{0.23395qB}{c_t m^*} \tag{2.59}$$

$$C_A = 5.456\frac{m}{m^*}e^{2.303\frac{P_{1hr}-P_{INT}}{m}} \tag{2.60}$$

Once the value of C_A is obtained from Eq. (2.60), the reservoir geometry can be obtained from **Table 2.1** by using the closest tabulated value ("exact for t_{DA}") and confront with the time to develop pseudosteady-state regime, $(t_{DA})_{pss}$ which is found from:

$$(t_{DA})_{pss} = 0.1833\frac{m^*}{m}t_{pss} \tag{2.61}$$

t_{pss} can be read from the Cartesian plot. However, this reading is inexact; therefore, it is recommended to plot the Cartesian pressure derivative and to find the exact point at which this becomes flat.

2.2.4 Tiab's direct synthesis (*TDS*) technique

TDS technique is the latest methodology for well test interpretation. Its basis started in 1989 [70]. *TDS*' creator was Tiab [71], who provided analytical and practical solutions for reservoir characterization using characteristic points or features—called by him "fingerprints"—read from a log-log plot of pressure and pressure derivative [15], versus time. Since the introduction of *TDS* in 1995, several scenarios, reservoir geometries, fluid types, well configurations, and operation conditions. For instance, extension of *TDS* technique to elongated systems can be found in [13, 14, 16–19, 23, 24, 28, 30, 31]. Some applications of conventional analysis in long reservoirs are given in [20, 29, 38, 54]. For vertical and horizontal gas wells with and without use of pseudotime, refer to [22, 36, 39]. Special cases of horizontal wells are found in [12, 47]. For transient rate analysis, refer to [27, 35, 49]. Applications on heavy oil (non-Newtonian fluids) can be found in [32, 34, 41, 42, 45, 52, 56, 62, 64]. For cases on shales reservoirs, refer to [49, 51, 56, 78]. Well test analysis by the *TDS* technique on secondary and tertiary oil recovery is presented by [25, 33, 60, 79]. For multirate testing in horizontal and vertical wells, refer, respectively, to [65, 67]. References [43, 46] are given for conductive faults. For deviated and partially penetrated wells, refer to [37, 64], respectively. *TDS* technique extended to multiphase flow was presented by [26]. Wedged and T-shaped reservoirs can be found in [48] and coalbed-methane reservoirs with bottom water drive are given in [53]. *TDS* technique is excellent for interpreting pressure test in hydraulically fractured vertical wells since unseen flow regimes can be generated [50, 69, 80]. The first publications on horizontal wells in naturally fractured and anisotropic media are given in [81, 82]. The threshold pressure gradient is dealt by [57, 72]. For vertical wells in double porosity and double permeability formations, refer, respectively, to [41, 83]. A book published by Escobar [56] presents the most recent topics covered by the *TDS* technique, and a more comprehensive state-of-the-art on *TDS* technique is given by [58]. This book revolves around this methodology; therefore, practically, the whole content of [71]—pioneer paper of *TDS* technique—will be brought here:

The starting point is the definition of the dimensionless pressure derivative from Eq. (1.89);

$$t_D * P_D{}' = \frac{kh(t * \Delta P')}{141.2 q\mu B} \tag{2.62}$$

By looking at Eqs. (2.17) and (2.18), we can conclude the difficulty of using hand mathematical operations with them. Instead of using these general solutions, Tiab [71] obtained partial solutions to the differential equation for each flow regime or time period. For instance, during early pseudosteady-state, the governing equation reduces to:

$$P_D = \frac{t_D}{C_D} \tag{2.63}$$

Combination of Eqs. (1.94) and (2.15) results in:

$$\frac{t_D}{C_D} = \left(2.95 \times 10^{-4}\frac{h}{\mu}\right)\frac{t}{C} \tag{2.64}$$

Replacing Eq. (1.89) in the above expression yields;

$$\left(\frac{kh}{141.2q\mu B}\right)\Delta P = \left(2.95 \times 10^{-4}\frac{kh}{\mu}\right)\frac{t}{C} \tag{2.65}$$

Solving for C;

$$C = \left(\frac{qB}{24}\right)\frac{t}{\Delta P} \tag{2.66}$$

The pressure derivative curve also has a straight line of unitary slope at early times. The equation of this line is obtained by taking the derivative of Eq. (2.63) with respect to the natural logarithm of t_D/C_D. So:

$$\left(\frac{t_D}{C_D}\right)P_D{}' = \frac{t_D}{C_D} \tag{2.67}$$

Where the derivative of the dimensionless pressure is:

$$P_D{}' = dP_D/dt_D = \left(\frac{kh}{141.2q\mu B}\right)dP \Big/ \left(\frac{0.0002637k}{\phi\mu c_t r_w^2}\right)dt \tag{2.68}$$

Rearranging;

$$P_D{}' = \left(\frac{26.856 r_w^2 \phi c_t h}{qB}\right)\Delta P' \tag{2.69}$$

Converting to dimensional form, the left-hand side of Eq. (2.67) by using the definitions given by Eqs. (2.64) and (2.68):

$$\left(\frac{t_D}{C_D}\right)P_D{}' = 0.00792252\left(\frac{kh}{q\mu B}\right)\left(\frac{\phi c_t h r_w^2}{C}\right)(t * \Delta P') \tag{2.70}$$

Multiplying and dividing by 0.8935;

$$\left(\frac{t_D}{C_D}\right)P_D{}' = 0.007087\left(\frac{kh}{q\mu B}\right)\left(\frac{\phi c_t h r_w^2}{0.8935C}\right)(t * \Delta P') \tag{2.71}$$

Recalling Eq. (2.15), the above becomes:

$$\left(\frac{t_D}{C_D}\right)P_D{}' = 0.007087\left(\frac{kh}{q\mu B}\right)\left(\frac{1}{C_D}\right)(t * \Delta P') \tag{2.72}$$

Since the unit slope is one, then $C_D = 1$, thus;

$$\left(\frac{t_D}{C_D}\right)P_D{}' = \frac{kh(t * \Delta P')}{141.2q\mu B} \tag{2.73}$$

From looking at **Figure 2.6**, both pressure and pressure derivative curves display a unitary slope at early times. Replacing Eqs. (2.64), (2.73) in (2.67) and solving for C will result:

$$C = \left(\frac{qB}{24}\right)\frac{t}{t * \Delta P'} \tag{2.74}$$

As seen in **Figure 2.6**, the infinitely acting radial flow portion of the pressure derivative is a horizontal straight line with intercept of 1/2. The governing equation is:

$$\left[\left(\frac{t_D}{C_D}\right)P_D{}'\right]_r = \frac{1}{2} \tag{2.75}$$

Combining the above equation with Eq. (2.73) results the best expression to estimate reservoir permeability:

$$k = \frac{70.6q\mu B}{h(t * \Delta P')_r} \tag{2.76}$$

Subscript r stands for radial flow line. A customary use of *TDS*, as established by Tiab [71], is to provide suffices to identify the different flow regimes. For instance, *pss* stands for pseudosteady state, i stands for either initial or intercept, etc. In terms of pressure, the equation of this line is:

$$P_{Dr} = 0.5\left\{ \ln\left(\frac{t_D}{C_D}\right)_r + 0.80907 + \ln\left[C_D e^{2s}\right]\right\} \tag{2.77}$$

It is recommended to draw a horizontal line throughout the radial flow regime and choose one convenient value of $(t^*\Delta P')_r$ falling on such line.

Tiab [71] also obtained the start time of the infinite line of action of the pressure derivative is:

$$\left(\frac{t_D}{C_D}\right)_{sr} = 10\ \log\left(C_D e^{2s}\right)^{10} \tag{2.78}$$

Replacing Eqs. (1.92) and (2.15) in the above equation will yield:

$$t_{sr} = \frac{\mu\, C}{6.9\ \times\ 10^{-5} kh}\left[\ln\left(\frac{0.8935C}{\phi c_t h r_w^2}\right) + 2s\right] \tag{2.79}$$

A better form of Eq. (2.78) was given by [84];

$$\left(\frac{t_D}{C_D}\right)_{sr} = \frac{1}{\alpha}\left[\ln\left(C_D e^{2s}\right) + \ln\left(\frac{t_D}{C_D}\right)_{SR}\right] \tag{2.80}$$

Setting a = 0.05 in the above equation and solving for C:

$$C = 0.056\phi c_t h r_w^2\left(\frac{t_{Dsr}}{2s + \ln t_{Dsr}}\right) \tag{2.81}$$

t_{Dsr} is calculated with Eq. (1.94) letting $t = t_{sr}$.

The point of intersection, i, between the early time unit-slope line defined by Eqs. (2.63) and (2.67) and the late-time infinite-acting line of the pressure derivative, defined by Eq. (2.75), is given by:

$$\left(\frac{t_D}{C_D} P_D{}'\right)_i = 0.5 \tag{2.82}$$

$$\left(\frac{t_D}{C_D}\right)_i = 0.5 \tag{2.83}$$

where i stands for intersection. After replacing the definitions given by Eqs. (1.94), (2.15), and (2.72) will, respectively, provide:

$$(t * \Delta P')_i = \frac{70.6 q\mu B}{kh} \tag{2.84}$$

$$k = \frac{1695\mu C}{h t_i} \tag{2.85}$$

For the unit-slope line, the pressure curve is the same as for the pressure derivative curve. Then, at the intersection point:

$$(\Delta P)_i = (t * \Delta P)_i = (t * \Delta P')_r \tag{2.86}$$

Tiab [71] correlated for $C_D e^{2s}$ > 100 permeability, wellbore storage coefficient, and skin factor with the coordinates of the maximum point—suffix x—displayed once the "hump" observed once wellbore storage effects start diminishing. These correlations are given as follows:

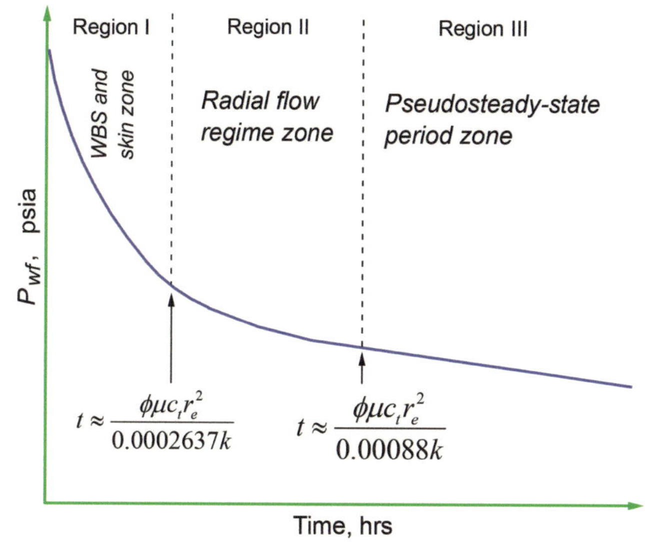

Figure 2.8.
Characteristics found in the Cartesian graph, taken from [68].

$$\left(\frac{t_D}{C_D}P_D{'}\right)_x = 0.35717\left(\frac{t_D}{C_D}\right)_x - 0.50 \tag{2.87}$$

$$\log\left(C_D e^{2s}\right) = 0.35\left(\frac{t_D}{c_D}\right)^{1.24} \tag{2.88}$$

$$\log\left(C_D e^{2s}\right) = 1.71\left(\frac{t_D}{C_D}P_D{'}\right)^{1.24} \tag{2.89}$$

Replacing Eqs. (2.64) and (2.73) into Eq. (2.87) leads to:

$$(t * \Delta P')_x = \left(0.015\frac{qB}{C}\right)t_x - 0.42\left(\frac{141.2q\mu B}{kh}\right) \tag{2.90}$$

Either formation permeability or wellbore storage coefficient can be determined using the coordinates of the peak, t_x and $(t^*\Delta P')_x$. Solving for both of these parameters from Eq. (2.90) results:

$$k = \left(\frac{70.6q\mu B}{h}\right)\frac{1}{(0.014879qB/C)\,t_x - (t * \Delta P')_x} \tag{2.91}$$

$$C = \frac{0.014879qBt_x}{(t * \Delta P')_x + (t * \Delta P')_r} \tag{2.92}$$

The constants in Eqs. (2.91) and (2.92) are slightly different as those in [58]. These new unpublished versions were performed by *TDS*' creator.

Eq. (2.91) is so helpful to find reservoir permeability in short test when radial flow is absent which is very common in fall-off tests. Once permeability is found from Eq. (2.91), solved for $(t^*\Delta P')_r$ from Eq. (2.76) and plot on a horizontal line throughout this value. Then, compare with the actual derivative plot and use engineering criterion to determine if the permeability value is acceptable. This means, if the straight line is either lower or higher than expected. Otherwise, new coordinates of the peak ought to be read for repeating the calculations since the hump should look some flat.

Substitution of Eqs. (2.64) and (2.73) in Eqs. (2.88) and (2.89) allows obtaining two new respective correlations for the determination of the mechanical skin factor:

$$s = 0.171\left(\frac{t_x}{t_i}\right)^{1.24} - 0.5\ \ln\left(\frac{0.8935C}{\phi\, hc_t r_w^2}\right) \tag{2.93}$$

$$s = 0.921\left(\frac{(t*\Delta P')_x}{(t*\Delta P')_i}\right)^{1.1} - 0.5\ \ln\left(\frac{0.8935C}{\phi hc_t r_w^2}\right) \tag{2.94}$$

Sometimes, the reading of the peak coordinates may be wrong due to the flat appearance of it. Then, it should be a good practice to estimate the skin factor using both Eqs. (2.93) and (2.94). These values should match each other.

Divide Eq. (2.87) by Eq. (2.75); then, in the result replace Eqs. (2.64) and (2.73) and solve for both permeability and wellbore storage:

$$k = 4745.36\frac{\mu C}{ht_x}\left\{\frac{(t*\Delta P')_x}{(t*\Delta P')_r} + 1\right\} \tag{2.95}$$

$$C = \frac{0.014879qBt_x}{(t*\Delta P')_x + (t*\Delta P')_r} \tag{2.96}$$

This last expression is useful to find wellbore storage coefficient when the early unitary slope line is absent.

TDS technique has a great particularity: for a given flow regime, the skin factor equation can be easily derived from dividing the dimensionless pressure equation by the dimensionless derivative equation of such flow regime. Then, the division of Eq. (2.77) by Eq. (2.75) leads to the below expression once the dimensionless parameters given by Eqs. (1.89), (1.94), and (2.73) are replaced in the resulting quotient. Solving for s from the final replacement leads to:

$$s = 0.5\left(\frac{\Delta P_r}{(t*\Delta P')_r} - \ln\left[\frac{k\, t_r}{\phi\mu c_t r_w^2}\right] + 7.43\right) \tag{2.97}$$

being t_r any convenient time during the infinite-acting radial flow regime throughout which a horizontal straight line should have been drawn. Read the ΔP_r corresponding to t_r. Comparison between Eqs. (2.38) and (2.76) allows concluding:

$$m = 2.303(t*\Delta P')_r = \ln(10)(t*\Delta P')_r \tag{2.98}$$

which avoids the need of using the semilog plot if the skin pressure drop is needed to be estimated by Eq. (2.40), otherwise, Eq. (2.40) becomes:

$$\Delta P_s = |2(t*\Delta P')_r|s, \quad \begin{cases} \text{if } s>0 \rightarrow \Delta P_s>0 \\ \text{if } s<0 \rightarrow \Delta P_s<0 \end{cases} \tag{2.99}$$

For the determination of well-drainage area, Tiab [69] expressed Eq. (2.75) as:

$$(t_D * P_D')_r = \frac{1}{2} \tag{2.100}$$

Also, Tiab [69] differentiated the dimensionless pressure with respect to dimensionless time in Eq. (2.57), so:

$$t_D * P_D' = 2\pi t_{DA} \tag{2.101}$$

Then, Tiab [69] based on the fact that two given flow regime governing equations can be intersected each other, regardless the physical meaning of such intersection, and solving for any given parameter, intercepted Eqs. (2.100) with (2.101), then, replaced in the resulting expression the dimensionless quantities given by Eqs. (2.92), (2.97), and (2.62) and solved for the area given in ft^2:

$$A = \frac{kt_{rpssi}}{301.77\phi\mu c_t} \tag{2.102}$$

Furthermore, Chacon et al. [85] replaced the dimensionless time given by Eq. (1.100) and the dimensionless pressure derivative of Eq. (2.62) into Eq. (2.102) and also solved for the area in ft^2:

$$A = \frac{0.234qBt_{pss}}{\phi c_t h\ (t * \Delta P')_{pss}} \tag{2.103}$$

The above expression uses any convenient point, t_{pss} and $(t^*\Delta P')_{pss}$, during the late time pseudosteady-state period. Because of noisy pressure derivative data, the readings of several arbitrary points may provide, even close, different area values. Therefore, it is convenient to use an average value. To do so, it is recommended to draw the best late-time unit-slope line passing through the higher number of pressure derivative points and extrapolate the line at the time of 1 h and read the pressure derivative value, $(t^*\Delta P')_{pss1}$. Under these circumstances, Eq. (2.103) becomes:

$$A = \frac{0.234qB}{\phi c_t h\ (t * \Delta P')_{pss1}} \tag{2.104}$$

Eqs. (2.102) through (2.104) apply only to closed-boundary reservoirs of any geometrical shape. For constant-pressure reservoirs, the works by Escobar et al. [28, 54] for *TDS* technique (summary given in **Table 2.2**) and for conventional analysis are used for well-drainage area determination in circular, square, and elongated systems.

TDS technique has certain step-by-step procedures which not necessarily are to be followed since the interpreter is welcome to explore and use *TDS* as desired. Then, they are not provided here but can be checked in [69, 71].

Example 2.3

Taken from [68]. The pressure and pressure derivative data given in **Table 2.3** corresponds to a drawdown test of a well. Well, fluid, and reservoir data are given below:

Find permeability, skin factor, drainage area, and flow efficiency by conventional analysis. Find permeability, skin factor, and three values of drainage area using *TDS* technique:

r_w = 0.267 ft	q = 250 BPD	μ = 1.2 cp	
c_t = 26.4 × 10^{-5}psi^{-1}	h = 16 pies	ϕ = 18%	B = 1.229 bbl/BF

Solution

Conventional analysis. Figure 2.9 and **2.10** present the semilog and Cartesian plots, respectively, to be used in conventional analysis. From **Figure 2.9**, the semilog slope, m, is of 18 psia/cycle and P_{1hr} = 2308 psia. Permeability and skin factor are calculated using Eqs. 2.38 and 2.39, respectively, thus:

Constant, Ξ	Equation	Reservoir geometries	Equation number
301.77	$A = \frac{kt_{rpssi}}{\Xi\phi\mu c_t}$	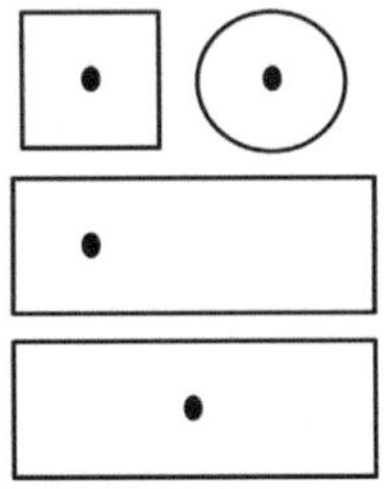	(2.100)
283.66			(2.101)
4066	$A = \sqrt[3]{\frac{kt_{ssri} Y_E^4}{\Xi\phi\mu c_t}}$	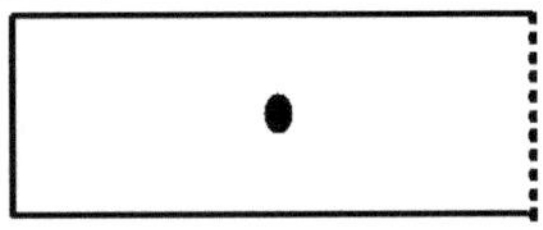	(2.102)
482.84		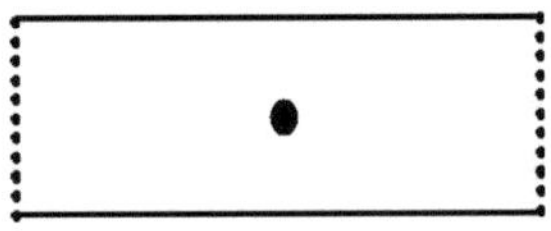	
7584.2		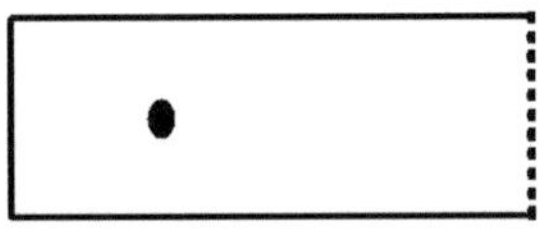	
2173.52	$A = \left(\frac{kt_{ss1ri,ss2ri}}{\Xi\phi\mu c_t}\right)^{2/3} \frac{Y_E^{5/3}}{b_x}$	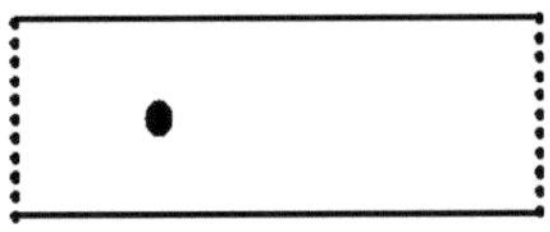	(2.103)
6828.34		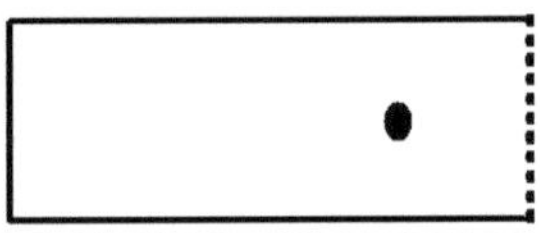	
41.82	$X_E = \frac{1}{\Xi}\left(\frac{kt_x}{\phi\mu c_t}\right)^{0.5}$	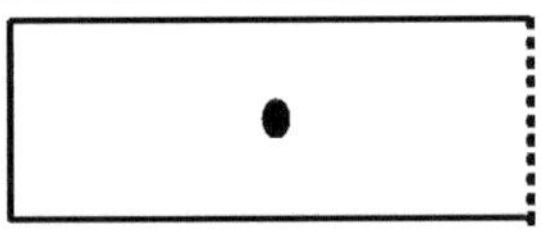	(2.104)
20.91		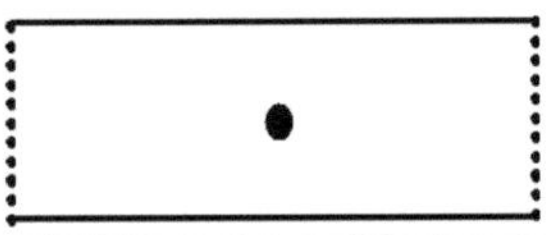	

Table 2.2.
Summary of equations, taken from [28].

$$k = \left| \frac{162.6\,(250)(1.2)(1.229)}{(-18)(16)} \right| = 208 \text{ md}$$

$$s = 1.1513 \left[\frac{2308 - 2733}{-18} - \log\left(\frac{208}{0.18(1.2)(26.4 \times 10^{-5})(0.267)^2} \right) + 3.23 \right] = 22.15$$

Find the pressure loss due to skin factor with Eq. (2.40);

t, h	P_{wf}, psia	D*P*, psia	*t**D*P*′, psia/h	*t*, h	P_{wf}, psia	D*P*, psia	*t**D*P*′, psia/h
0.00	2733	0		5	2312	421	65.42
0.10	2703	30	31.05	7	2293	440	35.32
0.20	2672	61	58.95	9.6	2291	442	5.86
0.30	2644	89	84.14	12	2290	443	5.85
0.40	2616	117	106.30	16.8	2287	446	7.63
0.65	2553	180	129.70	33.6	2282	451	7.99
1.00	2500	233	135.15	50	2279	454	7.94
1.50	2440	293	151.90	72	2276	457	10.50
2.00	2398	335	127.26	85	2274	459	12.18
3.00	2353	380	102.10	100	2272	461	13.36
4.00	2329	404	81.44				

Table 2.3.
Pressure and pressure derivative versus time data for example 2.3.

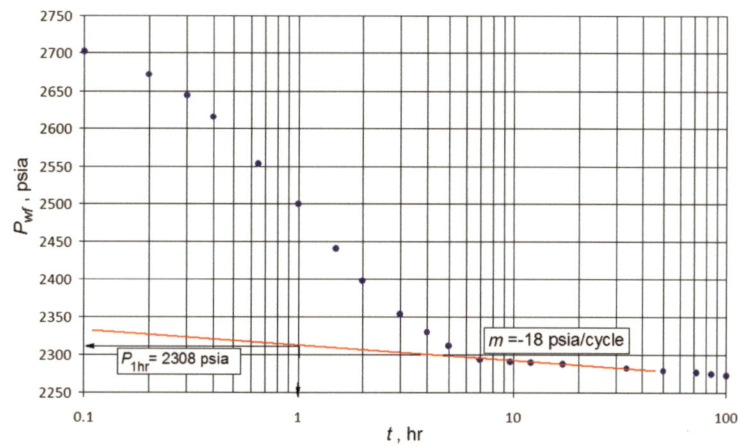

Figure 2.9.
Semilog plot for example 2.3.

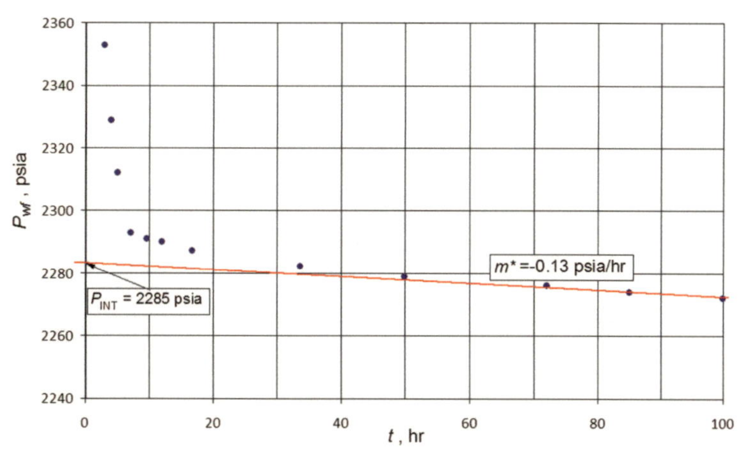

Figure 2.10.
Cartesian plot for example 2.3.

$$\Delta P_s = |0.87(-18)|22.15 = 346.7 \text{ psia}$$

Since the average reservoir pressure is not reported, then, the initial pressure value is taken instead. Eq. (2.51) allows estimating the flow efficiency.

$$FE = 1 - \frac{346.9}{2733 - 2272} = 24.75\,\%$$

From the Cartesian plot, **Figure 2.10**, is read the following data:

$m^* = -0.13$ psia/h	$P_{INT} = 2285$ psia	$t_{pss} \approx 50$ h

Use Eq. (2.59) to find well drainage area:

$$A = -\frac{0.234qB}{\phi h c_t m^*} = -\frac{0.234(250)(1.229)}{(0.18)(16)(26.4\times10^{-5})(-0.13)} = 727391.1 \text{ ft}^2 = 16.7 \text{ Ac}$$

Find the Dietz shape factor with Eq. (2.60);

$$C_A = 5.456\frac{-18}{-0.13}e^{\left[\frac{2.303(2308-2285)}{-18}\right]} = 39.82$$

As observed in **Table 2.1**, there exist three possible well drainage area geometry values (hexagon, circle, and square) close to the above value. To discriminate which one should be the appropriate system geometry find the dimensionless time in which pseudosteady-state period starts by using Eq. (2.61):

$$(t_{DA})_{pss} = 0.1833\frac{-0.13}{-18}50 = 0.066 \approx 0.1$$

TDS technique. The following are the characteristic points read from **Figure 2.11**:

$(t^*\Delta P')_r = 7.7$ psia	$t_r = 33.6$ h	$\Delta P_r = 451$ psia
$t_{pss} = 85$ h	$(t^*\Delta P')_{pss} = 12.18$ psia	$t_{rpi} = 58$ h
$(t^*\Delta P')_{pss1} = 0.14$ psia		

Find permeability and skin factor with Eqs. (2.76) and (2.97), respectively:

$$k = \frac{70.6q\mu B}{h(t*\Delta P')_r} = \frac{70.6(250)(1.2)(1.229)}{(16)(7.7)} = 211.3 \text{ md}$$

$$s = 0.5\left[\frac{451}{7.7} - \ln\left(\frac{211.3(33.6)}{0.18(1.2)(26.4\times10^{-5})(0.267^2)}\right) + 7.43\right] = 22.4$$

Determine the well drainage area with Eqs. (2.102) and (2.103), thus;

$$A = \frac{211.3(58)}{301.77(0.18)(1.2)(26.4\times10^{-5})(43560)} = 16.35 \text{ Ac}$$

$$A = \frac{0.234(250)(1.229)(85)}{(0.18)(26.4\times10^{-5})(16)(12.18)(43560)} = 15.15 \text{ Ac}$$

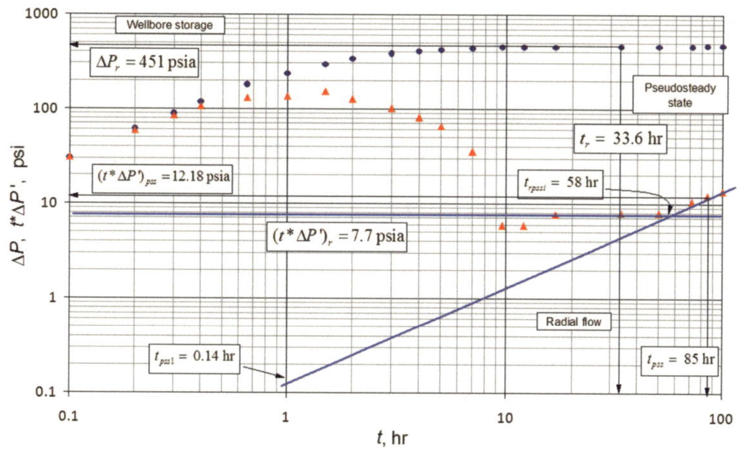

Figure 2.11.
Pressure and pressure derivative plot for example 2.3.

$$A = \frac{0.234(250)(1.229)}{(0.18)(26.4 \times 10^{-5})(16)(0.14)(43560)} = 15.5 \text{ Ac}$$

Even, more parameters can be reestimated with *TDS* technique for verification purposes but it will not be performed for saving-space reasons. However, the reader is invited to read the coordinates of the peak and the intersection point of the wellbore storage and radial flow lines. Then, estimate formation permeability with Eqs. (2.84), (2.85), and (2.91). Also, find the wellbore storage coefficient using Eqs. (2.74), (2.81), (2.92), and (2.96) and skin factor with Eqs. (2.93) and (2.94).

Example 2.4

Taken from [68] with the data from the previous example, Example 2.3, determine t_{SSL} and find if the well fluid level is increasing or decreasing in the annulus if the well has a drill pipe with 2 in external diameter inside a liner with 5 in of inner diameter including joint gaskets. The density of the wellbore fluid is 42.5 lbm/ft^3.

Solution

From **Figure 2.11**, a point is chosen on the early unit-slope line. This point has coordinates: D*P* = 59 psia and *t* = 0.2 h. Wellbore storage coefficient is found with Eq. (2.21):

$$C = \frac{(250)(1.229)}{24}\frac{0.2}{59} = 0.0434 \text{ bbl/psia}$$

Solving for annulus capacity from Eq. (2.5);

$$V_u = \left(\frac{\rho}{144}\right)C = \left(\frac{42.5}{144}\right)0.0434 = 0.0128 \text{ bbl/ft}$$

The theoretical capacity is found with Eq. (2.45), so:

$$V_u = 0.0009714(5^2 - 2^2) = 0.0204 \text{ bbl/ft}$$

This leads to the conclusion that the annular liquid is falling.

2.3. Multiphase flow

According to Perrine [86], the single fluid flow may be applied to the multiple fluid flow systems when the gas does not dominate the pressure tests, it means liquid production is much more relevant than gas flow. Under this condition, the diffusivity equation, Eq. (1.27), will result and the total fluid mobility is determined by Eq. (1.24). We also mentioned in Chapter 1 that Martin [63] provided some tips for a better use of Perrine method. Actually, Perrine method works very well in liquid systems.

The semilog equations for drawdown and build tests are, respectively, given below:

$$P_{wf} = P_i - \frac{162.6q_t}{\lambda_t h}\left(\log\frac{\lambda_t t}{1688\phi c_t r_w^2} + 0.869s\right) \tag{2.105}$$

$$P_{ws} = P_i - \frac{162.6q_t}{\lambda_t h}\log\left(\frac{t_p + \Delta t}{\Delta t}\right) \tag{2.106}$$

The flow rate is estimated by:

$$q_t = \left(q_o B_o + (q_g - q_o R_s/1000)B_g + q_w B_w\right)/B_o \tag{2.107}$$

Eq. (2.107) is recommended when oil flow dominates the test. It is removed from the denominator, otherwise. It advised to use consistent units in Eq. (2.107) meaning that the gas flow rate must be in Mscf/D and the gas volume factor bbl/SCF.

Once the semilog slope has been estimated, the total mobility, the phase effective permeabilities, and the mechanical skin factor are found from:

$$\lambda_t = -\frac{162.6q_t}{mh} \tag{2.108}$$

$$k_L = -\frac{162.6q_L B_L \mu_L}{mh};\ L = \text{water or oil} \tag{2.109}$$

$$k_g = -\frac{162.6\left(q_g - q_o R_s/1000\right)B_g \mu_g}{mh} \tag{2.110}$$

$$s = 1.1513\left(\frac{P_{wf} - P_{1hr}}{m} - \log\left(\frac{\lambda_t}{\phi c_t r_w^2}\right) + 3.23\right) \tag{2.111}$$

The best way of interpreting multiphasic flow tests in by using biphasic and/or triphasic pseudofunctions. Normally, well test software uses empirical relationships to estimate relative permeability data. The accuracy of the following expression is sensitive to the relative permeability data:

$$m(P) = \int_{P_0}^{P} \frac{k_{ro}}{\mu_o B_o} dP \tag{2.112}$$

The expressions used along this textbook for reservoir characterization may apply for both single fluid and multiple fluid production tests. Single mobility has to be changed by total fluid mobility and individual flow rate ought to be replaced by the total fluid rate. Just to cite a few of them, Eqs. (2.66), (2.76), (2.85), (2.91), (2.92), and (2.97) become:

$$C = \left(\frac{q_t}{24}\right)\frac{t}{\Delta P} \tag{2.113}$$

$$\lambda_t = \frac{k}{\mu}\bigg|_t = \frac{70.6q_t}{h(t*\Delta P')_r} \tag{2.114}$$

$$\left(\frac{k}{\mu}\right)_t = \frac{1695C}{ht_i} \tag{2.115}$$

$$\lambda_t = \left(\frac{70.6q_t}{h}\right)\frac{1}{(0.014879q_t/C)t_x - (t*\Delta P')_x} \tag{2.116}$$

$$C = \frac{0.014879q_t t_x}{(t*\Delta P')_x + (t*\Delta P')_r} \tag{2.117}$$

$$s = 0.5\left[\frac{\Delta P_r}{(t*\Delta P)_r} - \ln\left(\frac{\lambda_t t_r}{\phi c_t r_w^2}\right) + 7.43\right] \tag{2.118}$$

Also, the effective liquid permeabilities are found using the individual viscosity, rate, and volume factor. Then, Eq. (2.76) applied to oil and water will yield:

$$k_o = \frac{70.6q_o\mu_o B_o}{h(t*\Delta P')_r} \tag{2.119}$$

$$k_w = \frac{70.6q_w\mu_w B_w}{h(t*\Delta P')_r} \tag{2.120}$$

However, from a multiple fluid test, it is a challenge to find the reservoir absolute permeability. Several methods have been presented. For instance, Al-Khalifah et al. [87] presented a sophisticated method applied to either drawdown or multiple rate tests. Their method even includes the estimating of the saturation change respect to pressure. However, we presented the method by Kamal and Pan [88] which applies well for liquid fluid. Relative permeabilities must be known for its application. Once effective permeabilities are found, let us say from Eqs. (2.119) and (2.120), estimate the permeability ratio k_o/k_w and find the water saturation from the relative permeability curves as schematically depicted in **Figure 2.12** (left). Then, using the estimated water saturation value, enter **Figure 2.12** (right) and read a value from a relative permeability curve. Use the most dominant flow curve. The dominant phase is assumed to be oil for the example in **Figure 2.12**. Since both phase effective permeability and phase relative permeability are known, the absolute permeability is found from the definition of relative permeability:

$$k = \frac{k_o}{k_{ro}} \tag{2.121}$$

Further recommendations for handling multiphase flow tests are presented by Al-Khalifah et al. [89] and are also reported by Stanislav and Kabir [7].

2.4. Partial penetration and partial completion

When a well penetrates a small part of the formation thickness, hemispherical flow takes place. See **Figure 2.13** top. When the well is cased above the producer

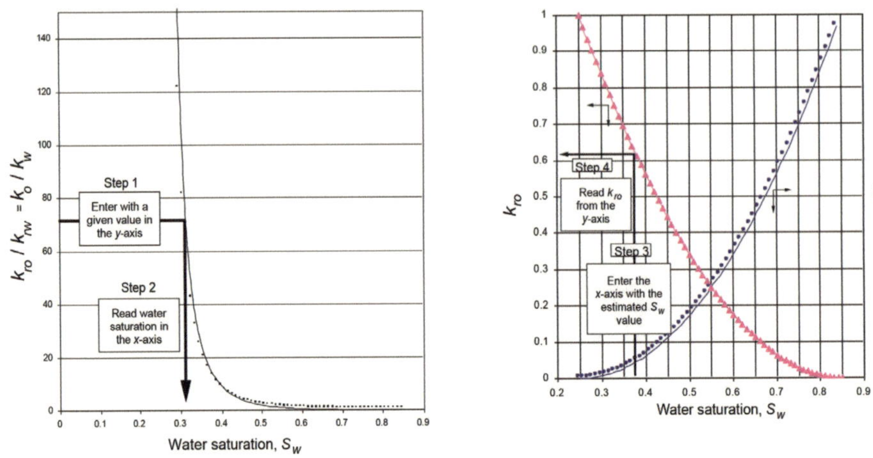

Figure 2.12.
Determination of absolute permeability as outlined by Kamal and Pan [88].

range and only a small part of the casing is perforated, spherical flow occurs in the region near the face of the well. See **Figure 2.13** bottom. As the transient moves further into depth of the formation, the flow becomes radial, but if the test is short, the flow will be spherical. Both types of flow are characterized by a slope of −1/2 in the log-log plot of pressure derivative versus time [90, 91]. Theoretically, before either hemispherical or spherical flow takes place, there exists a radial flow regime occurring by fluids withdrawn from the formation thickness that is close in height to the completion interval. This represents the transmissibility of the perforated interval. Actually, this flow regime is unpractical to be seen mainly because of wellbore storage effects. We will see further in this chapter that there are especial

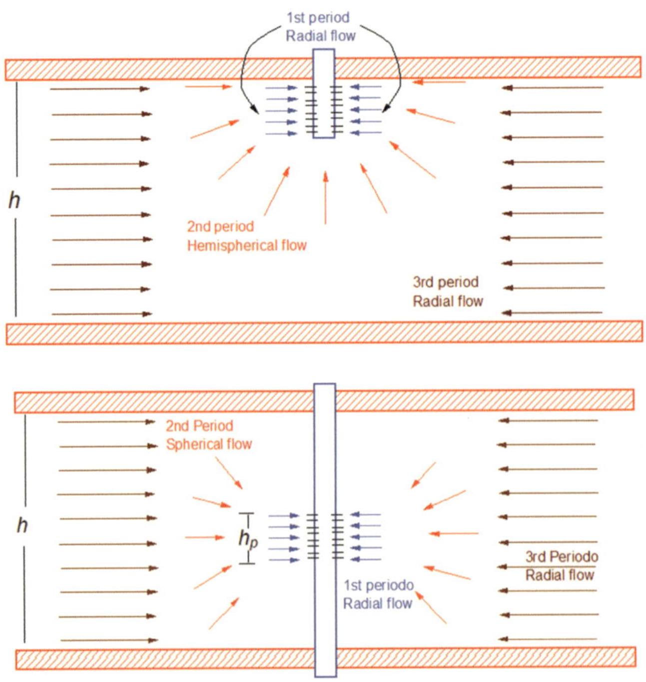

Figure 2.13.
Ideal flow regimes in partial penetration (top) and partial completion (bottom) systems, after [66].

conditions for hemispherical/spherical flow to be observed which occur later that the completion-interval-limited radial flow regime. Both hemispherical and spherical flow vanished when the top and bottom boundaries have been fully reached by the transient wave; the true radial flow is developed throughout the full reservoir thickness.

The apparent skin factor, s_a, obtained from pressure transient analysis is a combination of several "pseudoskin" factors such as [91]:

$$s_a = s + s_p + s_\theta + s_{cp} + \ldots\ldots \tag{2.122}$$

where s is the true damage factor caused by damage to the well portion, s_p is the pseudoskin factor due to the restricted flow entry, s_q is the pseudoskin factor resulting from a well deviation angle, and s_{cp} is the pseudoskin due to a change in permeability near the face of the well. s_p can be estimated from [92]:

$$s_p = \left(\frac{h}{h_p} - 1\right) \ln h_D \tag{2.123}$$

h_p = length of perforated or open interval. The equations of dimensionless thickness, h_D, for hemispherical and spherical flow, respectively:

$$h_D = \frac{h}{r_w}\sqrt{\frac{k_h}{k_z}} \tag{2.124}$$

$$h_D = \frac{h}{2r_w}\sqrt{\frac{k_h}{k_z}} \tag{2.125}$$

where k_h is the horizontal permeability, $k_z = k_v$ is the vertical permeability. The contribution of the pseudoskin of an inclined well is given by Cinco et al. [92]:

$$\psi' = \tan^{-1}\left(\sqrt{\frac{k_z}{k_h}}\tan\psi\right) \tag{2.126}$$

$$s_\theta = -\left(\frac{\psi'}{41}\right)^{2.06} - \left(\frac{\psi'}{56}\right)^{1.865} \times \log\left(\frac{h}{100r_w}\right) \tag{2.127}$$

According to Cinco et al. [92], the above equation is valid for $0° \leq q \leq 75°$, $h/r_w > 40$, and $t_D > 100$. Note that Eq. (2.127) could provide a negative value. This is because the deviation at the face of the well increases the flow area or presents reservoir pseudothickness. The pseudoskin responding for permeability changes near wellbore is given by [93]:

$$s_{cp} = \frac{h}{h_p}\left[1 - 0.2\left(\frac{r_s - r_w}{h_p}\right)\right]\left(\frac{k - k_s}{k_s}\right)\ln\frac{r_s}{r_w} \tag{2.128}$$

Example 2.4

Taken from [91]. A directional well which has an angle to the vertical of 24.1° has a skin factor $s = -0.8$. The thickness of the formation is 100 ft, the radius of the wellbore is 0.3 ft, and the horizontal to vertical permeability ratio is 5. Which portion of the damaged corresponds to the deviation of the well?

Solution

The deviation angle affected by the anisotropy is estimated with Eq. (2.126);

$$\psi' = \tan^{-1}\left(\sqrt{5}\tan(24.1)\right) = 45°$$

The pseudoskin factor caused by well deviation is found from Eq. (2.127):

$$s_\theta = -\left(\frac{45}{41}\right)^{2.06} - \left(\frac{45}{56}\right)^{1.865}\log\left(\frac{100}{100(0.3)}\right) = -1.56$$

From Eq. (2.122);

$$s_a = s + s_\theta = -0.8 - 1.56 = -2.36$$

Therefore, 66.1 % of the skin factor is due to the well deviation.

2.4.1 Conventional analysis for spherical flow

The diffusivity equation for spherical flow assuming constant porosity, compressibility, and mobility is given by Abbott et al. [90]:

$$\frac{1}{r^2}\frac{\partial}{\partial r}\left(r^2\frac{\partial P}{\partial r}\right) = \frac{\phi\mu c_t}{k_{sp}}\frac{\partial P}{\partial t} \tag{2.129}$$

where k_{sp} is the spherical permeability which is defined as the geometrical mean of the vertical and horizontal permeabilities:

$$k_{sp} = \sqrt[3]{k_v k_h^2} = k_{hs} \tag{2.130}$$

The physical system is illustrated in **Figure 2.14**, right. This region is called a "spherical sink." r_{sw} is given by:

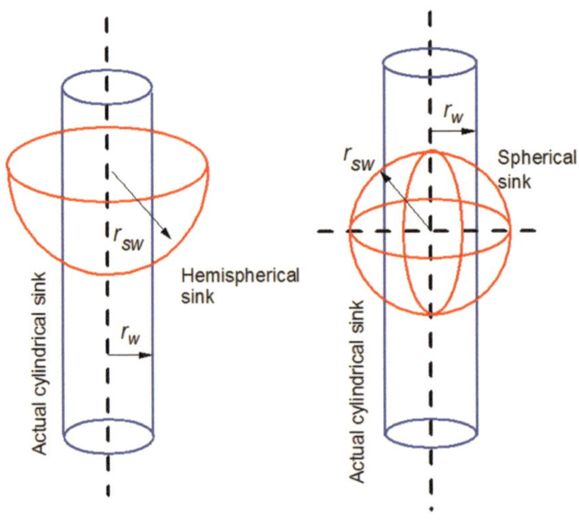

Figure 2.14.
Cylindrical, hemispherical, and spherical sinks, after [66].

$$r_{sw} = \frac{h_p}{2\ln\left(\frac{h_p}{r_w}\right)} \tag{2.131}$$

The spherical flow equations for pressure drawdown and pressure buildup when the flow time is much longer than the shut-in time were presented by [94]:

$$P_{wf} = P_i - \frac{70.6q\mu B}{k_{sp}r_{sw}}(1+s_{sp}) + \frac{2453q\mu B}{k_{sp}^{3/2}}\sqrt{\phi\mu c_t}\frac{1}{\sqrt{t}} \tag{2.132}$$

$$P_{ws} = P_{wf} + \frac{70.6q\mu B}{k_{sp}r_{sw}}(1+s_{sp}) - \frac{2453q\mu B}{k_{sp}^{3/2}}\sqrt{\phi\mu c_t}\left[\frac{1}{\sqrt{t_p}} + \frac{1}{\sqrt{\Delta t}} - \frac{1}{\sqrt{t_p+\Delta t}}\right] \tag{2.133}$$

The spherical pressure buildup equation when the flow time is shorter than the shut-in time:

$$P_{ws} = P_i - \frac{2453q\mu B}{k_{sp}^{3/2}}\sqrt{\phi\mu c_t}\left[\frac{1}{\sqrt{\Delta t}} - \frac{1}{\sqrt{t_p+\Delta t}}\right] + \frac{70.6q\mu B}{k_{sp}r_{sw}}s_{sp} \tag{2.134}$$

Then, from a Cartesian plot of P_{wf} as a function of $t^{-1/2}$, for drawdown, or P_{ws} as a function of either $[t_p^{-1/2} + \Delta t^{-1/2} - (t_p + \Delta t)^{-1/2}]$ or $[\Delta t^{-1/2} - (t_p + \Delta t)^{-1/2}]$ for buildup, we obtain a line which slope, m, and intercept, I, can be used to estimate tridimensional permeability and geometrical (spherical) skin factor.

$$k_{sp} = \left(-\frac{2453q\mu B}{m}\sqrt{\phi\mu c_t}\right)^{2/3} \tag{2.135}$$

$$s_{sp} = \frac{(I - P_{wf})k_{sp}r_{sw}}{70.6q\mu B} - 1 \tag{2.136}$$

Once the spherical permeability is known, we solve for the vertical permeability from Eq. (2.130), and then, estimate the value of skin effects due to partial penetration [94]:

$$s_c = \left(\frac{1}{b} - 1\right)[\ln h_D - G] \tag{2.137}$$

where $b = h_p/h$. h_D can be estimated from Eq. (2.125), and G is found from [94]:

$$G = 2.948 - 7.363b + 11.45b^2 - 4.576b^3 \tag{2.138}$$

2.4.2 Conventional analysis for hemispherical flow

The model for hemispheric flow is very similar to that of spherical flow [94]. The difference is that a boundary condition considers half sphere. **Figure 2.14** (left) outlines the geometry of such system. The drawdown and pressure equations are given below [94]:

$$P_{wf} = P_i - \frac{141.2q\mu B}{k_{sp}r_{sw}}(1+s_{sp}) + \frac{4906q\mu B}{k_{sp}^{3/2}}\sqrt{\phi\mu c_t}\frac{1}{\sqrt{t}} \tag{2.139}$$

$$P_{ws} = P_{wf} + \frac{141.2q\mu B}{k_{sp}r_{sw}}(1+s_{sp}) - \frac{4906q\mu B}{k_{sp}^{3/2}}\sqrt{\phi\mu c_t}\left[\frac{1}{\sqrt{t_p}} + \frac{1}{\sqrt{\Delta t}} - \frac{1}{\sqrt{t_p+\Delta t}}\right] \tag{2.140}$$

$$P_{ws} = P_i - \frac{4906q\mu B}{k_{sp}^{3/2}}\sqrt{\phi\mu c_t}\left[\frac{1}{\sqrt{\Delta t}} - \frac{1}{\sqrt{t_p+\Delta t}}\right] + \frac{141.2q\mu B}{k_{sp}r_{sw}}s_{sp} \tag{2.141}$$

As for the spherical case, from a Cartesian plot of P_{wf} as a function of $t^{-1/2}$, for drawdown, or P_{ws} as a function of either $[t_p^{-1/2} + \Delta t^{-1/2} - (t_p + \Delta t)^{-1/2}]$ or $[\Delta t^{-1/2} - (t_p + \Delta t)^{-1/2}]$ for buildup, we obtain a line which slope, m, and intercept, I, can be used to estimate spherical permeability and geometrical (spherical) skin factor.

$$k_{sp} = \left(-\frac{4906q\mu B}{m}\sqrt{\phi\mu c_t}\right)^{2/3} \tag{2.142}$$

$$s_{sp} = \frac{(I - P_{wf})k_{sp}r_{sw}}{141.2q\mu B} - 1 \tag{2.143}$$

2.4.3 *TDS* for spherical flow

Moncada et al. [66] presented the expressions for interpreting both pressure drawdown or buildup tests in either gas or oil reservoirs using the *TDS* methodology. Spherical permeability is estimated by reading the pressure derivative at any arbitrary time during which spherical flow can calculate spherical permeability and the spherical skin factor also uses the pressure reading at the same chosen time:

$$k_{sp} = \left(\frac{1227qB\mu}{(t*\Delta P')_{sp}}\sqrt{\frac{\phi\mu c_t}{t_{sp}}}\right)^{2/3} \tag{2.144}$$

$$s_{sp} = 34.74\sqrt{\frac{\phi\mu c_t r_{sw}^2}{k_{sp}t_{sp}}}\left[\frac{(\Delta P)_{sp}}{2(t*\Delta P')_{sp}} + 1\right] - 1 \tag{2.145}$$

The total skin, s_t, is defined as the sum of all skin effects at the well surroundings:

$$s_t = \frac{s}{b} + s_c + s_{sp} \tag{2.146}$$

If the radial flow were seen, the horizontal permeability can be estimated from:

$$k_H = k = \frac{70.6qB\mu}{h_p(t*\Delta P')_{r1}} \tag{2.147}$$

The suffix $r1$ implies the first radial flow.

Moncada et al. [66] observed that the value of the derivative for the late radial flow in spherical geometry is equivalent to 0.0066 instead of 0.5 as of the radial system. In addition, the slope line −½ corresponding to the spherical flow and the late radial flow line of the curve of the dimensionless pressure derivative in spherical symmetry intersect in:

$$(t_D * P_D')_i = \frac{1}{2\sqrt{\pi}}t_{Dsp}^{-1/2} = 0.0066 \tag{2.148}$$

Replacing the dimensionless time results:

$$t_i = 6927748.85\frac{\varphi\mu c_t r_{sw}^2}{k_{sp}} \tag{2.149}$$

In the above equation, suffix i denotes the intersection between the spherical flow and the late radial flow. If the radial flow is not observed, this time can give an initial point to draw the horizontal line corresponding to the radial flow regime, from which horizontal permeability is determined. This point can also be used to verify spherical permeability, k_{sp}. Another equation defining the mentioned dimensionless time can be found from the intersection of the slope line −½ (spherical flow) with the radial line of late radial flow but in radial symmetry, knowing that:

$$(t_D * P_D')_i = \frac{k^{3/2}h}{4k_{sp}^{3/2}\sqrt{\pi r_{sw}^2}}\frac{1}{\sqrt{t_D}} = 0.5 \tag{2.150}$$

Replacing the dimensionless time will give:

$$t_i = 301.77\frac{k^2h^2\phi\mu c_t}{k_{sp}^3} \tag{2.151}$$

Combining Eqs. (2.149) and (2.151), an expression to find the spherical wellbore radius, r_{sw}:

$$r_{sw} = 0.0066\frac{kh}{k_{sp}} \tag{2.152}$$

2.4.4 *TDS* for hemispherical flow

Here the same considerations are presented in Section 2.4.3. Using a pressure and a pressure derivative value reading at any time during hemispherical flow allows finding hemispherical permeability and partial penetration skin [66],

$$k_{hs} = \left(\frac{2453qB\mu}{(t * \Delta P')_{hs}}\sqrt{\frac{\phi\mu c_t}{t_{hs}}}\right)^{2/3} \tag{2.153}$$

$$s_{hs} = 34.74\sqrt{\frac{\phi\mu c_t r_{sw}^2}{k_{hs}t_{hs}}}\left[\frac{(\Delta P)_{hs}}{2(t * \Delta P')_{hs}} + 1\right] - 1 \tag{2.154}$$

Moncada et al. [66] also found that the derivative in spherical geometry of the late radial flow corresponds to 0.0033 instead of 0.5 as of the radial system. This time the line of radial flow and hemispheric flow, in hemispherical symmetry, intersect in:

$$(t_D * P_D')_i = \frac{1}{2\sqrt{\pi t_{Dsh}}} = 0.0033 \tag{2.155}$$

From where,

$$t_i = 27710995.41\frac{\phi\mu c_t r_{sw}^2}{k_{hs}} \tag{2.156}$$

As for the spherical case, there exists an expression to define the intersection time of the −½ slope line of the hemispherical flow regime pressure derivative and the late radial flow line pressure derivative but, now, in radial symmetry:

$$(t_D * P_D')_i = \frac{k^{3/2}h}{2k_{hs}^{3/2}\sqrt{\pi r_{sw}^2}}\frac{1}{\sqrt{t_D}} = 0.5 \tag{2.157}$$

$$t_i = 1207.09\frac{k^2h^2\phi\mu c_t}{k_{hs}^3} \tag{2.158}$$

This point of intersection in radial symmetry gives the following equation:

$$r_{sw} = 0.0033\frac{kh}{k_{hs}} \tag{2.159}$$

Skin factors are estimated in a manner similar to Section 2.4.3.

2.4.5 Wellbore storage and perforation length effects on hemispherical/spherical flow

It is important to identify the range of WBS values, which can influence the interpretation of the spherical and hemispheric flow regime. **Figure 2.15** is a plot of P_D vs. t_D providing an idea of the storage effect. As can be seen, the pressure response for several C_D values can be distinguished when storage is low (<10), whereas for larger C_D values, the response is almost identical. For C_D < 10, the slope of −½ that characterizes both spherical and hemispherical flow is well distinguished. For values of 10 < C_D < 100, the slope of −½ is more difficult to identify. For values of C_D > 100, the spherical flow regime has been practically masked by storage, which makes it impossible to apply the technique presented above to estimate the vertical permeability. Then, to ensure there is no C_D masking, it should be less than 10 [66].

The length of the completed interval or the length of the partial penetration, h_p, plays an important role in defining the spherical/hemispherical flow. The presence of spherical or hemispheric flow is characterized by a slope of −½. This characteristic slope of −½ is absent when the penetration ratio, $b = h_p/h$, is greater than 20% [66], as shown in **Figure 2.16**.

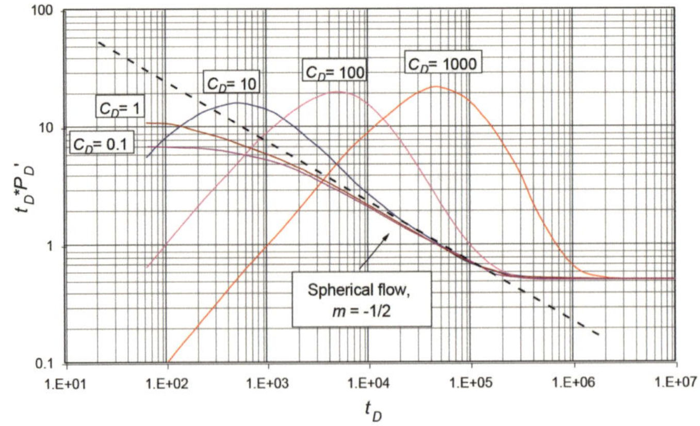

Figure 2.15.
Pressure derivative spherical source solution for a single well in an infinite system including WBS and no skin, after [66].

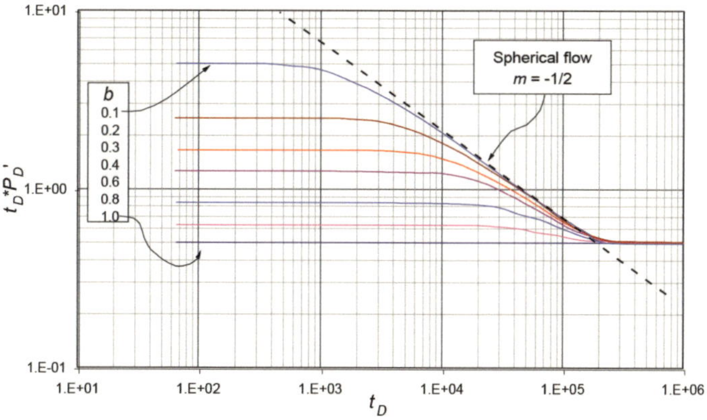

Figure 2.16.
Pressure derivative behavior for a single well in an infinite reservoir with different partial penetration lengths (C_D = 0, s = 0), *after [66].*

Example 2.5

Abbott et al. [90] presented pressure-time data for a pressure drawdown test. Well no. 20 is partially completed in a massive carbonate reservoir. The well was shut-in for stabilization and then flowed to 5200 BOPD for 8.5 h. The pressure data are given in **Table 2.4** and reservoir and fluid properties are given below:

h = 302 ft	r_w = 0.246 f	P_i = 2298 psia
h_p = 20 ft	q = 5200 BPD	B = 1.7 bbl/STB
φ = 0.2	μ = 0.21 cp	$c_t = 34.2 \times 10^{-6}$ psia^{-1}

Solution by conventional analysis

Using the slope value of −122 psia/cycle read from the semilog plot of **Figure 2.17**, the reservoir permeability is calculated with Eq. (2.38);

$$k = \frac{162.6qB\mu}{mh} = \left|\frac{162.6(5200)(1.7)(0.21)}{(-122)(302)}\right| = 8.19 \text{ md}$$

The mechanical skin factor is determined with Eq. (2.39) once the intercept of 2252 psia is read from **Figure 2.17**.

$$s = 1.1513\left[\frac{2252 - 2298}{-122} - \log\left(\frac{8.19}{(0.2)(0.21)(34.2\times10^{-6})(0.246)^2}\right) + 3.2275\right] = -5.03$$

Figure 2.18 contains a Cartesian graph of P_{wf} as a function of $t^{-1/2}$. From there, the observed slope is m = 250 psia (h$^{-1/2}$) and intercept, I = 2060 psia, spherical permeability, and spherical skin factors are calculated using Eqs. (2.239) and (2.240), respectively:

$$k_{sp} = \left(\frac{2453q\mu B}{m}\sqrt{\phi\mu c_t}\right)^{2/3} = \left(\frac{\frac{2453(5200)(0.21)(1.7)}{250}}{\sqrt{(0.2)(0.21)(34.2\times 10^{-6}}}\right)^{2/3} = 7.81 \text{ md}$$

t, h	$t^{-0.5}$, h$^{-0.5}$	P_{wf}, psia	ΔP, psia	$t^*\Delta P'$, psia
0.0		2266	0	
0.5	1.414	2255	11	11.5
1.0	1.000	2243	23	24.5
1.6	0.791	2228	38	40.0
2.0	0.707	2218	48	45.0
2.5	0.632	2208	58	52.5
3.0	0.577	2197	69	69.0
3.5	0.535	2185	81	66.5
4.0	0.500	2178	88	60.0
4.5	0.471	2170	96	56.3
5.5	0.426	2161	105	46.8
6.0	0.408	2157	109	48.0
6.5	0.392	2153	113	52.0
7.0	0.378	2149	117	49.0
7.5	0.365	2146	120	52.5
8.0	0.354	2142	124	48.0
8.5	0.343	2140	126	

Table 2.4.
Pressure and pressure derivative versus time data for example 2.5.

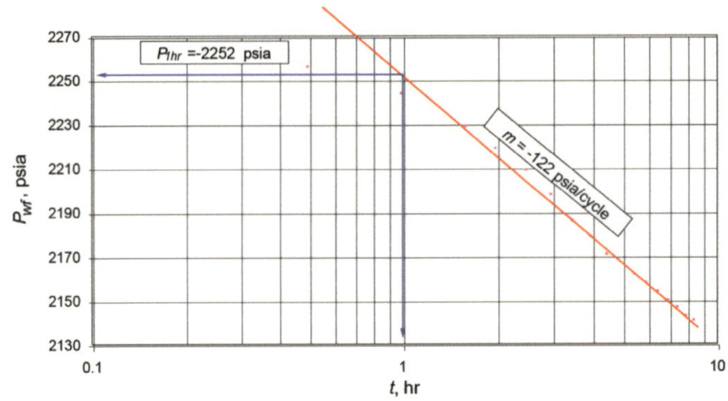

Figure 2.17.
Semilog plot for well no. 20.

$$s_{sp} = \frac{(P_i - I)k_{sp}r_{sw}}{70.6q\mu B} - 1 = \frac{(2298 - 2060)(7.81)(9.69)}{70.6(5200)(0.21)(1.7)} - 1 = -0.86$$

Vertical permeability and spherical wellbore radius are found with Eq. (2.130) and (2.131), respectively,

$$k_v = \frac{k_{sp}^3}{k_h^2} = \frac{7.81^3}{8.19^2} = 7.1 \text{ md}$$

$$r_{sw} = \frac{b}{2\ln\left(\frac{b}{r_w}\right)} = \frac{120}{2\ln\left(\frac{120}{0.246}\right)} = 9.69 \text{ ft}$$

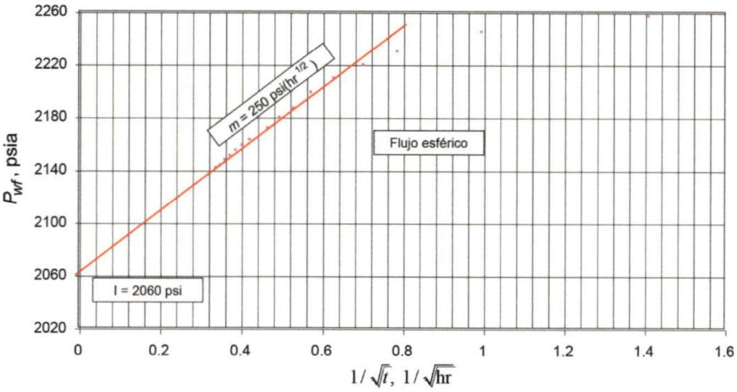

Figure 2.18.
Cartesian spherical plot for well no. 20.

With the value of the vertical permeability, it is possible to estimate the skin factor caused by partial penetration with Eqs. (2.125), (2.138), and (2.137), thus:

$$h_D = \sqrt{\frac{k_h}{k_v}}\left(\frac{h}{r_w}\right) = \sqrt{\frac{8.26}{7.1}}\left(\frac{302}{0.246}\right) = 1324.1$$

$$G = 2.948 - 7.363b + 11.45b^2 - 4.675b^3$$

$$= 2.948 - 7.363\left(\frac{120}{302}\right) + 11.45\left(\frac{120}{302}\right)^2 - 4.675\left(\frac{120}{302}\right)^3 = 1.57$$

$$s_c = \left(\frac{1}{h_p/h} - 1\right)[\ln\ h_D - G] = \left(\frac{1}{120/302} - 1\right)[\ln\ 1318.5 - 1.57] = 8.51$$

Solution by *TDS* technique

The following data points were read from **Figure 2.19**.

t_N = 1 h	ΔP = 23 psia	
$(t^*\Delta P')_{sp}$ = 56.25 psia	ΔP_s = 96 psia	t_{sp} = 4.5 h
$(t^*\Delta P')_{r2}$ = 52.5 psia	ΔP_{r2} = 96 psia	t_{r2} = 7.5 h

Wellbore storage coefficient is found from Eq. (2.66)

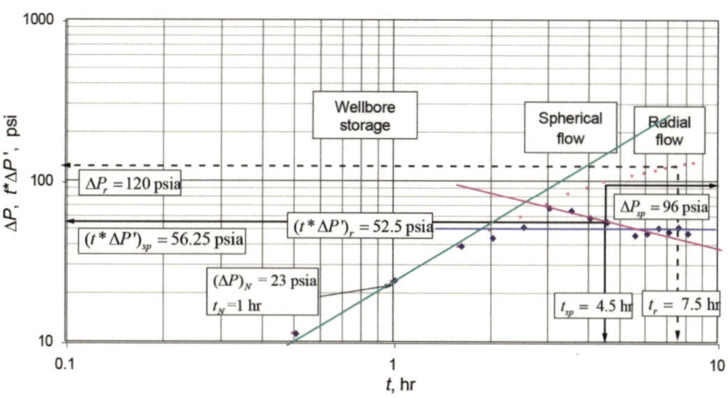

Figure 2.19.
Pressure and pressure derivative versus time log-log plot for well no. 20.

Parameter	Conventional	*TDS*
k_{sp}, md	7.01	8.05
s_{sp}	−0.86	−0.93
$\bar{k}$, md	8.19	8.26
s_r	−5.03	−5.53
k_v, md	7.10	7.65

Table 2.5.
Comparison of results.

$$C = \left(\frac{qB}{24}\right)\frac{t_N}{(\Delta P)_N} = \left(\frac{(5200)(1.7)}{24}\right)\frac{1}{23} = 16.01 \text{ bbl/psi}$$

From the spherical flow pressure derivative line, $m = -1/2$, the spherical permeability and mechanical spherical skin factor are, respectively, estimated by Eqs. (2.144) and (2.145);

$$k_{sp} = \left(\frac{1227qB\mu}{(t*\Delta P')_{sp}}\sqrt{\frac{\phi\mu c_t}{t_{sp}}}\right)^{2/3} = \left(\frac{1227\frac{(5200)(1.7)(0.21)}{56.25}}{\sqrt{\frac{(0.2)(0.21)(34.2\times10^{-6})}{4.5}}}\right)^{2/3} = 8.05 \text{ md}$$

$$s_{sp} = 34.74\sqrt{\frac{\phi\mu c_t r_{sw}^2}{k_{sp}t_{sp}}}\left[\frac{(\Delta P_w)_{sp}}{2(t*\Delta P')_{sp}}\right] - 1$$

$$s_{sp} = 34.74\sqrt{\frac{(0.2)(0.21)(34.2 \text{ x } 10^{-6})(9.69^2)}{(8.05)(4.5)}}\left[\frac{(96)}{2(56.25)}\right] - 1 = -0.93$$

The horizontal permeability and mechanical skin are found during the late radial flow using Eqs. (2.76) and (2.97), respectively;

$$k_r = \frac{70.6qB\mu}{h(t*\Delta P')_r} = \frac{70.6(5200)(1.7)(0.21)}{(302)(52.5)} = 8.26 \text{ md}$$

$$s = 0.5\left[\frac{\Delta P_r}{(t*\Delta P')_r} - \ln\left(\frac{k_r t_r}{\phi\mu c_t r_w^2}\right) + 7.43\right] = 0.5\left[\frac{120}{52.5} - \ln\left(\frac{(8.26)(7.5)}{(0.2)(0.21)(34.2\times10^{-6})(0.246^2)}\right) + 7.43\right] = -5.33$$

Vertical permeability is determined from Eq. (2.130);

$$k_v = \frac{k_{sp}^3}{k_h^2} = \frac{8.05^3}{8.26^2} = 7.65 \text{ md}$$

Table 2.5 presents the comparison of the results obtained by the conventional method and *TDS* technique.

2.5. Multirate testing

So far, the considerations revolve around a single flow test, meaning the production rate is kept constant for the application of the solution of the diffusivity equation. However, there are cases in which the flow rate changes; in such cases, the use of the solution to the diffusivity equation requires the application of the time superposition principle already studied in Section 1.14.2. Some reasons for the use of multirate testing are outlined as follows:

- It is often impractical to keep a constant rate for a long time to perform a complete pressure drawdown test.
- When the well was not shut-in long enough to reach the static pressure before the pressure drawdown test started. It implies superposition effects.
- When, it is not economically feasible shutting-in a well to run a pressure buildup test.

Whether the production rates are constant or not during the flow periods, there are mainly four types of multirate tests:

1. Uncontrolled variable flow rate;
2. Series of constant flow rates;
3. Variable flow rate while keeping constant bottom-hole pressure, Pwf. This test is common in gas wells producing very tight formations and more recently applied on testing of shale formations;
4. Pressure buildup (fall-off) tests.

Actually, a holistic classification of transient well testing is given in **Figure 1.4**. It starts with PTA which is known in the oil argot as pressure transient analysis. As seen in the figure, it is divided in single well tests, normally known as drawdown (flow) tests for production cases or injection tests for injection fluid projects. Our field of interest focuses on more than one rate operation (multirate testing) which includes all the four types just above described. It is worth to mention types 3 and 4. Type three is also known as rate transient analysis (RTA) which has been dealt with in a full chapter by this book's author in reference [56]. As far as case 4 is concerned, pressure buildup testing is the most basic multirate test ever existed since it comprises two flow rates: (1) one time period at a given q value different than zero and (2) another time period with a zero flow rate. This is because when a well is shut-in, the flow stops at surface by the formation keeps still providing fluid to the well due to inertia.

2.5.1 Conventional analysis

Considering the sketch of **Figure 2.20**, application of the superposition principle [2–4, 6, 7, 11, 27, 44, 56, 60, 65, 67, 95, 96, 97] leads to:

$$P_{wf}(t) = P_i - \frac{141.2\mu B}{kh}\left\{\begin{array}{l} q_1[P_D(t_D)+s] + (q_2-q_1)[P_D([t-t_1]_D)+s]+ \\ (q_3-q_2)[P_D([t-t_2]_D)+s] + (q_4-q_3)[P_D([t-t_3]_D)+s] \\ +\ldots. + (q_N-q_{N-1})[P_D([t-t_N]_D)+s] \end{array}\right\} \tag{2.160}$$

Rearranging;

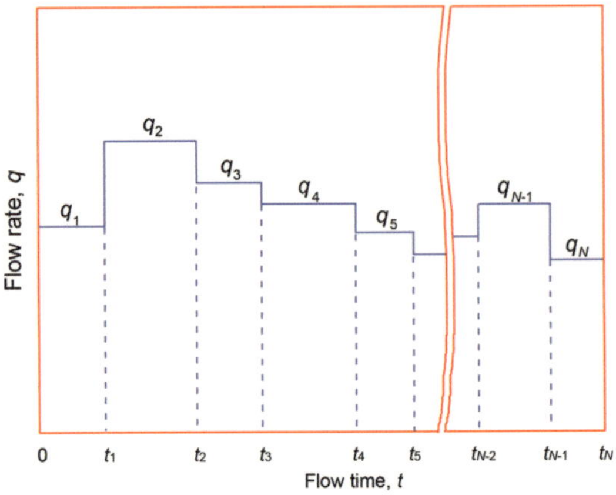

Figure 2.20.
Schematic representation of a multirate test (typ. 1).

$$P_{wf}(t) = P_i - \frac{141.2\mu B}{kh}\left\{\begin{array}{l} q_1\{P_D(t_D) - P_D([t-t_1]_D\} + q_2\{P_D([t-t_1])_D - \\ P_D([t-t_2])_D\} + \ldots + q_{N-1}\{P_D([t-t_{N-2}])_D - \\ P_D([t-t_{N-1}])_D\} + +q_N\{P_D([t-t_{N-1}])_D\} + s \end{array}\right\} \quad (2.161)$$

Next step is to replace P_D by an appropriate diffusivity equation solution which depends upon the flow regime dealt with. **Figure 2.21** presents the most typical superposition functions applied to individual flow regimes. The normal case is to use radial flow, top function in **Figure 2.21**. However, Escobar et al. [44] presented the inconvenience of not applying the appropriate superposition function for a given flow regime. They found, for instance, that if the radial superposition is used, instead of the linear, for characterization of an infinite-conductivity hydraulic fracture, the estimated half-fracture length would be almost three times longer than the actual one.

Coming back to Eq. (2.161), the assumed superposition function to be used is the radial one; then, this equation becomes:

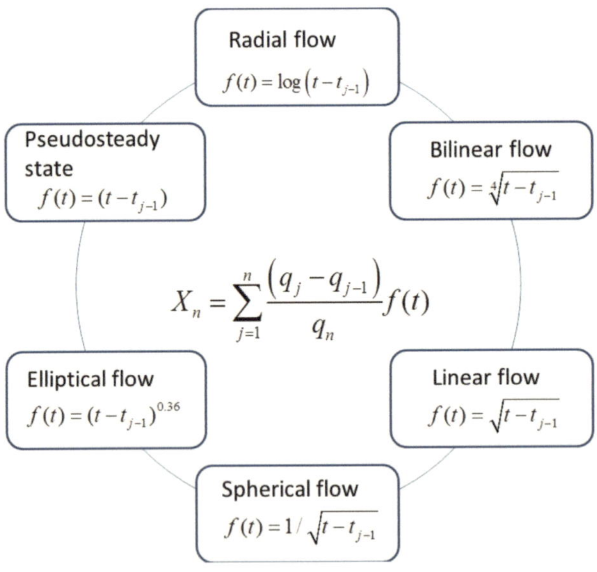

Figure 2.21.
Flow regime superposition functions.

$$P_{wf}(t)=P_i-\frac{70.6\mu B}{kh}\left\{\begin{array}{l} q_1\ln\left(\dfrac{t}{t-t_1}\right)+q_2\ln\left(\dfrac{t-t_1}{t-t_2}\right)+q_3\ln\left(\dfrac{t-t_2}{t-t_3}\right)+ \\ q_{N-1}\ln\left(\dfrac{t-t_{N-2}}{t-t_{N-1}}\right)+q_N\{\ln(t-t_{N-1})\}+ \\ \ln\dfrac{k}{\phi\mu c_t r_w^2}-7.4316+2s \end{array}\right\} \tag{2.162}$$

Since it is uneasy to find natural log paper in the stationary shops, then, dividing for the natural log of 10 is recommended to express Eq. (2.162) in decadic log; then,

$$P_{wf}(t)=P_i-\frac{162.6\mu B}{kh}\left\{\begin{array}{l} \sum\limits_{j=1}^{N-1} q_j\log\left(\dfrac{t-t_{j-1}}{t-t_j}\right)+q_N\{\log(t-t_{N-1})\}+ \\ \log\dfrac{k}{\phi\mu c_t r_w^2}-3.2275+0.8686s \end{array}\right\} \tag{2.163}$$

Simplifying;

$$\frac{P_i-P_{wf}(t)}{q_N}=\frac{162.6\mu B}{kh}\left\{\begin{array}{l} \sum\limits_{j=1}^{N}\left(\dfrac{q_j-q_{j-1}}{q_N}\right)\log(t-t_{j-1})+\log\dfrac{k}{\phi\mu c_t r_w^2} \\ -3.2275+0.8686s \end{array}\right\} \tag{2.164}$$

Let;

$$s'=\log\frac{k}{\phi\mu c_t r_w^2}-3.23+0.87\cdot s \tag{2.165}$$

$$m'=\frac{162.6\mu B}{kh} \tag{2.166}$$

Solving for skin factor from Eq. (2.165);

$$s=1.1513\left[\frac{b'}{m'}-\log\frac{k}{\phi\mu c_t r_w^2}+3.23\right] \tag{2.167}$$

$$X_n=\sum_{i=1}^{n}\left(\frac{q_i-q_{i-1}}{q_n}\right)\log(t-t_{i-1}) \tag{2.168}$$

Plugging Eqs. (2.165), (2.166), and (2.168) into Eq. (2.164) will lead to:

$$\frac{P_i-P_{wf}(t)}{q_n}=m'X_n+m's' \tag{2.169}$$

which indicates that a Cartesian plot of $\Delta P/q_n$ against the superposition time, X_n, provides a straight line which slope, m', and intercept, $m'b'$ allows finding reservoir permeability and skin factor using Eqs. (2.166) and (2.167), respectively. However, it is customary for radial flow well interpretation to employ a semilog plot instead of a Cartesian plot. This issue is easily solved by taking the antilogarithm to the superposition function resulting into the equivalent time, t_{eq}. Under this situation, Eq. (2.169) becomes:

$$\frac{P_i - P_{wf}(t_n)}{q_n} = m_n \log t_{eq} + b_n \tag{2.170}$$

And the equivalent time is then defined by,

$$\frac{P_i - P_{wf}(t_n)}{q_n} = m_n \log t_{eq} + b_n \tag{2.171}$$

$$t_{eq} = \prod_{i=1}^{n} (t_n - t_{i-1})^{(q_i - q_{i-1})/q_n} = 10^{X_n} \tag{2.172}$$

For a two-rate case, Russell [96] developed the governing well-flowing pressure equation, as follows:

$$P_{wf} = m'_1 \left[\log\left(\frac{t_1 + \Delta t}{\Delta t}\right) + \frac{q_2}{q_1} \log(\Delta t) \right] + P_{INT} \tag{2.173}$$

Therefore, the slope, m'_1, and intercept, P_{INT}, of a Cartesian plot of P_{wf} versus $\log[(t_1+\Delta t)/\Delta t] + (q_2/q_1)\log(\Delta t)$ allows finding permeability and skin factor from the following relationships:

$$k = -\frac{162.6 q_1 \mu B}{m_1' h} \tag{2.174}$$

$$s = 1.1513 \left[\frac{q_1}{q_1 - q_2} \left(\frac{P_{wf}(\Delta t = 0) - P_{1hr}}{m'_1} \right) - \log \frac{k}{\phi \mu c_t r_w^2} + 3.23 \right] \tag{2.175}$$

In general, the lag time, t_{lag}, transition occurred during the rate change, is shorter when there is a rate reduction than a rate increment, i.e., if $q_2 < q_1$, then the t_{lag} will be short and if $q_2 > q_1$, then the t_{lag} will be longer due to wellbore storage effects.

The pressure drop across the damage zone is:

$$\Delta P_s(q_1) = -0.87(m'_1)s \tag{2.176}$$

$$\Delta P_s(q_2) = -0.87 \frac{q_2}{q_1} (m'_1)s \tag{2.177}$$

And;

$$P^* = P_{\text{int}} - \frac{q_1}{(q_2 - q_1)} \left[P_{wf}(\Delta t = 0) - P_{1hr} \right] \tag{2.178}$$

P^* is known as “false pressure” and is often used to estimate the average reservoir pressure which is treated in Chapter 3.

2.5.2 *TDS* technique

The mathematical details of the derivation of the equations are presented in detail by Perrine [86]. Application of *TDS* technique requires estimating the following parameters:

$$\Delta P_q = \frac{P_i - P(t_n)}{q_n} \tag{2.179}$$

$$t_n = t_{n-1} + \Delta t \tag{2.180}$$

And equivalent time, t_{eq}, estimation is achieved using Eq. (2.172). Mongi and Tiab [67] suggested for moderate flow rate variation, to use real time rather than equivalent time with excellent results. In contrast, sudden changes in the flow rate provide unacceptable results. However, it is recommended here to always use equivalent time as will be demonstrated in the following exercise where using equivalent time the pressure derivative provides a better description. Mongi and Tiab [67] also recommended that test data be recorded at equal intervals of time to obtain smoother derivatives. However, it is not a practical suggestion since derivative plot is given in log coordinates. *TDS* is also applicable to two-rate tests and there is also a *TDS* technique where there is a constant flow rate proceeded by a variable flow rate. For variable injection tests, refer to [60].

With the equivalent time, Eq. (2.172) determines the pressure derivative, $t_{eq}{}^{*}$ $(DP/q)'$, and plot the derivative in a similar fashion as in Section 2.2.4; wellbore storage coefficient can be obtained by taking any point on the early-time unit-slope line by:

$$C = \left(\frac{B}{24}\right)\left(\frac{t}{\Delta P_q}\right) \tag{2.181}$$

Permeability and mechanical skin factor are estimated from:

$$k = \frac{70.6\mu B}{h\left(t_{eq} * \Delta P'_q\right)_r} \tag{2.182}$$

$$s = 0.5\left[\frac{(\Delta P_q)_r}{\left(t_{eq} * \Delta P_q'\right)_r} - \ln\left(\frac{k(t_{eq})_r}{\phi\mu c_t r_w^2}\right) + 7.43\right] \tag{2.183}$$

Once again, rigorous time instead of equivalent time can be used in Eqs. (2.182) and (2.183); however, a glance to **Figure 2.23** and **2.24** tells us not to do so.

Example 2.6

Earlougher and Kersch [8] presented an example to estimate permeability using a Cartesian plot of flowing pressure, P_{wf}, versus superposition time, X_n, and demonstrated the tedious application of Eq. (2.168). A slope of 0.227 psia/(BPD/cycle) was estimated which was used in Eq. (2.166) to allow finding a permeability value of 13.6. We determined an intercept of 0.5532 psia/(BPD/cycle) which led us to find a skin factor of −3.87 with Eq. (2.167).

Use semilog conventional analysis and *TDS* technique to find reservoir permeability and skin factor, as well. Pressure and rate data are given in **Table 2.6** along another parameters estimated here. Reservoir, fluid, and well parameters are given below:

P_i = 2906 psia	B = 1.27 bbl/STB	μ = 0.6 cp	
h = 40 ft	r_w = 0.29 ft	φ = 11.2%	$c_t = 2.4 \times 10^{-6}$1/psia

Solution by semilog conventional analysis

Figure 2.22 is a semilog graph of $[P_i - P_{wf}(t)]/q_n$ versus t and t_{eq}. The purpose of this graph is to compare between the rigorous analysis using equivalent time, t_{eq}, and analysis using the real time of flow, t. Note that during the first cycle, the graphs of t

n	t, h	q, BPD	P_{wf}, psia	ΔP, psia	$\Delta P/q$, psia/BPD	X_n	t_{eq}, h	$t^*(\Delta P/q)'$, psia/BPD	$t_{eq}{}^*(\Delta P/q)'$, psia/BPD
		0	2906						
1	1	1580	2023	883	0.559	0.000	1.000	0.559	0.261
1	1.5	1580	1968	938	0.594	0.176	1.500	0.594	0.131
1	1.89	1580	1941	965	0.611	0.276	1.890	0.611	0.102
1	2.4	1580							
2	3	1490	**1892**	**1014**	0.681	0.519	3.306	0.681	0.099
2	3.45	1490	**1882**	**1024**	0.687	0.569	3.707	0.687	0.103
2	3.98	1490	**1873**	**1033**	0.693	0.624	4.208	0.693	0.099
2	4.5	1490	1867	1039	0.697	0.673	4.712	0.697	0.095
2	4.8	1490							
3	5.5	1440	1853	1053	0.731	0.787	6.124	0.731	0.104
3	6.05	1440	1843	1063	0.738	0.819	6.596	0.738	0.111
3	6.55	1440	1834	1072	0.744	0.849	7.056	0.744	0.120
3	7	1440	1830	1076	0.747	0.874	7.481	0.747	0.128
3	7.2	1440							
4	7.5	1370	1827	1079	0.788	0.974	9.412	0.788	0.148
4	8.95	1370	1821	1085	0.792	1.009	10.212	0.792	0.154
4	9.6	1370							
5	10	1300	1815	1091	0.839	1.124	13.311	0.839	0.192
5	12	1300	1797	1109	0.853	1.153	14.239	0.853	0.188
6	14.4	1260							
7	15	1190	1775	1131	0.950	1.337	21.746	0.950	0.205
7	18	1190	1771	1135	0.954	1.355	22.662	0.954	0.206
7	19.2	1190							
8	20	1160	1772	1134	0.978	1.423	26.457	0.978	0.208
8	21.6	1160							
9	24	1137	1756	1150	1.011	1.485	30.553	1.011	0.208
10	28.8	1106							
11	30	1080	1751	1155	1.069	1.607	40.426	1.069	0.248
11	33.6	1080							
12	36	1000							
13	36.2	983	1756	1150	1.170	1.788	61.414	1.170	0.447
13	48	983	1743	1163	1.183	1.799	63.020	1.183	0.463

**The three last columns are not given in [8].*

Table 2.6.
*Pressure and rate data for example 2.6, after [8].**

and t_{eq} are practically the same. By regression for the real-time case gave a slope m' = 0.2411 psia/BPD/cycle and intercept $\Delta P/q_{(1hr)}$ = 0.553 psia/BPD/cycle. Permeability and skin factors are calculated with the Eqs. (2.166) and (2.167), respectively:

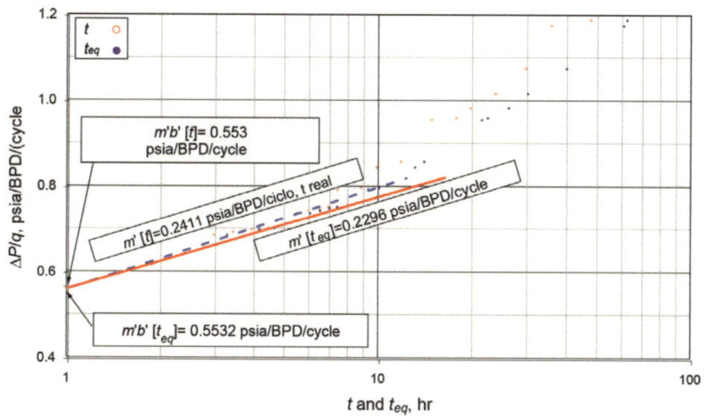

Figure 2.22.
Semilog of normalized pressure versus actual and equivalent time for example 2.6.

$$k = \frac{162.6\mu B}{m'h} = \frac{162.6(1.27)(0.6)}{0.2411(40)} = 12.84 \text{ md}$$

$$s = 1.1513\left[\frac{0.553}{0.2411} - \log\left(\frac{12.84}{(0.112)(0.6)(2.4\times10^{-6})(0.29^2)}\right) + 3.23\right] = -3.98$$

The straight line with t_{eq} has a slope m' = 0.2296 psia/BPD/cycle, and intercept $\Delta P/q(_{1\text{hr}})$ = 0.5532 psia/BPD/cycle. Then, permeability and skin factor estimated by Eqs. (2.166) and (2.167) are 13.49 md and −3.87, respectively.

Solution by *TDS* technique

The derivative of normalized pressure is also reported in **Table 2.6**. **Figure 2.23** illustrates a log-log plot of ΔP_q versus t_{eq} and $(t^*\Delta P'_q)$ and $(t_{eq}{}^*\Delta P'_q)$ versus t and t_{eq}. Both derivatives were estimated with a smooth value of 0.5. During the first cycle, the two sets of data have roughly the same trend; also the flow regimes are quite different. Also, the equivalent normalized pressure derivative suggests a faulted system and possibly the pseudosteady-state period has been reached. This last situation is unseen in the normalized pressure derivative. From this graph, the following values are read:

$(t^*\Delta P'_q)_r$ = 0.097 psia/BPD/cycle	$(\Delta P_q)_r$ = 0.693 psia/BPD/cycle	$(t_{eq})_r$ = 4.208 h

Permeability and skin factor are estimated, respectively, using Eqs. (2.182) and (2.183):

$$k = \frac{70.6\mu B}{h(t*\Delta P'_q)_r} = \frac{(70.6)(1.27)(0.6)}{0.097(40)} = 13.86 \text{ md}$$

$$s = 0.5\left[\frac{0.693}{0.097} - \ln\left(\frac{(13.86)(4.208)}{(0.112)(0.6)(2.4\times10^{-6})(0.29^2)}\right) + 7.43\right] = -3.804$$

The comparison of the results obtained by the different methods is summarized in **Table 2.7**. The permeability absolute deviation with respect to arithmetic mean is less than 5% using actual time. Note that all results agree well. Even though, when Earlougher and Kersch [8] written, pressure derivative function was still in diapers; then, it was not possible to differentiate the second straight-line which for Earlougher and Kersch [8] corresponded to pseudosteady-state period instead of a fault as clearly seen in **Figure 2.23**. Also, the absolute deviation of the flow rate (referred to the first value) is less than 10% during radial flow regime. However,

Methodology	***k*, md**	***s***
Superposition time, Cartesian plot	13.6	−3.87
Equivalent time, semilog plot	13.49	−3.87
Actual time, semilog plot	12.84	−3.98
TDS	13.86	−3.794
Average	13.45	−3.88

Table 2.7.
Comparison of estimated results of example 2.6.

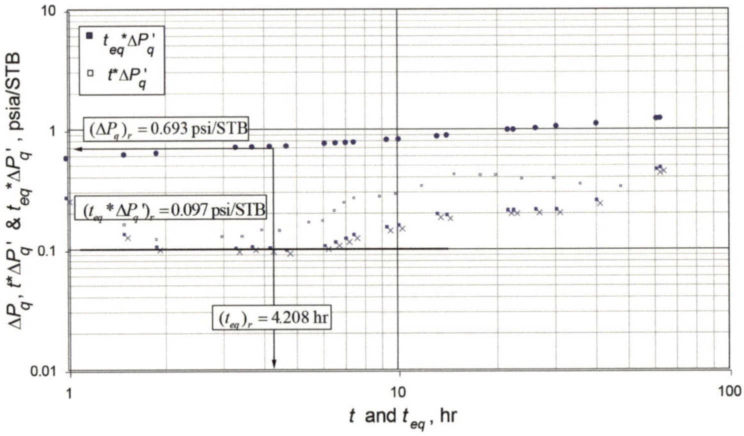

Figure 2.23.
Normalized pressure and pressure derivative versus time and equivalent time log-log plot for example 2.6.

when using real time, the radial flow regime is different; then, the recommendation is to always use equivalent time.

2.6. Pressure drawdown tests in developed reservoirs

Slider [11, 98, 99] suggested a methodology to analyze pressure tests when there are no constant conditions prior to the test. **Figure 2.24** schematizes a well with the shutting-in pressure declining (solid line) before the flow test started at a time t_1. The dotted

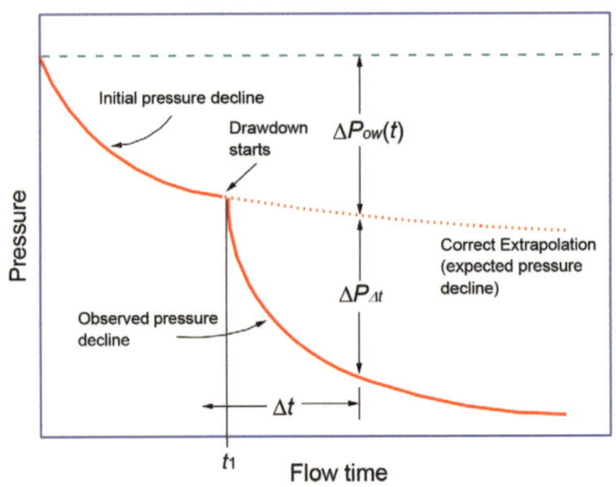

Figure 2.24.
Behavior of a declination test in a depleted well, after [11, 21].

line represents future extrapolation without the effect of other wells in the reservoir. The production starts at t_1 and the pressure behaves as shown by the solid line [11].

2.6.1 Conventional analysis

The procedure suggested by Slider [11, 99] to correctly analyze such tests is presented below:

a. Extrapolate the shutting-in pressure correctly (dotted line in **Figure 2.24**).

b. Estimate the difference between the observed well-flowing pressure and the extrapolated pressure, $\Delta P_{\Delta t}$.

c. Graph $\Delta P_{\Delta t}$ vs. Log Δt. This should give a straight line which slope and intercept can be used for estimation of permeability and skin factor using Eqs. (2.38) and (2.39), respectively. For this particular case, Eq. (2.39) is rewritten as:

$$s = 1.1513\left[\frac{(\Delta P_{\Delta t})_{1hr}}{m} - \log\left(\frac{k}{\phi\mu c_t r_w^2}\right) + 3.23\right] \quad (2.184)$$

However, this analysis could be modified as follows [8, 11, 21, 98, 99]. Consider a shut-in developed with other wells in operation. There is a pressure decline in the shut-in well resulting from the production of the other wells (superposition). After the test, well has been put into production at time t_1, its pressure will be:

$$P_{wf} = P_i - \frac{141.2q\mu B}{kh}[P_D(\Delta t_D,\ r_D = 1, \dots) + s] - \Delta P_{ow}(t) \quad (2.185)$$

According to **Figure 2.24**, $\Delta P_{wo}(t)$ is the pressure drop referred to P_i caused by other wells in the reservoir and measured at a time $t = t_1 + \Delta t$. $\Delta P_{wo}(t)$ can be estimated by superposing by:

$$\Delta P_{ow}(t) = P_i - P_w(t) = \frac{141.2\mu}{kh}\sum_{j=2}^{n} q_j B_j P_D(t_D,\ r_{Dj}\dots) \quad (2.186)$$

Eq. (2.186) assumes that all wells start to produce at $t = 0$. This is not always true. Including wells that start at different times require a more complex superposition. If the other wells in the reservoir operate under pseudosteady-state conditions, as is usually the case, Eq. (2.152) becomes:

$$\Delta P_{ow}(t) = b - m^* t \quad (2.187)$$

The slope, m^*, is negative when $\Delta P_{wo}(t)$ vs. t is plotted. Instead, it is positive, if P_w vs. t is plotted. m^* is estimated before the test well is opened in production at the pressure decline rate:

$$m^* = \frac{dP_{ws}}{dt} = \frac{(P_{ws})_2 - (P_{ws})_1}{t_2 - t_1} \quad (2.188)$$

If pressure data is available before the test, m^* can be easily estimated. Also, it can be estimated by an equation resulting from replacing Eq. (2.57) in (2.186):

$$m^* = \frac{-0.23395}{\phi c_t h A}\sum_{j=2}^{n} q_j B_j \tag{2.189}$$

The reservoir volume is given in ft^3. Combining Eq. (1.106) with r_D = 1, (1.94), (2.185), and (2.187), results:

$$P_{wf} - m^*\Delta t = m \log \Delta t + \Delta P_{1hr} \tag{2.190}$$

Eq. (2.190) indicates that a graph of P_{wf}–$m^*\Delta t$ vs. log Δt gives a straight line of slope m and intercept ΔP_{1hr} at Δt = 1 h. The permeability can be found from Eq. (2.38). The skin is estimated from an arrangement of Eq. (2.39):

$$s = 1.1513\left[\frac{\Delta P_{1hr} - P_{ws}(\Delta t = 0)}{m} - \log\left(\frac{k}{\phi \mu c_t r_w^2}\right) + 3.23\right] \tag{2.191}$$

2.6.2 *TDS* technique

TDS technique for developed reservoirs was extended by Escobar and Montealegre [21]. Escobar and Montealegre [21] showed that the technique could be applied taking the derivative to the pressure in a rigorous way, that is to say, without considering the effect of the production of other wells. As it will be seen in the example 2.7, this is not recommended since the derivative is not correctly defined and, therefore, the results could include deviations above 10%. In this case, it is advisable to correct or extrapolate the pressure by means of Eq. (2.192) and, then, take the extrapolated pressure derivative and apply the normal equations of the *TDS* technique given in Section 2.2.4. Needless to say that any *TDS* technique equation can also be used once the pressure derivative has been properly estimated with the extrapolated pressure:

$$P_{ext} = P_{wf} - m^*\Delta t \tag{2.192}$$

Example 2.7

Escobar and Montealegre [21] presented a simulated pressure test of a square-shaped reservoir with an area of 2295.7 acres having a testing well 1 in the center and another well 2 at 1956 ft north of well 1. Well 2 produced at a rate of 500 BPD during 14000 h. After 4000 h of flow, well 1 was opened at a flow rate of 320 BPD to run a pressure drawdown test which data are reported in **Table 2.8** and **Figure 2.26**. The data used for the simulation were:

r_w = 0.3 pie	μ = 3 cp	c_t = 3 × 10^{-6} psia^{-1}	h = 30 pies
ϕ = 10%	B = 1.2 bbl/BF	k = 33.33 md	s = 0

Interpret this test using conventional and *TDS* techniques considering and without considering the presence of well 2.

Solution by conventional analysis

A pressure change is observed in well 1 up to a time of 4000 h, after which it is put into production for the declination test, as shown in **Figure 2.25. Figure 2.26** presents a plot of P_{wf} vs. log Δt obtained with the information in **Table 2.9**. Hence, the slope and intercept are, respectively, −230 psia/cycle and 3330.9 psia. Permeability and skin factor are, respectively, estimated from Eqs. (2.38) and (2.191):

t, h	P_{wf}, psia	t, h	P_{wf}, psia	t, h	P_{wf}, psia
0	5000	4000.00	4278.93	7091.28	2007.41
4.51	5000.0001	4000.10	4134.44	7511.28	1899.99
10.10	4999.98	4000.20	4015.56	7931.28	1792.61
56.79	4991.08	4000.40	3830.82	8351.28	1685.19
100.98	4970.97	4000.64	3676.32	8771.28	1577.72
201.48	4926.98	4001.13	3478.40	9191.28	1470.25
319.33	4887.16	4001.80	3345.40	9611.28	1362.85
402.02	4864.59	4005.06	3166.11	10031.28	1255.45
506.11	4840.13	4017.96	3039.90	10451.28	1148.04
637.15	4813.27	4090.00	2891.65	10871.28	1040.57
802.13	4782.99	4201.48	2807.85	11291.28	933.06
1009.82	4747.74	4402.02	2720.00	11711.28	825.63
1271.28	4705.41	4637.15	2644.70	12131.28	718.26
1551.28	4661.10	5009.82	2542.16	12551.28	610.85
2111.28	4573.46	5411.28	2437.61	12971.28	503.40
2671.28	4486.12	5831.28	2329.70	13391.28	395.95
3091.28	4420.63	6251.28	2222.26	13811.28	288.51
3511.28	4355.13	6671.28	2114.85	14000.00	240.23

Table 2.8.
Pressure data of a developed reservoir in example 2.7, after [21].

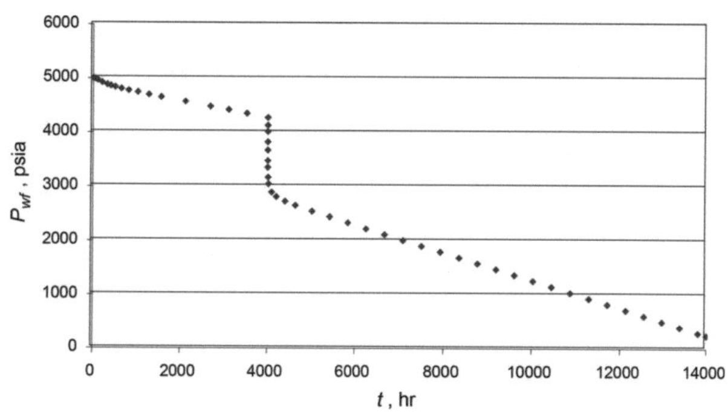

Figure 2.25.
Cartesian plot of pressure versus time data simulated for well 1, after [21].

$$k = \frac{162.2q\mu B}{hm} = \frac{162.6(320)(3)(1.2)}{30(230)} = 27.15 \text{ md}$$

$$s = 1.1513\left[\frac{3330.9 - 4278.93}{-230} - \log\left(\frac{27.15}{(0.1)(3)(3\times10^{-6})(0.3^2)}\right) + 3.23\right] = -1.35$$

Table 2.9 also reports the data of P_{wf}–$m^*\Delta t$. **Figure 2.26** presents, in addition, the plot of P_{wf}–$m^*\Delta t$ vs. log Δt. Now, the slope and intercept are, respectively, 193.9 psia/cycle and 3285.9 psia. A permeability of 32.2 md is found from Eq. (2.38) and a skin factor of −0.28 is estimated from Eq. (2.191).

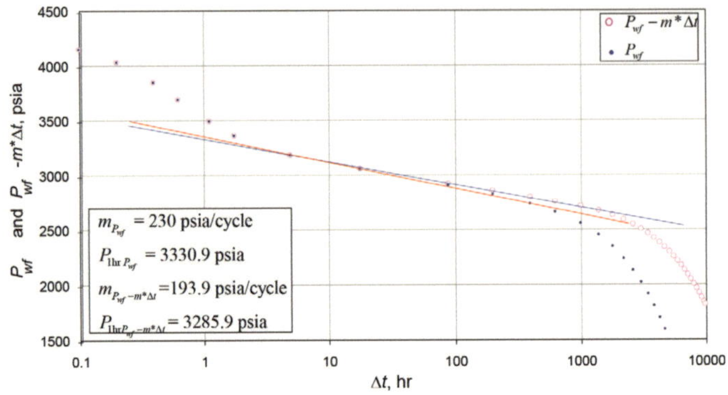

Figure 2.26.
Semilog plot for example 2.7, after [21].

From the derivative plot, **Figure 2.27**, we can observe that the pseudosteady-state period has been perfectly developed; as a consequence, we can obtain the Cartesian slopes performing a linear regression with the last 10 pressure points, namely: m^* (P_{wf} vs. Δt) = -0.256 psia/h and m^* (P_{ext} vs. Δt) = -0.0992 psia/h. Eq. (2.59) allows obtaining the well drainage area of well 2:

$$A_{(P_{wf})} = -\frac{0.23395qB}{\phi c_t h m^*} = \frac{0.23395(320)(1.2)}{(0.1)(3\times10^{-6})(30)(0.256)(43560)} = 895.2 \text{ Ac}$$

$$A_{(P_{ext})} = -\frac{0.23395qB}{\phi c_t h m^*} = \frac{0.23395(320)(1.2)}{(0.1)(3\times10^{-6})(30)(0.0992)(43560)} = 2310 \text{ Ac}$$

Solution by *TDS* technique

Application of *TDS*, the pressure derivative is initially taken to the well-flowing pressure data, see **Table 2.9**. Then, the derivative is taken to the corrected pressure, P_{wf}–$m^*\Delta t$. Both pressure derivatives are reported in **Figure 2.27**. For the uncorrected pressure, the following information was read from **Figure 2.27**:

t_r = 35.826 h	$(t^*\Delta P')_r$ = 90.4 psia	ΔP_r = 1301.7 psia

Permeability and skin factor are calculated with Eqs. (2.76) and (2.97);

$$k = \frac{70.6q\mu B}{h(t*\Delta P')_r} = \frac{70.6(320)(3)(1.2)}{30(90.4)} \cong 30 \text{ md}$$

$$s = 0.5\left[\frac{1301.7}{90.4} - \ln\left(\frac{(30)(35.826)}{(0.1)(3)(3\times10^{6})(0.3^2)}\right) + 7.43\right] = -0.74$$

Then, for the corrected pressure case, the following data were read from **Figure 2.27**;

t_r = 319.3321 h	$(t^*\Delta P')_r$ = 82.1177 psia	ΔP_r = 1477.3508 psia

With these data, Eq. (2.76) provided a permeability value of 33.07 md and Eq. (2.97) allows estimating a skin factor of -0.087. Eq. (2.102) is used to find the well drainage area using t_{rpi} = 376.6049 h (uncorrected pressure) and t_{rpi} = 800.5503 h (corrected pressure) read from **Figure 2.28**, then,

Δt, h	P_{wf}, psia	ΔP_{wf}, psia	$t^*\Delta P_{wf}'$, psia	P_{ext}, psia	ΔP_{ext}, psia	$t^*\Delta P_{ext}'$, psia
0.00	4278.93	0.00	0.00	4278.93	0.00	0.00
0.01	4263.17	15.76	16.17	4263.17	15.76	16.17
0.02	4247.80	31.13	31.50	4247.80	31.13	31.50
0.03	4232.77	46.16	46.15	4232.77	46.16	46.14
0.05	4203.65	75.28	73.43	4203.66	75.27	73.42
0.06	4189.53	89.40	86.24	4189.54	89.39	86.23
0.08	4162.11	116.83	110.14	4162.12	116.81	110.13
0.113	4118.69	160.24	145.70	4118.71	160.22	145.69
0.160	4062.04	216.89	187.89	4062.06	216.87	187.86
0.226	3989.51	289.42	234.79	3989.55	289.38	234.75
0.319	3899.81	379.12	281.46	3899.86	379.07	281.41
0.451	3793.92	485.01	319.72	3793.99	484.94	319.65
0.637	3676.32	602.61	339.60	3676.42	602.51	339.50
0.900	3555.47	723.47	333.33	3555.61	723.33	333.19
1.271	3442.11	836.82	300.48	3442.31	836.63	300.27
1.796	3345.40	933.53	250.08	3345.68	933.25	249.79
2.537	3269.25	1009.68	196.69	3269.65	1009.28	196.28
3.583	3211.35	1067.58	152.72	3211.91	1067.02	152.14
5.061	3166.11	1112.82	123.22	3166.90	1112.03	122.39
7.149	3128.10	1150.84	106.39	3129.21	1149.72	105.23
10.098	3093.70	1185.23	97.75	3095.28	1183.65	96.11
16.005	3050.46	1228.47	92.62	3052.96	1225.97	90.01
22.61	3018.94	1259.99	91.11	3022.47	1256.46	87.43
31.93	2987.69	1291.25	90.86	2992.67	1286.26	85.67
45.11	2956.32	1322.61	91.76	2963.37	1315.57	84.42
63.72	2924.46	1354.47	93.89	2934.41	1344.52	83.53
90.00	2891.65	1387.28	97.54	2905.70	1373.23	82.89
127.13	2857.33	1421.60	103.15	2877.17	1401.76	82.47
179.57	2820.74	1458.19	111.39	2848.77	1430.16	82.17
253.65	2780.81	1498.12	123.24	2820.41	1458.52	81.97
358.30	2736.24	1542.70	140.71	2792.17	1486.76	82.41
506.11	2684.75	1594.18	167.76	2763.76	1515.17	85.41
714.90	2622.34	1656.59	211.07	2733.94	1544.99	94.76
1009.82	2542.16	1736.77	279.69	2699.80	1579.13	115.40
1411.28	2437.61	1841.32	380.45	2657.93	1621.01	150.84
1831.28	2329.70	1949.23	489.81	2615.59	1663.35	191.86
2251.28	2222.26	2056.67	600.72	2573.71	1705.22	234.44
2811.28	2079.04	2199.90	749.48	2517.91	1761.03	292.10
3371.28	1935.78	2343.15	898.68	2462.07	1816.86	350.18
4071.28	1756.82	2522.11	1085.26	2392.38	1886.55	422.88

Δt, h	P_{wf}, psia	ΔP_{wf}, psia	$t^*\Delta P_{wf}'$, psia	P_{ext}, psia	ΔP_{ext}, psia	$t^*\Delta P_{ext}'$, psia
4771.28	1577.72	2701.21	1271.94	2322.57	1956.36	495.67
5611.28	1362.85	2916.08	1495.85	2238.83	2040.10	582.91
6451.28	1148.04	3130.89	1719.92	2155.15	2123.78	670.32
6591.28	1112.23	3166.71	1757.27	2141.19	2137.74	684.89
7711.28	825.63	3453.31	2055.97	2029.43	2249.50	801.38
8831.28	539.22	3739.71	2354.68	1917.88	2361.06	917.87
9951.28	252.69	4026.24	2653.47	1806.19	2472.74	1034.44
10000.00	240.23	4038.70	2666.47	1801.33	2477.60	1039.51

Table 2.9.
Data of P_{wf}, $P_{ext} = P_{wf} - m^*\Delta t$, $t^*\Delta P_{wf}'$, $t^*\Delta P_{ext}'$ for example 2.7, after [21].

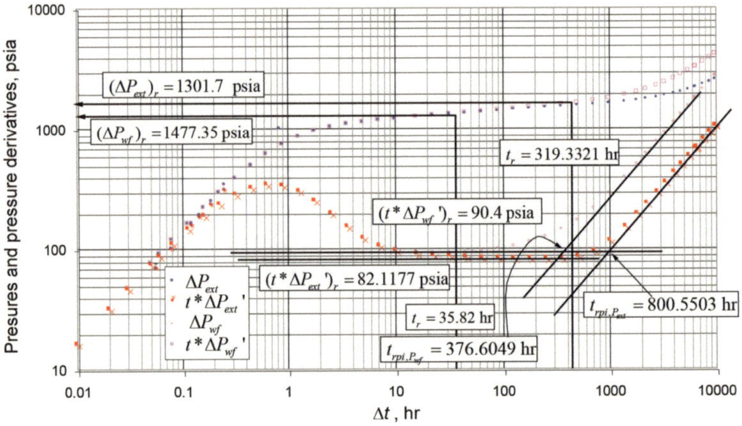

Figure 2.27.
Log-log plot of pressures and pressure derivatives versus time for example 2.7, after [21].

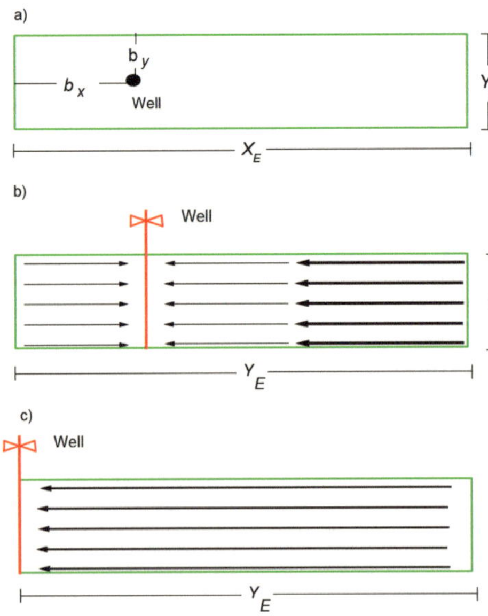

Figure 2.28.
Reservoir geometry and description of flow regimes. (a) Reservoir approximated geometry, (b) Dual linear flow, (c) Single linear or hemilinear flow.

Method	*k*, md	Abs. error, %	*s*	Abs. error, %
Simulation	33.33		0	
Semilog with P_{wf}	27.15	18.54	−1.35	135
Semilog with P_{ext}	32.2	3.39	−0.29	29
TDS with P_{wf}	30	9.99	−0.74	74
TDS with P_{ext}	33.07	0.78	−0.087	8.7

Table 2.10.
Permeability and skin factor results for example 2.7, after [21].

$$A_{P_{wf}} = \frac{kt_{rpssi}}{301.77\phi\mu c_t} = \frac{(30)(376.604)}{301.77(0.1)(3)(3\times10^{-6})}\frac{1}{43560} = 955 \text{ Ac}$$

$$A_{P_{ext}} = \frac{kt_{rpssi}}{301.77\phi\mu c_t} = \frac{(33.07)(800.5503)}{301.77(0.1)(3)(3\times10^{-6})}\frac{1}{43560} = 2237.8 \text{ Ac}$$

Figure 2.28 provides a comparison of the derivative of the flowing bottom pressure ignoring the effect of well 2 and the pressure derivative including the effect of well 2. It is noted there that the radial flow zone is shorter and less defined. On the other hand, the pseudosteady-state zone appears first when the effect of the adjacent well is not included, indicating that the well drainage area, and therefore, the reserves present therein will be substantially underestimated. **Table 2.10** shows all the permeability and skin factor values obtained for this example with their respective absolute errors with reference to the input simulation value. *TDS* when corrected pressure is taken gives the best results.

2.7. Elongated systems

These deposits can be approximated to the geometry described by **Figure 2.28**. They mainly result from fluvial depositions (deltaic), commonly called channels, terrace faulting, and carbonate reefs. The possible flow regimes when the well is completely off-center are presented in **Figure 2.28b** when the parallel reservoir boundaries are no-flow type (closed). Once radial flow vanishes, two linear flows take place at both sides of the reservoir. This flow regime is normally known as linear flow regime, see **Figure 2.27b**; actually, it consists of two linear flow regimes forming a 180° angle between each other. Therefore, Escobar et al. [19] named it dual-linear flow. Once the shorter reservoir boundary has been reached by the transient wave, only a unique linear flow is kept and lasts until the other boundary is reached. This unique flow is referred as single-linear flow by Escobar et al. [19]. However, since linear flow is taken on one side of the reservoir, it is also known as hemilinear flow regime. This is the only linear flow taken place in the system depicted in **Figure 2.28c**.

Both linear flows are characterized by a slope of 0.5 in the pressure derivative curve. **Figure 2.29** sketches the pressure derivative behavior of the mentioned systems.

2.7.1 *TDS* technique

The governing pressure and pressure derivative equations for the single-linear and dual-linear flow regimes are, respectively, given below [13, 16–20, 23, 24, 28, 29, 31, 35, 38, 55, 56]:

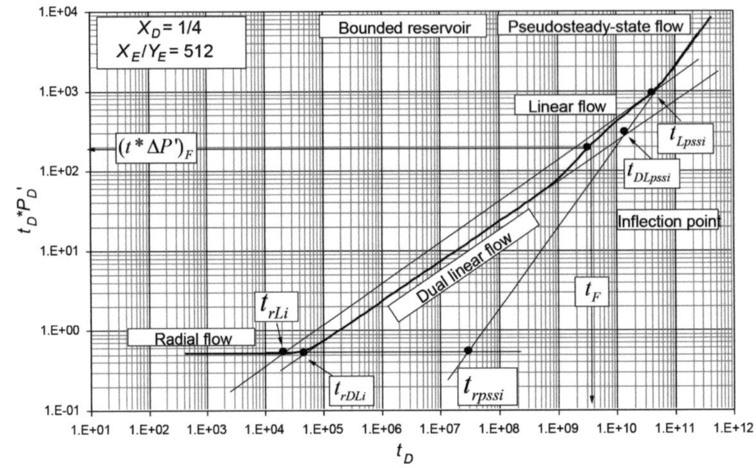

Figure 2.29.
Dimensionless well pressure derivative versus time behavior for a rectangular reservoir with the well located off-center, after [19].

$$(P_D)_L = 2\pi\sqrt{t_{D_L}} + s_L = \frac{2\pi\sqrt{t_D}}{W_D} + s_L \tag{2.193}$$

$$(t_D * P_D')_L = \frac{\pi\sqrt{t_D}}{W_D} \tag{2.194}$$

$$(P_D)_{DL} = \frac{2\sqrt{\pi t_D}}{W_D} + s_{DL} \tag{2.195}$$

$$(t_D * P_D')_{DL} = \frac{\sqrt{\pi t_D}}{W_D} \tag{2.196}$$

Being s_L is the geometrical skin factor caused by converging from either radial to linear flow regime (well located at one end of reservoir sides, **Figure 2.28c** or from dual-linear to linear flow (well off-center well). s_{DL} is the geometrical skin factor caused by converging from either radial to linear flow regime. The dimensionless parameters are defined by Escobar et al. [19] as:

$$W_D = \frac{Y_E}{r_w} \tag{2.197}$$

$$t_{DL} = \frac{t_D}{W_D^2} \tag{2.198}$$

The dimensionless distances are given by:

$$X_D = \frac{2b_x}{X_E} \tag{2.199}$$

$$Y_D = \frac{2b_y}{Y_E} \tag{2.200}$$

Variables b_x and b_y correspond to the nearest distances from the well to the reservoir boundaries in the directions x and y, respectively. See **Figure 2.28a**. Replacing Eqs. (1.94), (2.62) and (2.197) in Eq. (2.194) and solving for the root product of permeability by the reservoir width, Y_E, will yield:

$$\sqrt{k}Y_E = \frac{7.2034qB}{h(t*\Delta P')_L}\sqrt{\frac{t_L\mu}{\phi c_t}} \tag{2.201}$$

Since, *TDS* equations apply to either drawdown or buildup tests; then, when either t or Δt = 1 h, Eq. (2.200) becomes:

$$\sqrt{k}Y_E = \frac{7.2034qB}{h(t*\Delta P')_{L1}}\sqrt{\frac{\mu}{\phi c_t}} \tag{2.202}$$

The root product of permeability by the reservoir width can be also calculated from the dual-linear flow, *DL*. This can be performed by replacing also Eqs. (1.94), (2.62), and (2.197) into the dimensionless pressure derivative equation into Eq. (2.196) leading to:

$$\sqrt{k}Y_E = \frac{4.064qB}{h(t*\Delta P')_{DL}}\sqrt{\frac{t_{DL}\mu}{\phi c_t}} \tag{2.203}$$

Again at either t or Δt = 1 h, the above equation becomes:

$$\sqrt{k}Y_E = \frac{4.064qB}{h(t*\Delta P')_{DL1}}\sqrt{\frac{\mu}{\phi c_t}} \tag{2.204}$$

2.7.1.1 Intersection points

For long production times, the pseudosteady-state period is reached. Both pressure and pressure derivative are joined into a unit-slope line, we obtain a straight line. The governing pressure derivative equation at this time is given by Eq. (2.101). For the systems dealt with in this section, Eq. (2.102) which uses the point of intersection radial-pseudosteady state, Eq. (2.103) and (2.104) also apply. The straight line given by Eq. (2.101) also intersects the lines given by Eqs. (2.96) and (2.98); then, reservoir area can be found from such intersection times, thus, [13, 19]:

$$A = \sqrt{\frac{kt_{DLpssi}Y_E^2}{301.77\phi\mu c_t}} \tag{2.205}$$

$$A = \sqrt{\frac{kt_{Lpssi}Y_E^2}{948.047\phi\mu c_t}} \tag{2.206}$$

Likewise, the intersection times of the line of infinite radial behavior of the pressure derivative (horizontal straight line) with the hemilinear and dual-linear flow regimes lead to obtain reservoir width from:

$$Y_E = 0.05756\sqrt{\frac{kt_{rDLi}}{\phi\mu c_t}} \tag{2.207}$$

$$Y_E = 0.102\sqrt{\frac{kt_{rLi}}{\phi\mu c_t}} \tag{2.208}$$

As indicated by Tiab [71], the geometrical skin factors, or any skin factor, can be obtained by dividing the pressure equation by its derivative equation and solving for the skin factor. Following this, Escobar et al. [19] divided Eqs. (2.195) and (2.193) by Eqs. (2.196) and (2.194), respectively, after replacing the dimensionless quantities given by Eqs. (1.94), (1.89), (2.62), and (2.197) and solving for the geometrical skin factor will provide:

$$s_{DL} = \left(\frac{\Delta P_{DL}}{(t*\Delta P')_{DL}} - 2\right)\frac{1}{19.601 Y_E}\sqrt{\frac{k t_{DL}}{\phi \mu c_t}} - s \tag{2.209}$$

$$s_L = \left(\frac{\Delta P_L}{(t*\Delta P')_L} - 2\right)\frac{1}{34.743 Y_E}\sqrt{\frac{k t_L}{\phi \mu c_t}} - s_{DL} \tag{2.210}$$

where both t_{DL} and t_L are read at any convenient point during each respective linear flow regime. The pressure and pressure derivative values, ΔP_{DL}, $t^*\Delta P_{DL}'$, ΔP_L, and $t^*\Delta P_L'$, used in either Eqs. Eqs. (2.209) or (2.210) are read at these arbitrary times. The characteristic points used so far in this section are better explained in **Figure 2.29**.

In linear deposits, when the well is off-centered and there is a simultaneous action of the linear flow on one reservoir side and the steady state on the other side, a slope flow of −1/2 develops, which does not correspond to either spherical or hemispherical flows, see **Figure 2.30**. Given the isobaric geometry, this flow regime is called parabolic flow [19]. Although Sui et al. [100] called it dipolar flow, Escobar et al. [16, 17] performed numerical simulation and plotted the isobaric lines and found that the closest geometry shape corresponds to a parabola. The governing equations of this flow regime are:

$$P_D = -(W_D)(X_D)^2\left(\frac{X_E}{Y_E}\right)^2 t_D^{-0.5} + s_{PB} \tag{2.211}$$

$$t_D * P_D' = \frac{W_D}{2}(X_D)^2\left(\frac{X_E}{Y_E}\right)^2 t_D^{-0.5} \tag{2.212}$$

Once the division of the pressure equation by the pressure derivative equation is attained and the appropriate dimensionless expressions are replaced in the resulting division, the parabolic skin factor equation is obtained:

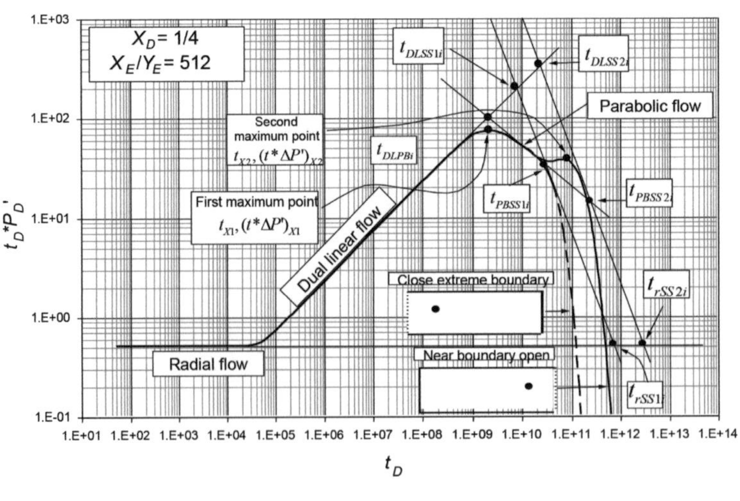

Figure 2.30.
Dimensionless pressure derivative versus time behavior for a well displaying parabolic flow regime, after [19].

$$s_{PB} = \left(\frac{\Delta P_{PB}}{(t * \Delta P')_{PB}} + 2\right)\frac{123.16\, b_x}{Y_E}\sqrt{\frac{\phi \mu c_t}{k t_{PB}}} - s_{DL} \tag{2.213}$$

Also, by substituting the dimensionless quantities into Eq. (2.212), the following equation is derived:

$$\frac{k^{1.5} Y_E}{b_x^2} = 17390\left[\frac{q\mu B}{h(t * \Delta P')_{PB}}\right]\left[\frac{\phi \mu c_t}{t_{PB}}\right]^{0.5} \tag{2.214}$$

In the above two equations, the pressure and pressure derivative values are read to a convenient or arbitrary point, t_{PB}.

The total skin factor for this type of reservoir is evaluated according to the flow regimes that are presented:

- *Well near a closed boundary*. In this case, radial, dual-linear, and hemilinear flows are presented.

$$s = s_r + s_{DL} + s_L \tag{2.215}$$

- *Well near an open boundary.* In this case, radial, dual-linear, and parabolic flows are presented.

$$s_t = s + s_{DL} + s_{PB} \tag{2.216}$$

If dual-linear is unseen, as presented in **Figure 2.28c**, Eq. (2.215) reduces to;

$$s_t = s + s_L \tag{2.217}$$

Escobar and Montealegre [18] performed a detailed analysis of the geometrical skin factor causes.

The points of intersection, see **Figures 2.30** and **2.31**, found between the different lines of the pressure derivative curve allows developing the following equations:

$$b_x = \frac{1}{65.41}\sqrt{\frac{k t_{DLPBi}}{\phi \mu c_t}} \tag{2.218}$$

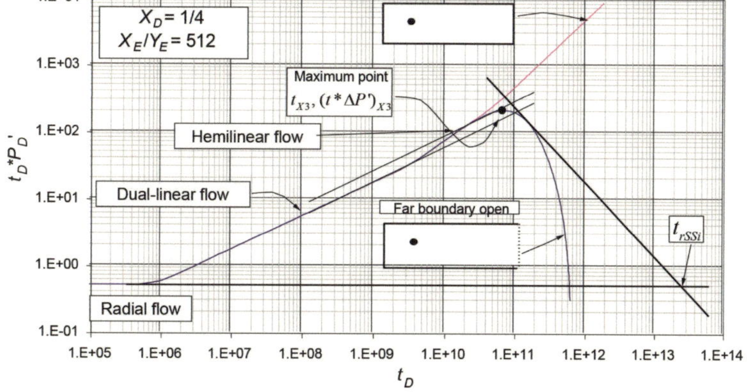

Figure 2.31.
Dimensionless pressure derivative versus time behavior for an off-centered well near a no-flow boundary and the far boundary is either at constant pressure or no flow, after [19].

$$b_x = \sqrt{\left(\frac{Y_E}{246.32}\right) * \left(\frac{kt_{rPBi}}{\phi\mu c_t}\right)^{0.5}} \quad (2.219)$$

For steady-state cases, a negative unit-slope line, *SS*1, tangent to the pressure derivative curve during late time is drawn. This occurs when the far boundary is at constant pressure. Its intercept with the parabolic flow straight line makes it possible to estimate the length of the reservoir, see **Figure 2.30**.

$$X_E^3 = \frac{1}{77.9}\left(\frac{kt_{PBSS1i}}{\phi\mu c_t}\right)b_x \quad (2.220)$$

Several scenarios arise for cases of lateral constant-pressure boundaries:

- Intersection of the −1-slope line with the dual-linear flow regime line:

$$X_E^3 = \left(\frac{1}{1.426 \times 10^9}\right)\left(\frac{kt_{DLSS1i}}{\phi\mu c_t}\right)^3\left(\frac{1}{b_x^3}\right) \quad (2.221)$$

- Intersection of the −1-slope line with the radial flow regime line:

$$X_E^3 = \left(\frac{1}{4.72 \times 10^6}\right)\left(\frac{kt_{rSS_1i}}{\phi\mu c_t}\right)^2\left(\frac{Y_E^2}{b_x^3}\right) \quad (2.222)$$

- Intersection of the −1-slope line with the parabolic flow regime line:

$$X_E^3 = \frac{1}{77.9}\left(\frac{kt_{PBSS1i}}{\phi\mu c_t}\right)b_x \quad (2.223)$$

Again, a negative unit-slope line, *SS*2, tangent to the pressure derivative curve during late time is drawn. This takes place when a no-flow far boundary exists. Its intercept with the dual-linear, radial, and parabolic flow straight lines can provide equations to estimate the length of the reservoir, see **Figure 2.30**.

- Intersection of the −1-slope line with the dual-linear flow regime line:

$$X_E^3 = \left(\frac{1}{1.42 \times 10^{10}}\right)\left(\frac{kt_{DLSS2i}}{\phi\mu c_t}\right)^3\left(\frac{1}{b_x^3}\right) \quad (2.224)$$

- Intersection of the −1-slope line with the radial flow regime line:

$$X_E^3 = \left(\frac{1}{4.66 \times 10^7}\right)\left(\frac{kt_{rSS2i}}{\phi\mu c_t}\right)^2\left(\frac{Y_E^2}{b_x^3}\right) \quad (2.225)$$

- Intersection of the −1-slope line with the parabolic flow regime line:

$$X_E^3 = \frac{1}{768.4}\left(\frac{kt_{PBSS2i}}{\phi\mu c_t}\right)b_x \quad (2.226)$$

From the inflection point between linear and dual-linear flow, the position of the well can be obtained by any of the following relationships:

$$b_x = \sqrt{\frac{kt_F}{5448.2\phi\mu c_t}} \tag{2.227}$$

$$b_x = \frac{khY_E(t * \Delta P')_F}{415.84q\mu B} \tag{2.228}$$

2.7.1.2 Maximum points

As seen in **Figure 2.30**, when the well is located near a constant-boundary pressure but the far boundary has no-flow, both parabolic flow regime and a maximum point, $X1$ (between dual linear and parabolic) are observed. If the far boundary is at constant pressure, another maximum, $X2$, can be developed. The first maximum is governed by:

$$(t_D * P_D')_{X1} = \frac{2}{3}\frac{\sqrt{\pi}}{W_D}t_{DX1}^{0.5} \tag{2.229}$$

$$\frac{X_E}{Y_E} = \frac{2}{3}\left(\frac{\sqrt{\pi}}{W_D X_D}\right)t_{DX1}^{0.5} \tag{2.230}$$

$$\frac{X_E}{Y_E} = \frac{2}{3}\left(\frac{\sqrt{\pi}}{W_D X_D}\right)t_{DX1}^{0.5} \tag{2.231}$$

From which it is obtained:

$$b_x = \left(\frac{1}{58.8}\right)\sqrt{\left(\frac{kt_{X1}}{\phi\mu c_t}\right)} \tag{2.232}$$

$$b_x = \frac{khY_E(t * \Delta P')_{X1}}{159.327q\mu B} \tag{2.233}$$

The second maximum has a governing equation given by:

$$(t_D * P_D')_{X2} = \frac{\sqrt{\pi}}{W_D}(X_D^2)t_{DX2}^{0.5} \tag{2.234}$$

$$\frac{X_E}{Y_E} = \left(\frac{\pi}{2W_D}\right)t_{DX2}^{0.5} \tag{2.235}$$

$$\frac{X_E}{Y_E} = \left(\frac{\sqrt{\pi}}{2X_D^2}\right)(t_D * P_D')_{X2} \tag{2.236}$$

From which is obtained:

$$X_E = 637.3\left(\frac{b_x^2}{Y_E}\right)\left(\frac{q\mu B}{kh}\right)\left(\frac{1}{(t * \Delta P')_{X2}}\right) \tag{2.237}$$

$$X_E = \frac{1}{39.2}\left(\frac{kt_{X2}}{\phi\mu c_t}\right)^{0.5} \tag{2.238}$$

When a rectangular reservoir has mixed boundaries and the well is near the no-flow boundary, see **Figure 2.31**, another maximum point, $X3$, can be displayed on the pressure derivative once the constant-pressure boundary is felt. The governing equation for this maximum point is:

$$\frac{X_E}{Y_E} = \frac{\pi^{1.5}}{4}\left(\frac{1}{W_D}\right)t_{DX3}^{0.5} \tag{2.239}$$

After replacing the dimensionless parameters and solving for the reservoir length, it will result:

$$X_E = \frac{1}{44.24}\sqrt{\frac{kt_{X3}}{\phi\mu c_t}} \tag{2.240}$$

Another steady-state period is depicted in **Figure 2.31** when the well is near a no-flow boundary and the farther one is at constant pressure. Again, one negative-unit-slope line is drawn tangent to the pressure derivative curve. In this case, both dual-linear flow and single linear flow regimes are developed. This is followed by a maximum. The governing equation of the mentioned negative slope line is:

$$t_D * P_D' = \left(\frac{X_E}{Y_E}\right)^3 \frac{W_D^2}{t_D} \tag{2.241}$$

Equating Eq. (2.75) with Eq. (2.241), an equation will be obtained that uses the radial and steady-state intercept to find reservoir length:

$$X_E = \sqrt[3]{\frac{kt_{rSSi}Y_E}{7584.2\phi\mu c_t}} \tag{2.242}$$

If it is assumed that the area is obtained from the product of the width by the length of the reservoir, $A = X_E Y_E$, then,

$$A = \sqrt[3]{\frac{kt_{rSSi}Y_E^4}{7584.2\phi\mu c_t}} \tag{2.243}$$

When the well is centered along the rectangular reservoir, different behavior occurs if one or both boundaries are at constant pressure as seen in **Figure 2.32**. The

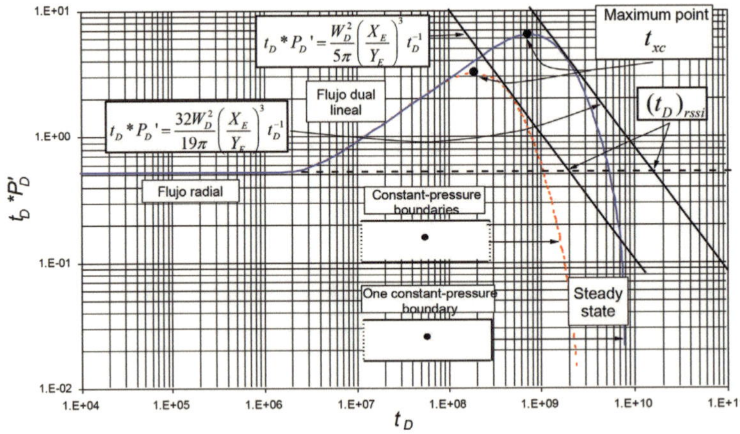

Figure 2.32.
Dimensionless pressure derivative versus time behavior for a centered well when one or both boundaries are at constant pressure, after [28].

equations of the straight line with unit slope passing tangentially to the pressure derivative curve are, respectively, given by the following expressions:

$$t_D * P_D' = \frac{32W_D^2}{19\pi}\left(\frac{X_E}{Y_E}\right)^3 t_D^{-1} \tag{2.244}$$

$$t_D * P_D' = \frac{W_D^2}{5\pi}\left(\frac{X_E}{Y_E}\right)^3 t_D^{-1} \tag{2.245}$$

The equations to estimate the drainage area is obtained from the intercept of Eqs. (2.140) and (2.141) with Eq. (2.75). After replacing the dimensionless parameters and assuming perfect rectangular geometry, we, respectively, have:

$$A = \sqrt[3]{\frac{kt_{rSSi}Y_E^4}{4066\phi\mu c_t}} \tag{2.246}$$

$$A = \sqrt[3]{\frac{kt_{rSSi}Y_E^4}{482.84\phi\mu c_t}} \tag{2.247}$$

The maximum point when one of the two boundaries is at constant pressure is given by:

$$\frac{X_E}{Y_E} = \frac{15}{32}\pi\left(\frac{1}{W_D}\right)t_{Dxc}^{0.5} \tag{2.248}$$

And, for the other case, when both extreme boundaries are subjected to a constant pressure:

$$\frac{X_E}{Y_E} = \frac{15}{16}\pi\left(\frac{1}{W_D}\right)t_{Dxc}^{0.5} \tag{2.249}$$

After replacing the dimensionless quantities in Eqs. (2.248) and (2.249), it is possible to find expressions to determine reservoir length and area, respectively:

$$X_E = \frac{1}{41.82}\left(\frac{kt_{xc}}{\phi\mu c_t}\right)^{0.5} \tag{2.250}$$

$$A = \frac{Y_E}{41.82}\left(\frac{kt_{xc}}{\phi\mu c_t}\right)^{0.5} \tag{2.251}$$

$$A = \frac{Y_E}{41.82}\left(\frac{kt_{xc}}{\phi\mu c_t}\right)^{0.5} \tag{2.252}$$

$$X_E = \frac{1}{20.91}\left(\frac{kt_{xc}}{\phi\mu c_t}\right)^{0.5} \tag{2.253}$$

$$A = \frac{Y_E}{20.91}\left(\frac{kt_{xc}}{\phi\mu c_t}\right)^{0.5} \tag{2.254}$$

Escobar et al. [28] determined the pressure derivative governing equation for constant pressure both circular or square systems;

$$t_D * P_D' \approx \frac{\pi}{84t_{DA}} \tag{2.255}$$

which intercept with the radial flow pressure derivative equation, Eq. (2.75), allows finding the well-drainage area:

$$A = \frac{kt_{rSSi}}{283.66\phi\mu c_t} \tag{2.256}$$

Table 2.11 presents a summary of the different equations to determine the drainage area in constant-pressure systems, since Escobar et al. [28] showed that Eq. (2.102) hugely fails in constant-pressure systems.

Escobar et al. [23] presented *TDS* technique for long reservoirs when the width is known from another source, like seismic. Under this condition, the reservoir areal anisotropy and even the anisotropy angle can be determined. Later, Escobar et al.

Constant, Ξ	Equation	Reservoir geometries
301.77	$A = \frac{kt_{rpssi}}{\Xi\phi\mu c_t}$	
283.66		
4066	$A = \sqrt[3]{\frac{kt_{rSSi}Y_E^4}{\Xi\phi\mu c_t}}$	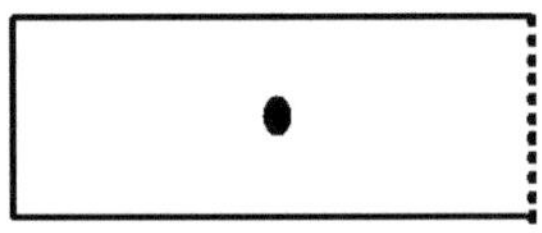
482.84		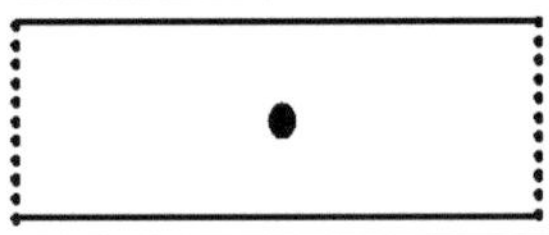
7584.2		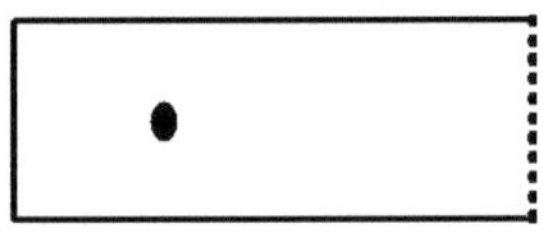
2173.52	$A = \left(\frac{kt_{rSS1i,\,rSS2i}}{\Xi\phi\mu c_t}\right)^{2/3}\frac{Y_E^{5/3}}{b_x}$	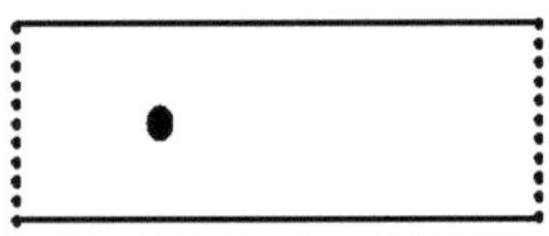
6828.34		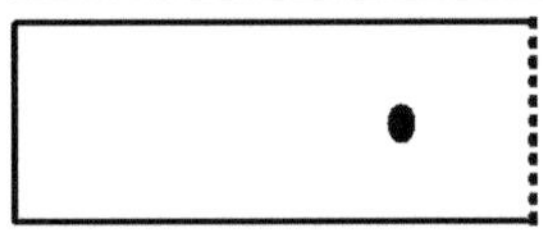
41.82	$X_E = \frac{1}{\Xi}\left(\frac{kt_x}{\phi\mu c_t}\right)^{0.5}$	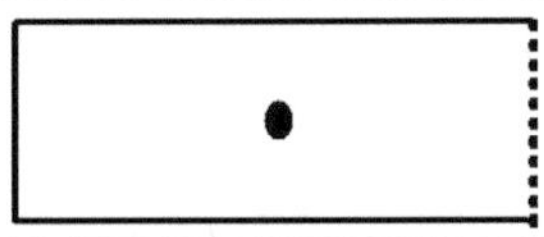
20.91		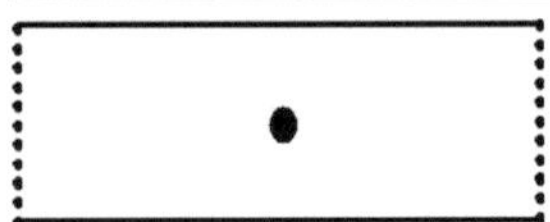

Table 2.11.
Equations for area determination in constant-pressure systems, after [28].

[29, 30] presented conventional analysis and *TDS* technique, respectively, when changes in either reservoir width or facies are seen in elongated systems.

2.7.2 Conventional method

The dimensional pressure governing equation for dual-linear flow regime is [20, 24]:

$$\Delta P = \frac{8.1282}{Y_E}\frac{qB}{kh}\left(\frac{\mu}{\phi c_t k}\right)^{0.5}\sqrt{t} + \frac{141.2q\mu B}{kh}(s_{DL} - s) \tag{2.257}$$

For pressure buildup tests, the superposition principle leads to find:

$$\Delta P = \frac{8.1282}{Y_E}\frac{qB}{kh}\left(\frac{\mu}{\phi c_t k}\right)^{0.5}\left(\sqrt{t_p + \Delta t} - \sqrt{\Delta t}\right) \tag{2.258}$$

Eqs. (2.257) and (2.258) indicate that a linear plot of pressure drop or pressure versus either $t^{0.5}$ (for drawdown tests) or $(t_p + \Delta t)^{0.5} - \Delta t^{0.5}$ (for buildup tests), tandem square root, will yield a straight line which slope, m_{DLF}, and intercept, b_{DLF}, are used to, respectively, find reservoir width, Y_E, and dual-linear (geometrical) skin factor, s_{DL};

$$Y_E = 8.1282\frac{qB}{m_{DLF}h}\left[\frac{\mu}{k\phi c_t}\right]^{0.5} \tag{2.259}$$

$$s_{DL} = \frac{khb_{DLF}}{141.2q\mu B} - s \tag{2.260}$$

Wong et al. [101] presented another version of the skin equation:

$$s_{DL} = \frac{1}{2}\left[\frac{khb_{DLF}}{141.2q\mu B} + \ln\left(\frac{r_w}{Y_E}\right)\right] \tag{2.261}$$

Escobar and Montealegre [20] found that Eq. (2.260) compared quite well with the results of [59, 102]. The governing equations for drawdown and buildup, respectively, for hemilinear flow regime are [20, 24, 28]:

$$\Delta P_{wf} = \frac{14.407}{Y_E}\frac{q\mu B}{kh}\sqrt{\frac{kt}{\phi\mu c_t}} + \frac{141.2q\mu B}{kh}s_L \tag{2.262}$$

$$\Delta P_{ws} = \frac{14.407}{Y_E}\frac{q\mu B}{kh}\sqrt{\frac{k}{\phi\mu c_t}}\left(\sqrt{t_p + \Delta t} - \sqrt{\Delta t}\right) \tag{2.263}$$

Similar to the dual-linear case, when plotting in Cartesian coordinates either P or ΔP versus either $t^{0.5}$ (for drawdown tests) or $(t_p + \Delta t)^{0.5} - \Delta t^{0.5}$ (for buildup tests), a straight line influenced by the linear flow will be obtained. Its slope, m_{LF}, and intercept, b_{LF}, are used, respectively, to estimate reservoir width, Y_E, and skin factor, s_L.

$$Y_E = \frac{14.407}{m_{LF}}\frac{qB}{h}\left(\frac{\mu}{\phi c_t k}\right)^{0.5} \tag{2.264}$$

$$s_L = \frac{khb_{LF}}{141.2q\mu B} - s_{DL} \qquad (2.265)$$

The governing equations for parabolic flow regime under drawdown and buildup conditions are given by [13, 19, 20, 24]:

$$\Delta P_{wf} = -\frac{34780.8qBb_x^2\sqrt{\phi c_t}}{hY_E}\left(\frac{\mu}{k}\right)^{1.5}\frac{1}{\sqrt{t}} + \frac{141.2q\mu B}{kh}s_{PB} \qquad (2.266)$$

$$\Delta P_{ws} = -\frac{34780.8qBb_x^2\sqrt{\phi c_t}}{hY_E}\left(\frac{\mu}{k}\right)^{1.5}\left(\frac{1}{\sqrt{t_p+\Delta t}} - \frac{1}{\sqrt{\Delta t}}\right) \qquad (2.267)$$

A straight line will be observed on a Cartesian plot of either P or ΔP versus either $1/t^{0.5}$ (for drawdown tests) or $1/(t_p+\Delta t)^{0.5} - 1/\Delta t^{0.5}$ (for buildup tests). Its m_{PB} and intercept, b_{PB}, lead to obtain well position along the x-direction, b_x, and parabolic skin factor, s_{PB}, respectively, from:

$$b_x = \sqrt{-\frac{m_{PB}hY_E}{34780.8qB\sqrt{\phi c_t}}\left(\frac{k}{\mu}\right)^{1.5}} \qquad (2.268)$$

$$s_{PB} = \frac{khb_{PB}}{141.2q\mu B} - s_{DL} \qquad (2.269)$$

Area, $A = X_E Y_E$, can be found from the Cartesian plot of pressure versus time using Eq. (2.59).

Example 2.8

Escobar et al. [19] presented a pressure test run in a reservoir in South America. Test data are given in **Table 2.12** and other relevant information is given below:

q = 1400 BPD	h = 14 ft	$c_t = 9 \times 10^{-6}$ psia^{-1}	P_i = 1326.28 psia
r_w = 0.51 pies	ϕ = 24%	B = 1.07 bbl/STB	μ = 3.5 cp

It is required to conduct the interpretation of the test by *TDS* and conventional analysis.

Solution by *TDS* technique

The following information was read from **Figure 2.33**;

$(t^*\Delta P')_r$ = 60 psia	ΔP_r = 122.424 psia	t_{DL} = 2 h
$(t^*\Delta P')_{DL}$ = 105.81 psia	ΔP_{DL} = 265.942 psia	t_{RDLi} = 0.7 h
t_{PB} = 10.157 h	$(t^*\Delta P')_{PB}$ = 132.873 psia	ΔP_{PB} = 458.466 psia
t_{PBDLi} = 6 h	t_{PBRi} = 50 h	t_{DLSS1i} = 7.5 h
t_{RSS1i} = 24 h	t_{PBSS1i} = 12 h	

Permeability is obtained from Eq. (2.76) and reservoir width with Eq. (2.203), respectively:

$$k = \frac{70.6q\mu B}{h(t*\Delta P')_r} = \frac{70.6(1400)(3.5)(1.07)}{(14)(60)} = 440.7 \text{ md}$$

t, h	ΔP, psia	t*ΔP′, psia	t, h	ΔP, psia	t*ΔP′, psia	t, h	ΔP, psia	t*ΔP′, psia
0.165	49.00	51.00	4.331	331.24	147.65	9.824	458.47	136.70
0.332	99.00	67.00	4.498	336.89	149.17	10.157	463.23	132.87
0.498	122.42	60.38	4.665	342.30	151.39	10.490	467.44	134.03
0.665	140.49	66.57	4.831	347.70	152.79	10.824	471.48	132.58
0.831	156.07	73.69	4.998	352.84	154.09	11.157	475.61	130.61
0.998	170.14	80.12	5.165	358.01	155.83	11.490	479.47	127.78
1.165	182.92	84.74	5.331	362.96	156.87	11.824	483.19	126.71
1.331	194.48	88.67	5.498	367.77	157.85	12.157	486.70	125.21
1.498	205.17	92.44	5.665	372.54	159.51	12.490	489.94	122.97
1.665	215.09	96.63	5.831	377.15	159.55	12.824	493.12	119.84
1.831	224.53	101.16	5.998	381.67	159.94	13.157	496.26	117.32
1.998	233.54	105.28	6.165	386.10	161.25	13.490	499.19	115.01
2.165	242.11	109.19	6.331	390.50	161.31	13.824	502.05	113.78
2.331	250.33	113.12	6.498	394.60	161.74	14.157	504.71	111.16
2.498	258.24	116.37	6.665	398.63	161.58	14.490	507.15	109.55
2.665	265.94	120.68	6.831	402.76	161.88	14.990	510.78	106.05
2.831	273.23	124.15	6.998	406.64	161.66	15.490	514.29	102.52
2.998	280.57	126.97	7.165	410.42	161.90	15.990	517.45	99.40
3.165	287.49	130.91	7.331	414.19	161.66	16.490	520.59	97.21
3.331	294.22	133.24	7.657	421.18	161.21	16.990	523.48	93.62
3.498	300.85	136.37	7.990	428.16	160.73	17.490	526.10	90.44
3.665	307.28	138.93	8.324	434.62	153.77	17.990	528.57	86.87
3.831	313.54	141.42	8.657	440.94	149.99	20.474	470.79	50.29
3.998	319.60	143.73	8.990	446.87	146.59	22.640	538.42	10.60
4.165	325.50	145.48	9.490	453.81	140.47			

Table 2.12.
Pressure and pressure derivative data versus time for example 2.8.

$$Y_E = \frac{4.064qB}{\sqrt{kh}(t*\Delta P')_{DL}}\sqrt{\frac{\Delta t_{DL}\mu}{\phi c_t}} = \frac{4.064(1400)(1.07)}{\sqrt{440.7}(14)(105.81)}\sqrt{\frac{(2)(3.5)}{(0.24)(9\times 10^{-6})}} = 352.4 \text{ ft}$$

Verify Y_E with Eq. (2.207):

$$Y_E = 0.05756\sqrt{\frac{kt_{RDLi}}{\phi\mu c_t}} = 0.05756\sqrt{\frac{(440.7)(0.7)}{(0.24)(3.5)(9\times 10^{-6})}} = 367.7 \text{ ft}$$

The well position along the reservoir is found with Eq. (2.214):

$$b_x = \sqrt{\frac{440.7^{1.5}(367.8)}{17390\left[\frac{1400(3.5)(1.07)}{14(132.873)}\right]\left[\frac{0.24(3.5)(9\times 10^{-6})}{10.157}\right]^{0.5}}} = 283.7 \text{ ft}$$

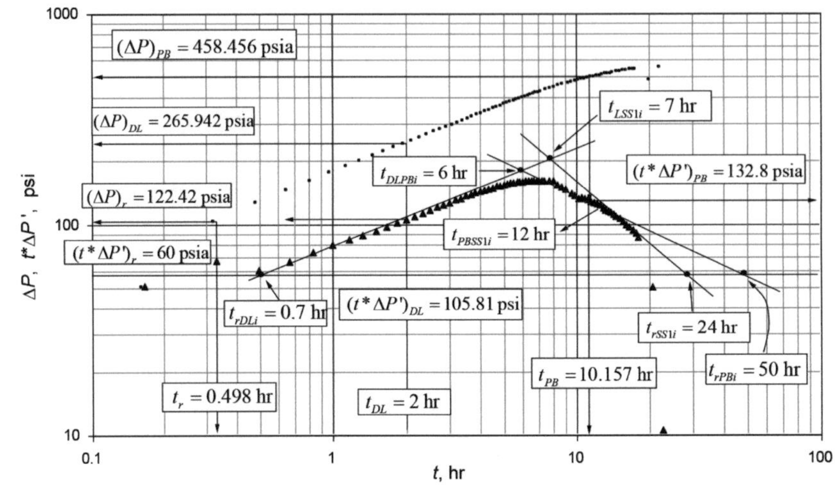

Figure 2.33.
Log-log plot of pressure and pressure derivative versus time for example 2.8, after [19].

Verify b_x of Eqs. (2.218) and (2.219):

$$b_x = \frac{1}{65.41}\sqrt{\frac{kt_{DLPBi}}{\phi\mu c_t}} = \frac{1}{65.41}\sqrt{\frac{440.7*6}{0.24(3.5)(9\times10^{-6})}} = 285.9 \text{ ft}$$

$$b_x = \sqrt{\left(\frac{Y_E}{246.32}\right)*\left(\frac{kt_{rPBi}}{\phi\mu c_t}\right)^{0.5}} = \sqrt{\left(\frac{367.7}{246.32}\right)\left(\frac{440.7(50)}{0.24(3.5)(9\times10^{-6})}\right)^{0.5}} = 283.9 \text{ ft}$$

Observe on the pressure derivative curve that once the parabolic flow is finished, before falling, it rises a little, from which it is inferred that the far boundary is of no flow, this maximum point is not observed with much clarity; then, Eqs. (2.224), (2.226), and (2.225), using the intersection of the −1-slope line with the dual linear, parabolic, and radial flow lines are used:

$$X_E^3 = \left(\frac{1}{1.41\times10^{10}}\right)\left(\frac{440.7\times7.5}{0.24\times3.5\times(9\times10^{-6})}\right)^3\left(\frac{1}{284^3}\right) = 637.2 \text{ ft}$$

$$X_E^3 = \left(\frac{1}{4.66\times10^{7}}\right)\left(\frac{440.7\times24}{0.24\times3.5\times(9\times10^{-6})}\right)^2\left(\frac{367.7^2}{284^3}\right) = 628.2 \text{ ft}$$

$$X_E = \left(\frac{1}{768.4}\left(\frac{440.7\times12}{0.24\times3.5\times(9\times10^{-6})}\right)\times284\right)^{1/3} = 637.1 \text{ ft}$$

Skin factor are found with Eqs. (2.97), (2.209), (2.210), and (2.213):

$$s = 0.5\left[\frac{122.424}{60} - \ln\left(\frac{440.7\times0.5}{0.24\times3.5\times9\times10^{-6}\times0.33^2}\right)+7.43\right] = -4.9$$

$$s_{DL} = \left(\frac{265.942}{105.81}-2\right)\frac{1}{34.743\times367.7}\sqrt{\frac{440.7\times2}{0.24\times3.5\times9\times10^{-6}}} = 0.4+4.9 = 5.3$$

$$s_{PB} = \left(\frac{458.466}{132.873}+2\right)\frac{123.16\,(283.7^2)}{352.4}\sqrt{\frac{(0.24)(3.5)(9\times10^{-6})}{(440.7)(10.157)}} = 6.3\text{-}5.3 = 1$$

The total skin factor is calculated from the sum of the partial skin factors, Eq. (2.216):

$$s = s_r + s_{DL} + s_{PB} = -4.9 + 5.3 + 1 = 1.4$$

Solution by conventional analysis

The following information was read from **Figure 2.34**, **2.35**, and **2.36**;

m = 140 psia/cycle	$m_{DLF} = -150.8$	b_{DLF} = 19.4
$m_{PB} = -851.6$	b_{PB} = 730.64	P_{1hr} = 1158 psia

Permeability is determined using the slope of the semilog plot, m, by means of Eq. (2.38) and mechanical skin factor with Eq. (2.39);

$$k = \left|\frac{162.6q\mu B}{hm}\right| = \left|\frac{162.6(1400)(3.5)(1.07)}{(14)(140)}\right| = 434.96 \text{ md}$$

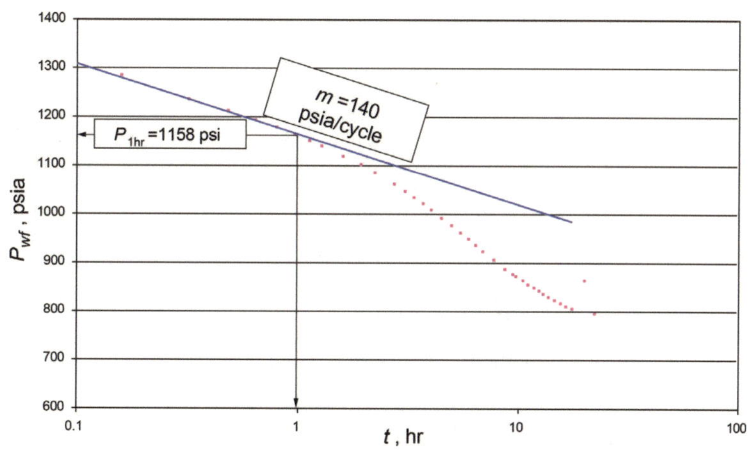

Figure 2.34.
Semilog plot for example 2.8, after [20].

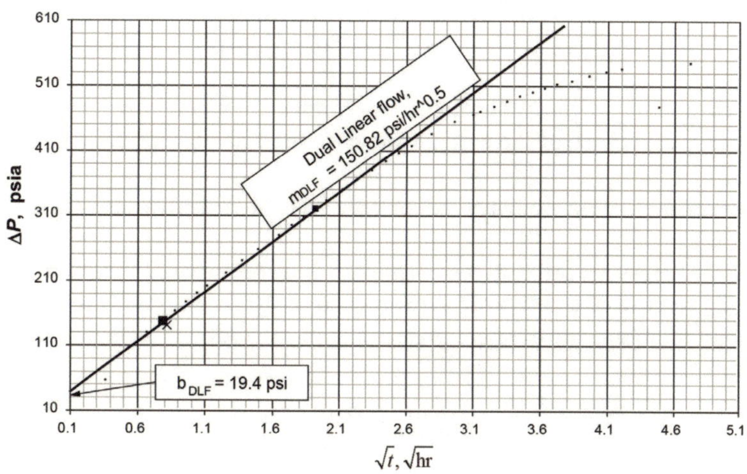

Figure 2.35.
Cartesian plot of ΔP vs. $t^{0.5}$ for example 2.8, after [20].

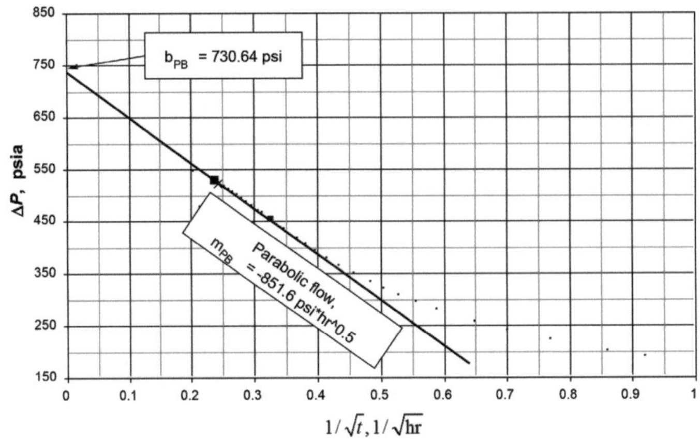

Figure 2.36.
Cartesian plot of ΔP vs. $1/t^{-0.5}$ for example 2.8, after [20].

$$s = 1.1513\left[\frac{1158}{-140} - \log\left(\frac{434.96}{(0.24)(3.5)(9\times10^{-6})(0.51)^2}\right) + 3.2275\right] = -4.6$$

Using m_{DLF} the reservoir width value is calculated Eq. (2.259);

$$Y_E = 8.1282\frac{(1400)(1.07)}{(150.8)(14)}\left[\frac{(3.5)}{(434.96)(0.24)(9\times10^{-6})}\right]^{0.5} = 350 \text{ ft}$$

b_x is found from Eq. (2.268):

$$b_x = \sqrt{-\frac{-851.6(14)(350)}{34780.8(1400)(1.07)\sqrt{(0.24)(9\times10^{-6})}}\left(\frac{434.96}{3.5}\right)^{1.5}} = 277.4 \text{ ft}$$

Geometrical skin factor is found from Eqs. (2.260) and (2.269), thus:

$$s_{DL} = \frac{khb_{DLF}}{141.2q\mu B} + 4.6 = \frac{(434.96)(14)(19.4)}{141.2(1400)(3.5)(1.07)} = 4.8$$

$$s_{PB} = \frac{khb_{PB}}{141.2q\mu B} - s_{DL} = \frac{(434.96)(14)(730.64)}{141.2(1400)(3.5)(1.07)} - 4.8 = 1.3$$

When comparing with the results of the simulation with those obtained by the *TDS* technique and those of the conventional method no greater difference is found.

2.8. Determination of average reservoir pressure from flow tests

Until 2010, pressure buildup tests, chapter 3, were the only means to determine the average pressure of a reservoir. However, in 2010, Agarwal [1] presented a methodology to obtain the average pressure from drawdown tests, using the following expression:

$$\overline{P} = P_{wf} + \frac{887.18q\,B\,\mu}{kh} \tag{2.270}$$

This does not consider the Dietz shape factor, but conditions that the well-flowing pressure is determined at the point where the late pseudosteady-state period develops. According to Agarwal [1], this point is determined using the arithmetic derivative (not multiplied by time). In the arithmetic derivative, the radial flow is presented with a slope of −1. The pseudosteady-state period (postradial) takes place when the slope of the arithmetic derivative becomes zero (flat). That is the right moment where the well-flowing pressure, P_{wf}, is read.

Nomenclature

A	area, ft^2
B_g	gas volume factor, ft^3/STB
B_o	oil volume factor, bbl/STB
B_w	oil volume factor, bbl/STB
b	fraction of penetration/completion
b_{DLF}	intercept of P vs $t^{0.25}$ plot during dual-linear flow, psia$^{0.5}$
b_{LF}	intercept of P vs $t^{0.5}$ plot during hemilinear flow, psia$^{0.5}$
b_{PB}	intercept of P vs $1/t^{0.5}$ plot during hemilinear flow, h^{-1}
b_x	distance from closer lateral boundary to well along the x-direction, ft
b_y	distance from closer lateral boundary to well along the y-direction, ft
c	compressibility, 1/psia
C	wellbore storage coefficient, bbl/psia
C_A	reservoir shape factor
c_t	total or system compressibility, 1/psia
DF	damage factor
DR	damage ratio
FE	flow index
$f(t)$	time function
h	formation thickness, ft
h_p	length of perforations, ft
I	intercept
J	productivity index, bbl/psia
k	permeability, md
k_g	gas effective permeability, md
k_o	oil effective permeability, md
k_w	water effective permeability, md
ID_{csg}	internal casing diameter, in
m	slope of P vs log t plot, psia/h/cycle
m^*	slope of P vs t plot, psia/h
m_{DLF}	slope of P vs $t^{0.25}$ plot during dual-linear flow, psia$^{0.5}$/h
m_{LF}	slope of P vs $t^{0.5}$ plot during hemilinear flow, psia$^{0.5}$/h
m_{PB}	slope of P vs $1/t^{0.5}$ plot during hemilinear flow, (psia$^{0.5}$/h)$^{-1}$
m'	slope of superposition or equivalent time plot, psia/BPD/cycle
$m'b'$	intercept of superposition or equivalent time plot, psia/BPD/cycle
$m(P)$	pseudopressure, psia/cp
OD_{csg}	external casing diameter, in
P	pressure, psia
$\overline{P}$	average reservoir pressure, psia
P_D'	dimensionless pressure derivative
P_D	dimensionless pressure
P_i	initial reservoir pressure, psia
PR	productivity ratio

P_t	shut-in casing pressure, psia
P_{wf}	well flowing pressure, psia
P_{ws}	well shut-in or static pressure, psia
P_{1hr}	intercept of the semilog plot, psia
P^*	false pressure, psia
ΔP_s	pressure drop due to skin conditions, psia
q	liquid flow rate, bbl/D
q_{sc}	gas flow rate, Mscf/D
r_D	dimensionless radius
r	radius, ft
r_e	drainage radius, ft
R_s	gas dissolved in oil, scf/STB
r_w	well radius, ft
s	skin factor
s_c	skin due to partial penetration
s_{cp}	skin due to a change in permeability
s_{DL}	geometrical skin factor converging from radial to dual-linear flow
s_L	geometrical skin factor converging from dual-linear to linear flow
s_p	skin factor due to the restricted flow entry
s_{PB}	geometrical skin factor converging from dual-linear to parabolic flow
s_t	total skin factor
s_θ	skin factor resulting from a well deviation angle
T	reservoir temperature, °R, Transmissivity, md-ft/cp
t	time, h
t_p	production (horner) time before shutting-in a well, h
t_D	dimensionless time based on well radius
t_{DA}	dimensionless time based on reservoir area
t^*DP'	pressure derivative, psia
V	volume, ft^3
V_u	wellbore volume/unit length, bbl/ft
X	distance in the x-direction
X_E	reservoir length, ft
X_N	superposition time, h
Y	distance in the y-direction
Y_E	reservoir width, ft
W_D	dimensionless reservoir width
Z	height, ft

Greek

Δ	change, drop
Δt	shut-in time, h
φ	porosity, fraction
λ	mobility, md/cp
ρ	fluid density, lbm/ft^3
θ	deviation angle, °
ψ	measured deviation angle, °
ψ'	corrected deviation angle, °
μ	viscosity, cp

Suffices

0	base conditions
$1hr$	time of 1 h
a	actual
d	drainage
D	dimensionless
DA	dimensionless with respect to area
DL	dual linear flow
$DL1$	dual linear flow at 1 h
$DLpssi$	intersection of pseudosteady-state line with dual-linear line
$DLSS1$	intercept between dual-linear line and the −1-slope line (SS1)
$DLSS2$	intercept between dual-linear line and the −1-slope line (SS2)
eq	equivalent
F	inflection
g	gas
h	horizontal
hs	hemispherical
i	intersection or initial conditions
$ideal$	ideal
INT	intercept
inv	investigation
L	linear or hemilinear flow
$L1$	linear flow at 1 h
lag	lag
$Lpssi$	intercept of linear and pseudosteady state lines
N	an arbitrary point during early pseudosteady-state period
max	maximum
o	oil
OP	oil price, US$/STB
p	production, porous
PB	parabolic flow
$PBSS1$	intercept between parabolic line and the −1-slope line (SS1)
$PBSS2$	intercept between parabolic line and the −1-slope line (SS2)
pss	pseudosteady state
$pss1$	pseudosteady state at 1 h
r	radial flow
$rDLi$	intercept of radial and dual linear lines
r_1	radial flow before spherical/hemispherical flow
rg	relative to gas
rLi	intercept of radial and linear lines
ro	relative to oil
$rpssi$	intersection of pseudosteady-state line with radial line
$rSSi$	intersection between the radial line and the −1-slope line
$rSS1i$	intersection between the radial line and the −1-slope line (SS1)
$rSS2i$	intersection between the radial line and the −1-slope line (SS2)
rw	relative to water
s	skin
sf	sandface
sp	spherical

SS	steady
SSL	start of semilog line (radial flow)
*SS*1	−1-slope line formed when the parabolic flow ends and steady-state flow regime starts. Well is near the open boundary and the far boundary is opened
*SS*2	−1-slope line formed when the parabolic flow ends and steady-state flow regime starts. Well is near the open boundary and the far boundary is closed
sw	spherical/hemispherical wellbore
w	well, water
wa	apparent wellbore
wb	wellbore
wD	dimensionless emphasizing at wellbore
wf	well flowing
ws	well shut-in
x	maximum point (peak) during wellbore storage
xc	maximum point for centered wells
*X*1	maximum point between dual linear and parabolic lines
*X*2	maximum point between parabolic and negative unit slope lines
*X*3	maximum point between hemilinear and negative unit slope lines
z	vertical direction

References

[1] Agarwal RG. Direct Method of Estimating Average Reservoir Pressure for Flowing Oil and Gas Wells. Society of Petroleum Engineers; 2010. DOI: 10.2118/135804-MS

[2] Earlougher RC Jr. Advances in Well Test Analysis. Monograph Series. Vol. 5. Dallas, TX: SPE; 1977

[3] Lee J, Rollins JB, Spivey JP. Pressure Transient Testing. Vol. 9. Richardson, TX: SPE Textbook; 2003. 336p

[4] Mathews CS, Russell DG. Pressure Buildup and Flow Tests in Wells. SPE Monograph Series. Vol. 1. Dallas, TX: Society of Petroleum Engineers of AIME; 1967. 163p

[5] Rhagavan R. Well Test Analysis. New Jersey: Prentice Hall; 1993. 558p

[6] Sabet M. Well Testing. Gulf Publishing Co; 1991. 460p

[7] Stanislav JF, Kabir CS. Pressure Transient Analysis. New Jersey: Prentice Hall; 1990. 320p

[8] Earlougher RC, Kersch KM. Analysis of short-time transient test data by type-curve matching. Journal of Petroleum Technology. Society of Petroleum Engineers. 1974. pp. 793-800. DOI: 10.2118/4488-PA

[9] Ramey HJ. Short-time well test data interpretation in the presence of skin effect and wellbore storage. Journal of Petroleum Technology. Society of Petroleum Engineers. 1970:97-104. DOI: 10.2118/2336-PA

[10] Escobar FH, Olaya DA, Medina MV. Approximation of the starting time of radial flow regime under variable wellbore storage conditions in vertical wells. Journal of Engineering and Applied Sciences. 2014;**9**(7): 1047-1052

[11] Slider HC. Wordlwide Practical Petroleum Reservoir Engineering Methods. Revised Edition. Tulsa, Oklahoma, USA: PennWell Publishing Co.; 1983. 826p

[12] Escobar FH, Muñoz OF, Sepúlveda JA. Horizontal permeability determination from the elliptical flow regime for horizontal wells. Revista CT&F—Ciencia Tecnología y Futuro. 2004;**2**(5):83-95

[13] Escobar FH, Saavedra NF, Hernandez CM, Hernandez YA, Pilataxi JF, Pinto DA. Pressure and Pressure Derivative Analysis for Linear Homogeneous Reservoirs without using Type-Curve Matching. Society of Petroleum Engineers; 2004. DOI: 10.2118/88874-MS

[14] Escobar FH, Muñoz OF, Sepulveda JA, Montealegre M, Hernández YA. Parabolic flow: A new flow regime found in long, narrow reservoirs. In: XI Congreso Colombiano del Petróleo; October; 2005. pp. 18-21

[15] Escobar FH, Navarrete JM, Losada HD. Evaluation of Pressure Derivative Algorithms for Well-Test Analysis. Society of Petroleum Engineers; 2004. DOI: 10.2118/86936-MS

[16] Escobar FH, Muñoz OF, Sepulveda JA, Montealegre M. New finding on pressure response in long, narrow reservoirs. CT&F—Ciencia, Tecnología y Futuro. 2005:131-140

[17] Escobar FH, Muñoz OF, Sepulveda JA, Montealegre-MM, Hernandez YA. Parabolic flow: A new flow regime found in long, narrow reservoirs. XI Congreso Colombiano del Petróleo (Colombian Petroleum Symposium); October; 2005. pp. 18-21

[18] Escobar FH, Montealegre-MM. Effect of well stimulation on the skin

factor in elongated reservoirs. CT&F—Ciencia, Tecnología y Futuro. 2006; **3**(2):109-119. ISSN 0122-5383

[19] Escobar FH, Hernández YA, Hernandez CM. Pressure transient analysis for long homogeneous reservoirs using TDS technique. Journal of Petroleum Science and Engineering. 2007;**58**(1–2):68-82

[20] Escobar FH, Montealegre M. A complementary conventional analysis for channelized reservoirs. CT&F—Ciencia, Tecnología y Futuro. 2007;**3**(3): 137-146

[21] Escobar FH, Montealegre-MM. Application of TDS technique to developed reservoirs. Journal of Petroleum Science and Engineering. 2007;**55**(3–4):252-258

[22] Escobar FH, Lopez AM, Cantillo JH. Effect of the pseudotime function on gas reservoir drainage area determination. CT&F—Ciencia, Tecnología y Futuro. 2007;**3**(3):113-124

[23] Escobar FH, Tiab D, Tovar LV. Determination of areal anisotropy from a single vertical pressure test and geological data in elongated reservoirs. Journal of Engineering and Applied Sciences. 2007;**2**(11):1627-1639

[24] Escobar FH. Recent advances in well test analysis for long and narrow reservoirs. In: Korin L. Montclaire, editor. Petroleum Science Research Progress. Nova Publisher; 2008

[25] Escobar FH, Martinez JA, Montealegre M. Pressure and pressure derivative analysis for injection tests with variable temperature without type-curve matching. CT&F—Ciencia, Tecnología y Futuro. 2008;**4**(4):83-91

[26] Escobar FH, Montealegre M. Application of TDS technique to multiphase flow. CT&F—Ciencia, Tecnología y Futuro. 2008;**4**(4):93-105

[27] Escobar FH, Sanchez JA, Cantillo JH. Rate transient analysis for homogeneous and heterogeneous gas reservoirs using the TDS technique. CT&F—Ciencia, Tecnología y Futuro. 2008;**4**(4):45-59

[28] Escobar FH, Hernandez YA, Tiab D. Determination of reservoir drainage area for constant-pressure systems using well test data. CT&F—Ciencia, Tecnología y Futuro. 2010;**4**(1):51-72

[29] Escobar FH, Montealegre M, Carrillo-Moreno D. Straight line methods for estimating permeability or width for two-zone composite channelized reservoirs. CT&F—Ciencia, Tecnología y Futuro. 2009;**3**(5): 107-124

[30] Escobar FH, Montealegre-MM, Carrillo-Moreno D. Pressure and pressure derivative transient analysis without type-curve matching for elongated reservoirs with changes in permeability or width. CT&F—Ciencia, Tecnología y Futuro. 2010;**4**(1):75-88

[31] Escobar FH, Hernandez DP, Saavedra JA. Pressure and pressure derivative analysis for long naturally fractured reservoirs using the TDS technique. Dyna. 2010;**77**(163):102-114

[32] Escobar FH, Martínez JA, Montealegre M. Pressure and pressure derivative analysis for a well in a radial composite reservoir with a non-Newtonian/Newtonian interface. CT& F. 2010;**4**(1):33-42

[33] Escobar FH, Martinez JA, Bonilla LF. Pressure and pressure derivative analysis different of type-curve matching for thermal recovery processes. CT&F. 2011;**4**(4):23-35

[34] Escobar FH. Transient pressure and pressure derivative analysis for non-Newtonian fluids. In: New Technologies in the Oil and Gas Industry. Rijeka: Intech; 2012. http://www.intechopen.

com/books/export/citation/Zotero/new-technologies-in-the-oil-and-gas-industry/transient-pressure-and-pressure-derivative-analysis-for-non-newtonian-fluids

[35] Escobar FH, Rojas MM, Bonilla LF. Transient-rate analysis for long homogeneous and naturally fractured reservoir by the *TDS* technique. Journal of Engineering and Applied Sciences. 2012;**7**(3):353-370

[36] Escobar FH, Martinez LY, Méndez LJ, Bonilla LF. Pseudotime application to hydraulically fractured vertical gas wells and heterogeneous gas reservoirs using the TDS technique. Journal of Engineering and Applied Sciences. 2012; **7**(3):260-271

[37] Escobar FH, Corredor CM, Gomez BE, Cantillo JH, Prent LA. Pressure and pressure derivative analysis for slanted and partially penetrating wells. Journal of Engineering and Applied Sciences. 2012;**7**(8):932-938

[38] Escobar FH, Rojas MM, Cantillo JH. Straight-line conventional transient rate analysis for long homogeneous and heterogeneous reservoirs. Dyna. 2012; **79**(172):153-163

[39] Escobar FH, Muñoz YEM, Cerquera WM. Pseudotime function effect on reservoir width determination in homogeneous and naturally fractured gas reservoir drained by horizontal wells. Entornos Journal. 2012;(24): 221-231

[40] Escobar FH, Vega J, Diaz MR. Pressure and pressure derivative analysis for double-permeability systems without type-curve matching. Journal of Engineering and Applied Sciences. 2012;**7**(10)

[41] Escobar FH, Vega LJ, Bonilla LF. Determination of well-drainage area for power-law fluids by transient pressure analysis. CT&F. 2012;**5**(1):45-55

[42] Escobar FH, Martinez JA, Bonilla LF. Transient pressure analysis for vertical wells with spherical power-law flow. CT&F. 2012;**5**(1):19-25

[43] Escobar F-H, Martínez JA, Montealegre M. Pressure transient analysis for a reservoir with a finite-conductivity fault. CT&F—Ciencia, Tecnología y Futuro. 2013;**5**:5-18

[44] Escobar FH, Alzate HD, Moreno L. Effect of extending the radial superposition function to other flow regimes. Journal of Engineering and Applied Sciences. 2013;**8**(8):625-634

[45] Escobar FH, Ascencio JM, Real DF. Injection and fall-off tests transient analysis of non-Newtonian fluids. Journal of Engineering and Applied Sciences. 2013;**8**(9):703-707

[46] Escobar FH, Martinez JA, Montealegre M. Pressure and pressure derivative analysis in a reservoir with a finite-conductivity fault and contrast of mobilities. Fuentes Journal. 2013;**11**(2): 17-25

[47] Escobar FH, Meneses AR, Losada LM. Straight-line conventional transient pressure analysis for horizontal wells with isolated zones. Dyna. 2014;**81**(185): 78-85

[48] Escobar FH, Polanco MP, Benavidez W. Pressure and pressure derivative analysis for a vertical well in wedged and T-shaped reservoirs. Journal of Engineering and Applied Sciences. 2014;**9**(6):845-851

[49] Escobar FH, Montenegro LM, Bernal KM. Transient-rate analysis for hydraulically-fractured gas shale wells using the concept of induced permeability field. Journal of Engineering and Applied Sciences. 2014;**9**(8):1244-1254

[50] Escobar FH, Ghisays-Ruiz A, Bonilla LF. New model for elliptical flow regime

in hydraulically-fractured vertical wells in homogeneous and naturally-fractured systems. Journal of Engineering and Applied Sciences. 2014;**9**(9):1629-1636

[51] Escobar FH, Bernal KM, Olaya-Marin G. Pressure and pressure derivative analysis for fractured horizontal wells in unconventional shale reservoirs using dual-porosity models in the stimulated reservoir volume. Journal of Engineering and Applied Sciences. 2014;**9**(12):2650-2669

[52] Escobar FH, Salcedo LN, Pinzon C. Pressure and pressure derivative analysis for fractal homogeneous reservoirs with power-law fluids. Journal of Engineering and Applied Sciences. 2015;**10**(11):4763-4772

[53] Escobar FH, Srivastava P, Wu X. A practical method to determine aquifer leakage factor from well test data in CBM reservoirs. Journal of Engineering and Applied Sciences. 2015;**10**(11): 4857-4863

[54] Escobar FH, Fahes M, Gonzalez R, Pinchao DM, Zhao YL. Determination of reservoir drainage area for constant-pressure systems by conventional transient pressure analysis. Journal of Engineering and Applied Sciences. 2015; **10**(12):5193-5199

[55] Escobar FH, Ghisays-Ruiz A, Srivastava P. Characterization of the spherical stabilization flow regime by transient pressure analysis. Journal of Engineering and Applied Sciences. 2015; **10**(14):5815-5822. ISSN 1819-6608

[56] Escobar FH. Recent Advances in Practical Applied Well Test Analysis. New York: Nova publishers; 2015. 422p

[57] Escobar FH, Zhao YL, Pournik M, Liu QG, Olaya-Marin G. Interpretation of pressure tests in uniform-flux fractured vertical wells with threshold pressure gradient in low permeability reservoirs. Journal of Engineering and Applied Sciences. 2015;**10**(20): 9364-9372

[58] Escobar FH, Jongkittnarukorn J, Hernandez CM. The power of TDS technique for well test interpretation: A short review. Switzerland AG: Journal of Petroleum Exploration and Production Technology. 2018. DOI: 10.1007/s13202-018-0517-5

[59] Ehlig-Economides C, Economides MJ. Pressure transient analysis in an elongated linear flow system. Society of Petroleum Engineers Journal. 1985: 839-847. DOI: 10.2118/12520-PA

[60] Hachlaf H, Tiab D, Escobar FH. Effect of Variable Injection Rate on Falloff and Injectivity Tests. Society of Petroleum Engineers; 2002. DOI: 10.2118/76714-MS

[61] Jongkittinarukorn K, Tiab D, Escobar FH. Interpretation of Horizontal Well Performance in Complicated Systems by the Boundary Element Method. Society of Petroleum Engineers; 1998. DOI: 10.2118/50437-MS

[62] Martinez JA, Escobar FH, Montealegre M. Vertical well pressure and pressure derivative analysis for bingham fluids in a homogeneous reservoirs. Dyna. 2011;**78**(166):21-28

[63] Martin JC. Simplified Equations of Flow in Gas Drive Reservoirs and the Theoretical Foundation of Multiphase Pressure Buildup Analyses. Society of Petroleum Engineers; 1959

[64] Martinez JA, Escobar FH, Cantillo JH. Application of the *TDS* technique to dilatant non-Newtonian/Newtonian fluid composite reservoirs. Ingeniería e Investigación journal. 2011;**31**(3): 130-134

[65] Merzouk K, Tiab D, Escobar FH. Multirate Test in Horizontal Wells.

Society of Petroleum Engineers; 2002. DOI: 10.2118/77951-MS

[66] Moncada K, Tiab D, Escobar FH, Montealegre-MM, Chacon A, Zamora RA, Nese SL. Determination of vertical and horizontal permeabilities for vertical oil and gas wells with partial completion and partial penetration using pressure and pressure derivative plots without type-curve matching. CT&F—Ciencia, Tecnología y Futuro. 2005;**2**(6)

[67] Mongi A, Tiab D. Application of Tiab's Technique to Multi-Rate Tests. Society of Petroleum Engineers; 2000. DOI: 10.2118/62607-MS

[68] Tiab D. PE-5553: Well test analysis. Lecture notes. Fall 1993. The University of Oklahoma

[69] Tiab D. Analysis of pressure derivative without type-curve matching: Vertically fractured wells in closed systems. Journal of Petroleum Science and Engineering. 1994;**11**:323-333. This paper was originally presented as Tiab D. Analysis of Pressure and Pressure Derivative without Type-Curve Matching—III. Vertically Fractured Wells in Closed Systems. Society of Petroleum Engineers; 1993. DOI: 10.2118/26138-MS

[70] Tiab D. Direct Type-Curve Synthesis of Pressure Transient Tests. Society of Petroleum Engineers; 1989. DOI: 10.2118/18992-MS

[71] Tiab D. Analysis of pressure and pressure derivative without type-curve matching: 1—Skin and wellbore storage. Journal of Petroleum Science and Engineering. 1995;**12**:171-181. Also Tiab D. Analysis of Pressure and Pressure Derivatives Without Type-Curve Matching: I—Skin and Wellbore Storage. Society of Petroleum Engineers; 1993. DOI: 10.2118/25426-MS

[72] Zhao YL, Escobar FH, Jamili A, Olaya-Marin G, Ghisays-Ruiz A. Effect of wellbore storage on the vertical well pressure behavior with threshold pressure gradient in low permeability reservoirs. Journal of Engineering and Applied Sciences. 2015;**10**(22): 10659-10664

[73] Bourdet D, Ayoub JA, Pirard YM. (June 1). Use of Pressure Derivative in Well Test Interpretation. SPEFE. Society of Petroleum Engineers; 1989. pp. 293-302. DOI: 10.2118/12777-PA

[74] Brons F, Miller WC. (August 1). A simple method for correcting spot pressure readings. Journal of Petroleum Technology; Transactions of the AIME. Society of Petroleum Engineers. 1961; **222**. pp. 803-805. DOI: 10.2118/1610-G-PA

[75] Dietz DN. Determination of average reservoir pressure from build-up surveys. Journal of Petroleum Technology. Transactions AIME. Society of Petroleum Engineers. 1965; **234**. pp. 955-959. DOI: 10.2118/1156-PA

[76] Jones P. Reservoir limit test. Oil and Gas Journal. 1956:184-196

[77] Ramey HR Jr, Cobb WM. A general buildup theory for a well in a closed drainage area. JPT. Society of Petroleum Engineers. 1971:1493-1505. DOI: 10.2118/3012-PA

[78] Bernal KM, Escobar FH, Ghisays-Ruiz A. Pressure and pressure derivative analysis for hydraulically-fractured shale formations using the concept of induced permeability field. Journal of Engineering and Applied Sciences. 2014;**9**(10):1952-1958

[79] Boussalem R, Djebbar T, Escobar FH. Effect of Mobility Ratio on the Pressure and Pressure Derivative of Wells in Closed Composite Reservoirs. Society of Petroleum Engineers; 2002. DOI: 10.2118/76781-MS

[80] Tiab D, Bettam Y. Practical Interpretation of Pressure Tests of Hydraulically Fractured Wells in a Naturally Fractured Reservoir. Society of Petroleum Engineers; 2007. DOI: 10.2118/107013-MS

[81] Engler T, Tiab D. Analysis of pressure and pressure derivative without type-curve matching, 5. Horizontal well tests in naturally fractured reservoirs. Journal of Petroleum Science and Engineering. 1996;**15**:139-151

[82] Engler T, Tiab D. Analysis of pressure and pressure derivative without type-curve matching, 6. Horizontal well tests in anisotropic media. Journal of Petroleum Science and Engineering. 1996;**15**:153-168

[83] Engler T, Tiab D. Analysis of pressure and pressure derivative without type curve matching, 4. Naturally fractured reservoirs. Journal of Petroleum Science and Engineering. 1996;**15**:127-138

[84] Vongvuthipornchai S, Raghavan R. A Note on the Duration of the Transitional Period of Responses Influenced by Wellbore Storage and Skin. SPE Formation Evaluation. Society of Petroleum Engineers; 1988. pp. 207-214. DOI: 10.2118/15273-PA.

[85] Chacon A, Djebrouni A, Tiab D. Determining the Average Reservoir Pressure from Vertical and Horizontal Well Test Analysis Using the Tiab's Direct Synthesis Technique. Society of Petroleum Engineers; 2004. DOI: 10.2118/88619-MS

[86] Perrine RL. Analysis of pressure-buildup curves. In: Drilling and Production Practices; American Petroleum Institute (API); 1956. pp. 482-509. ID API-56-482

[87] Al-Khalifah A-JA, Horne RN, Aziz K. In-Place Determination of Reservoir Relative Permeability Using Well Test Analysis. Society of Petroleum Engineers; 1987. DOI: 10.2118/16774-MS

[88] Kamal MM, Pan Y. Use of Transient Data To Calculate Absolute Permeability and Average Fluid Saturations. Society of Petroleum Engineers; 2010. DOI: 10.2118/113903-PA

[89] Al-Khalifah A-JA, Aziz K, Horne RN. A New Approach to Multiphase Well Test Analysis. Society of Petroleum Engineers; 1987. DOI: 10.2118/16743-MS

[90] Abbott WA, Collins T, Tippie DB. Practical Application of Spherical Flow Transient Analysis. Society of Petroleum Engineers; 1978. DOI: 10.2118/7485-MS

[91] Dykstra H, Kazemi H, Raghavan R, Gulati MS. Pressure Transient Testing Methods, SPE Reprint Series No. 14. Dallas, Texas: Society of Petroleum Engineers; 1967. 264p

[92] Cinco H, Miller FG, Ramey HJ. Unsteady-state pressure distribution created by a directionally drilled well. Journal of Petroleum Technology, Transactions of AIME. Society of Petroleum Engineers. 1975;**259**. pp. 1392-1400. DOI: 10.2118/5131-PA

[93] Jones LG, Watts JW. Estimating skin effect in a partially completed damaged well. Journal of Petroleum Technology, Transactions of AIME. Society of Petroleum Engineers. 1971; **251**:249-252. DOI: 10.2118/2616-PA

[94] Joseph JA. Unsteady-state cylindrical, spherical and linear flow in porous media [PhD Dissertation]. Rolla: University of Missouri; 1984

[95] Merzouk K, Tiab D, Escobar FH. Multirate Test in Horizontal Wells. Society of Petroleum Engineers; 2002. DOI: 10.2118/77951-MS

[96] Russell DG. Determination of formation characteristics from two-rate

flow tests. Journal of Petroleum Technology; Transactions of AIME. Society of Petroleum Engineers. 1963; **228**:1347-1355. DOI: 10.2118/645-PA

[97] Tracy GW, Coats KH, Kazemi H, Odeh AS, Lebourg M, Prats M, et al. Pressure Analysis Methods. SPE Reprint Series No. 9. Society of Petroleum Engineers of AIME; 1967

[98] Slider HC. Application of Pseudo-Steady-State Flow to Pressure-Buildup Analysis. Society of Petroleum Engineers; 1966. DOI: 10.2118/1403-MS

[99] Slider HC. A simplified method for pressure buildup analysis for a stabilized well. Journal of Petroleum Technology. 1971:1155-1160. Slider HC. A Simplified Method of Pressure Buildup Analysis for a Stabilized Well. Society of Petroleum Engineers; 1967. DOI: 10.2118/1765-MS

[100] Sui W, Mou J, Bi L, Deng J, Ehlig-Economides CA. New Flow Regimes for Well Near Constant Pressure Boundary. Society of Petroleum Engineers; 2007. DOI: 10.2118/106922-MS

[101] Wong DW, Mothersele CD, Harrington AG, Cinco-Ley H. Pressure Transient Analysis in Finite Linear Reservoirs Using Derivative and Conventional Techniques: Field Examples. Society of Petroleum Engineers; 1986. DOI: 10.2118/15421-MS

[102] Nutakki R, Mattar L. Pressure Transient Analysis of Wells in Very Long Narrow Reservoirs. Society of Petroleum Engineers; 1982. DOI: 10.2118/11221-MS

[103] Horne RN. Modern Well Test Analysis: A Computer-aided Approach. Palo Alto, CA: Petroway Inc.; 1990. 185p

[104] Jones P. Drawdown Exploration Reservoir Limit, Well and Formation Evaluation. Society of Petroleum Engineers; 1957. DOI: 10.2118/824-G

[105] Jones P. Reservoir limit test on gas wells. Journal of Petroleum Technology. Society of Petroleum Engineers. 1962: 613-619. DOI: 10.2118/24-PA

[106] Odeh AS, Jones LG. Pressure Drawdown Analysis, Variable-Rate Case. JPT; Transactions of AIME. Society of Petroleum Engineers. 1965; **234**:960-964. DOI: 10.2118/1084-PA

[107] Smith CR, Tracy GW. Applied Reservoir Engineering. Tulsa, OK: Oil & Gas Consultants, Inc.; 1987

Chapter 3

Pressure Buildup Testing

A pressure buildup test has been a very popular technique used in the hydrocarbon industry. Several reasons have become a very popular test, some of these are: (a) it does not require very detailed supervision and (b) permeability and skin factor can be determined from both pressure buildup and drawdown tests. However, as studied in Section 2.8, until 2010, a flow test did not allow estimating the average reservoir pressure, while a pressure test does [7–10, 25, 26, 28, 30].

Figure 3.1 shows a plot of an ideal pressure buildup test. In general terms, it requires shutting-in a producer well after it has produced for some time, t_p, with a stable flow rate. A pressure buildup test is run as follows:

1. Place the pressure sensors in the selected site. It is recommended as close as possible to the perforations.

2. Stabilize the well to a constant production rate, q.

3. Close the well and record the P_{wf} value (just before closing).

4. Read the well bottom-hole pressure, P_{ws}, at short time intervals of 15 s for the first few minutes (10–15 min), then I could be every 10 min for the first hour. During the next 10 h, hourly pressure readings should be taken. When the test progresses, the time intervals can be expanded to 5 h. With recently introduced pressure recorders, the readings can be taken at shorter intervals. It can start reading every second or less.

To run a pressure buildup test, the well produces a constant rate for a period of time, t_p. A pressure recorder is lowered to the well immediately before closing. t_p should not be too small to avoid problems associated with superposition and investigation radius [12].

3.1. Superposition principle

Suppose that after the well has produced a constant rate for a time period, t_p, it is decided to shut-in the well to obtain a pressure buildup test. Intuitively, fluid movement is expected at the reservoir after the well is shut-in, but at surface $q = 0$. A similar situation arises in fall-off testing, but injections takes place instead of production. An analogy is made to the fluid movement at the reservoir [10–12, 28, 37, 40] as follows: the well is allowed to produce indefinitely at a flow rate, q, and at the instant of shutting-in the well, the same flow rate, q, is injected into the same well, and then the pressure drop is added due to the production of q and same pressure data multiplied by -1 and displaced at the time the well is shut-in. This, however, is not easy to understand. The better way is to understand, refer to **Figure 3.1**, is to estimate the well pressure drop at a time, $t_p + \Delta t$, and then subtract the pressure drop during a time, Δt. Mathematically;

$$P_{Dws} = P_D(t_p + \Delta t)_D - P_D(\Delta t)_D \tag{3.1}$$

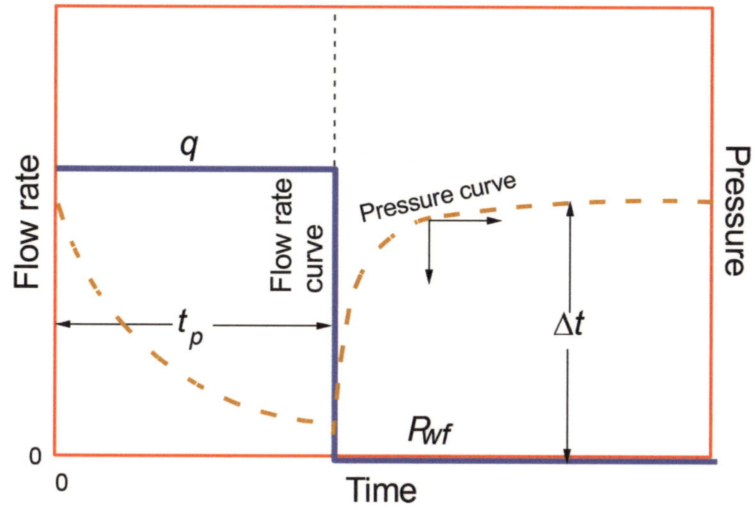

Figure 3.1.
Schematic representation of pressure restoration.

If t_p is not given, it can be estimated if cumulative production, Np, is known,

$$t_p = \frac{24Np}{q} \tag{3.2}$$

Assuming wellbore storage is neglected and the reservoir is of infinite size; then, Eq. (1.115) applies:

$$P_D(t_p + \Delta t)_D = \frac{1}{2}\left[\ln (t_p + \Delta t)_D + 0.80907\right] - s \tag{3.3}$$

$$P_D(\Delta t)_D = \frac{1}{2}\left[\ln (\Delta t)_D + 0.80907\right] - s \tag{3.4}$$

Combining Eqs. (3.3) and (3.4) in Eq. (3.1), then replacing in the resulting combination the dimensionless parameter given by Eqs. (1.89) and (1.94) yields:

$$P_{ws} = P_i - \frac{162.6 q\mu B}{kh} \log\left(\frac{t_p + \Delta t}{\Delta t}\right) \tag{3.5}$$

This is known as Horner equation. As a result of the application of the superposition principle is that the skin factor, s, disappears in the Horner's simplified equation. That means the slope of the Horner plot is not affected by the skin factor. However, the skin factor alters, even greater than in flow tests, the shape of the pressure buildup curve. The skin factor affects the buildup test more than the drawdown test because wellbore storage persists.

3.2. Buildup test methods

3.2.1 Horner method

Eq. (3.5) suggests that a semi-log plot of well-shut-in pressure versus $(t_p+\Delta t)/\Delta t$ will yield a straight line which slope allows finding the permeability from Eq. (2.34). Estimating the Horner time, $(t_p+\Delta t)/\Delta t$ was tedious before 1970 when computer power was limited which is not today's case. When superposition is overcome, the semi-log plot of P_{ws} versus Δt can be applied. In pressure buildup testing, this semi-log plot is rather known as MDH plot, as shown later.

Horner plot is generally not preferred, since it requires more work than MDH. It is strongly recommended to be used when $t_p < t_{pss}$ [12, 26]. This is because superposition effects make the semi-log straight line difficult to identify. Actually, Horner plot virtually increases about four times the length of the semi-log slope. If t_p is at least twice the size of t_{pss}, it is then justified to plot using t_{pss} instead of t_{pss} in finite systems [12, 26], since the Horner plot tends to prolong the semi-log line. Preparing a Horner plot with t_{pss} instead of t_p has meaning to minimize errors in the estimation of the average pressure. However, with the advent of the pressure derivative function, the identification of radial flow became easier [6, 13, 25].

Just to look alike a MDH plot, Horner plot uses inverted abscissa scale as shown in **Figure 3.2**. For long producing times, a slight variation of Eq. (2.34) is used to find skin factor when t_p> 1;

$$s = 1.1513\left[\frac{P_{1hr} - P_{wf}}{m} - \log\left(\frac{k}{\phi\mu c_t r_w^2}\right) + 3.23\right] \tag{3.6}$$

Here P_{wf} is used instead of P_i. P_{wf} is the registered pressure just before shutting-in the well. Finding $P_{1\text{hr}}$ requires using tp and adding one to that value. Use that estimated $(t_p+\Delta t)/\Delta t$ value and enter the Horner plot and read on the semi-log line the value of $P_{1\text{hr}}$. It is meaningless to estimate $(t_p+\Delta t)/\Delta t=1$. However, such reading will be used later to estimate the average reservoir pressure. When $t_p < 1$, the following expression ought to be used to find skin factor:

$$s = 1.1513\left[\frac{P_{1hr} - P_{wf}}{m} + \log\left(1+\frac{1}{t_p}\right) - \log\left(\frac{k}{\phi\mu c_t r_w^2}\right) + 3.2275\right] \tag{3.7}$$

Once the skin factor is estimated, the skin pressure drop and flow efficient can be found using Eqs. (2.35) and (2.46).

When the well is shut-in for a buildup test, the formation fluid keeps flowing, even though using downhole devise shutting. Again, this after-flow duration can be determined easily from the pressure derivative plot once radial flow starts. This was not the case before 1980s. This after-flow rate, q_{af}, due to the wellbore storage, has a significant influence on the pressure data. This occurs because the head pressure is not equal to the bottom shut-in pressure, and therefore the fluid continues to flow

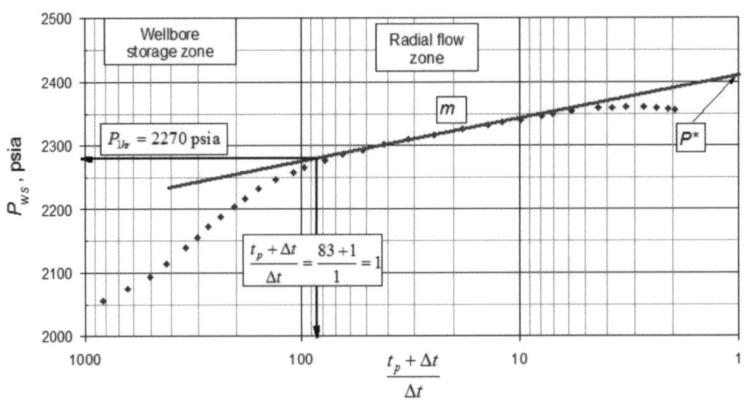

Figure 3.2.
Typical Horner plot, t_p = *83 hr.*

from the formation to the well. Then the pressure does not recover as fast as expected. As the flow rate tends to zero, the pressure increases rapidly. The semi-log graph is pronounced and linear in this period and can be confused with the semi-log slope [10, 12, 26, 28, 37, 40].

$$q_{af} = \frac{24CV_w}{B}\frac{dP_{ws}}{d(\Delta t)} \tag{3.8}$$

C is found from transient pressure analysis using Eq. (2.18). For producer and injector wells, respectively, the after-flow duration can be estimated from [40];

$$\Delta t_{af} = 204\left(\frac{C}{BJ}\right) \tag{3.9}$$

$$\Delta t_{af} \cong 204\left(\frac{CV_u}{BJ}\right) \tag{3.10}$$

where J is the productivity index, Eq. (2.44), B is the volume factor and, V_u, the wellbore capacity, Eq. (2.4), and C is the wellbore storage coefficient found from Eq. (2.18). When $q_{af}/q < 0.01$, it is concluded that wellbore storage does not affect the semi-log slope. In other words, after this time, WBS effects are negligible.

Because of superposition, skin and wellbore storage effects, the start time of the semi-log slope, Δt_{SSL}, is given by [12],

$$(\Delta t_D)_{SSL} = 50C_D e^{0.14s} \tag{3.11}$$

By taking a glance to Eq. (2.19), it is appreciated a higher effect of skin factor and wellbore storage in the above expression. After replacing the dimensionless parameters, Eqs. (2.14) and (1.94), in Eq. (3.11), it results:

$$\Delta t_{SSL} = \frac{170000\mu Ce^{0.14s}}{kh} \tag{3.12}$$

Eq. (3.5) applies to infinite-size reservoir. For finite reservoirs, Eq. (3.59), becomes [12, 40],

$$P_{ws} = P* - \frac{162.6q\mu B}{kh}\log\left(\frac{t_p+\Delta t}{\Delta t}\right) = P* - m\log\left(\frac{t_p+\Delta t}{\Delta t}\right) \tag{3.13}$$

However, this equation applies similar to Eq. (2.5). The false pressure, P^*, is read at a Horner time, $(t_p+\Delta t)/\Delta t=1$, and does not have physical meaning, but is useful to determine the average reservoir pressure [28].

3.2.2 Miller-Dyes-Hutchinson (MDH) method

This is based on the assumption that the production time, t_p, is long enough to reach the pseudo-steady-state period; then, it is more representative to use average pressure than initial pressure. The MDH method is preferred in old wells or depleted formations, which would make it difficult to obtain stabilization before shutting-in [40]. The Horner plot can be simplified [12, 28, 40], if $\Delta t \lll t_p$, then:

$$t_p + \Delta t \cong t_p \tag{3.14}$$

Then,

$$\log\left(\frac{t_p + \Delta t}{\Delta t}\right) \approx \log\, t_p - \log\, \Delta t \tag{3.15}$$

Combining Horner equation, Eq. (3.13) with Eq. (3.15), it yields [12, 41]:

$$P_{ws} = P * - m \log\, t_p + m \log\, \Delta t \tag{3.16}$$

If $P^* - m \log t_p$ = constant = intercept; then,

$$P_{ws} = P_{1hr} + \frac{162.6 q\mu\, B}{kh} \log\, \Delta t \tag{3.17}$$

This suggests that a semi-log plot of P_{ws} versus Δt will yield a straight line which slope, m, and intercept, $P_{1\text{hr}}$, are used to find reservoir permeability with Eq. (2.33) and skin factor with Eq. (3.6).

Some expressions and plots [12] can be used to estimate the end of the semi-log straight line. However, the use of the pressure derivative [6, 13, 25] avoids using them. Therefore, they are omitted in this chapter.

3.2.3 Extended Muskat method

It is a trial-and-error method that is more attractive in cases of constant pressure or water injection systems (filling) because in these cases, the straight line would be longer and, therefore, easier to identify [12, 32]. Muskat [32] proposed to build a potential plot (log(ΔP) versus Δt), see **Figure 3.3**. Cobb and Smith [8] and Ramey and Cobb [39] recommended using it only as a method of late-time analysis. For the application of the method, the average reservoir pressure is assumed as many times as a straight line results in the plot. This author found that changing the average reservoir pressure in a range of 30 psia above or below the target average pressure always provides a straight line. Permeability and skin factor are found from the intercept of such plot, ΔP_{Mint}, read at Δt = 0:

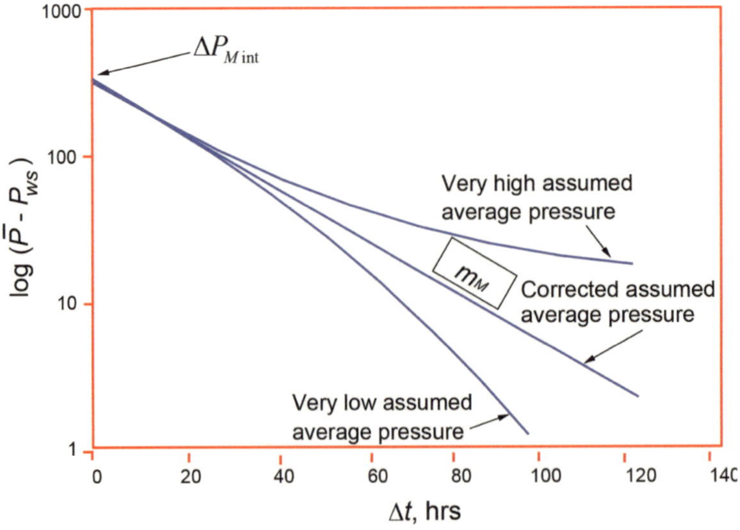

Figure 3.3.
Schematic representation of the Muskat plot for the analysis of pressure buildup tests, after [12].

$$k = \frac{141.2q\mu B}{h}\frac{P_D(t_{DA})_{\text{int}}}{\Delta P_{M\text{int}}} \tag{3.17.1}$$

$$s = \left(\frac{P_D(t_{DA})_{\text{int}}}{\Delta P_{\text{int}}}\right)\left[\overline{P} - P_{wf}(\Delta t = 0)\right] - \ln\frac{r_e}{r_w} + 0.75 \tag{3.18}$$

$P_D(t_{pDA})$ is normally found from plots [12]. However, this author fitted the curves of such plots to polynomials. For a well within a square shaped reservoir—constant pressure case:

$$P_{DM\text{ int}} = -0.0118157 + 1.3509395\left(1 - \exp\left(-21.692995\ t_{pDA}\right)\right) \tag{3.18.1}$$

For a well within a square geometry reservoir—no-flow boundary case.

$$P_{DM\text{ int}} = -0.02056 + 0.682297\left(1 - \exp\left(-50.7038508\ t_{pDA}\right)\right) \tag{3.19}$$

where

$$t_{pDA} = \left(\frac{0.0002637kt_p}{\phi\mu c_t A}\right) \tag{3.20}$$

The slope of the Muskat, m_M, plot can be used to find the drain area:

$$A = \left(\frac{k}{\phi\mu c_t m_M}\right)M_{SF} \tag{3.21}$$

For the values of $t_{pDA} > 1$, the Musk shape factor, M_{SF}, is 0.67, 1.34, and 0.84 for no-flow boundary square reservoirs, constant-pressure boundary square reservoirs and no-flow boundary circular reservoirs, respectively, with a unique well in the center of such systems [12].

If A is known, then,

$$\phi\, c_t\, h = S_t = \frac{kh}{43560\mu\, m_M A}M_{SF} = T\frac{M_{SF}}{43560 m_M A} \tag{3.22}$$

It can be concluded that MDH is generally preferred because it is easy to use. For short production times, it is recommended to use the Horner method since the semi-log line is longer than that provided by MDH. Earlougher [12] and Tiab [40] recommend the following aspects:

a. The Horner method could be used to analyze pressure buildup data, assuming t_p is known. However, MDH, and then Horner, are usually used as the first choice. If t_p is unknown, then use MDH.

b. Use MDH as the first test unless $t_p < t_{pss}$ (reservoir acting as infinity, then Horner is applied) or unless the well is in the center of a square shaped reservoir with open boundaries, such as an injection pattern of five points.

c. The Muskat method is used as a last option. It also provides the determination of the drainage area.

As for the MDH case, the starting and the end of the Muskat straight line can be estimated only for square shape reservoirs with the well at the center within it. For this reason and the help of the pressure derivative, it is not presented here. However, these procedures can be found in [12, 39].

3.2.4 Type-curve matching

Such type curves as the given in **Figure 2.4** and **2.6** and their accompanying equations also apply for buildup tests. However, superposition may cause trouble as can be seen in **Figure 3.4**. This was reported by Gringarten [23] when demonstrating the importance of deconvolution. Then, both pressure and pressure derivative must be corrected [3], before applying type-curve matching. To overcome this issue, Agarwal [1] introduced the equivalent time, given by:

$$\Delta t_e = \frac{t_p \Delta t}{\Delta t + t_p} \tag{3.23}$$

Eq. (3.23) is the most common equivalent-time equation. However, it was developed only for radial flow regime; therefore, it may fail providing good results if applied to other flow regimes. Then, the equivalent-time equations for bilinear [38], linear, [38], birradial (elliptical), and spherical/hemispherical/parabolic flow regimes are, respectively, given as follows:

$$\Delta t_{eBL} = \left[\sqrt[4]{t_p} + \sqrt[4]{\Delta t} - \sqrt[4]{t_p + \Delta t}\right]^4 \tag{3.24}$$

$$\Delta t_{eL} = \left[\sqrt{t_p} + \sqrt{\Delta t} - \sqrt{t_p + \Delta t}\right]^2 \tag{3.25}$$

$$\Delta t_{eBR} = \left[\sqrt[25]{t_p^9} + \sqrt[25]{\Delta t^9} - \sqrt[25]{(t_p + \Delta t)^9}\right]^{25/9} \tag{3.26}$$

$$\Delta t_{eSP} = \frac{1}{\left[\frac{1}{\sqrt{t_p}} + \frac{1}{\sqrt{\Delta t}} - \frac{1}{\sqrt{t_p + \Delta t}}\right]^2} \tag{3.27}$$

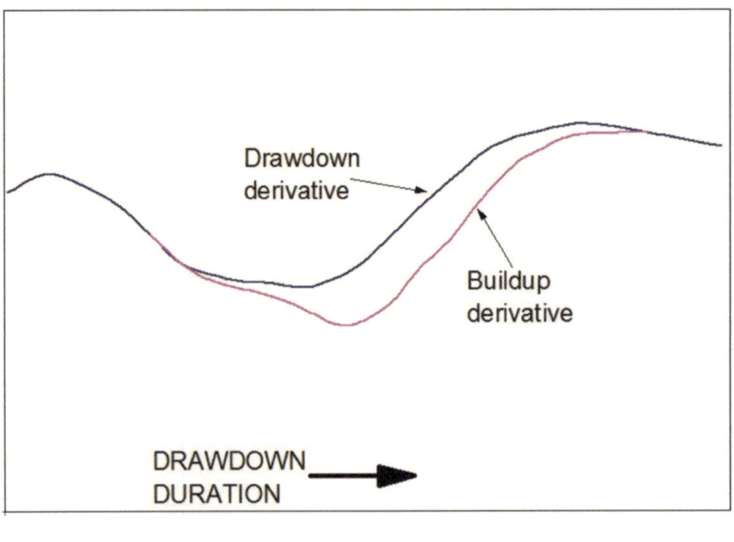

Figure 3.4.
Drawdown versus buildup log-log derivative shapes, after [23].

Once, the equivalent time is determined, then, the equivalent pressure derivative is estimated by:

$$\Delta t_e * \Delta P' = \Delta t_e \left[\frac{d(\Delta P)}{d(\Delta t_e)}\right] \tag{3.28}$$

Since Eq. (3.23) is normally used for time corrections, possibly, some recommendations given by [38] regarding the use of equivalent time were provided:

- t_{eq} is primarily useful for homogeneous infinite-acting radial flow systems when $t_p >> \Delta t$.
- t_{eq} is not recommended for fractured wells where linear flow dominates early time.
- t_{eq} should not be used if multiphase flow is dominant.
- Pressure data affected by boundaries are usually better plotted with Δt.

3.2.5 *TDS* technique

Good news! *TDS* technique applies to drawdown, buildup and, of course, drill stem tests. The equations already seen in Chapter 2 also apply here. Care must be taken while taking the pressure derivative. If superposition effects are observed; then, it is recommended to use equivalent time, Section 3.2.4, for the pressure derivative estimation. Drawdown pressure derivative may be taken, otherwise.

Once the derivative is estimated and the log-log of pressure and pressure derivative versus time is built, Equations provided in Chapter 2 also apply for pressure buildup test analysis. Just to name a few references [15–22, 31, 41, 42] also apply here.

3.3. Pressure buildup tests in developed reservoirs

The methods presented above may yield erroneous results when the test well produces under pseudo-steady-state conditions before shutting-in for a pressure buildup test or undergoes a pressure drawdown due to the production of adjacent wells in the reservoir. In such cases, it is better to use Eq. (3.1) in a more general way. Slider [34–37] has suggested a technique to treat the case of pressure tests in wells where the pressure drop contains the contribution of nearby wells. A procedure similar to that presented for the case of pressure drawdown, Section 2.6, is presented.

3.3.3 Conventional buildup analysis for developed reservoirs

It is required to extrapolate the well-flowing pressure over the pressure buildup period to estimate, $P_{w\ ext}$, see **Figure 3.5**. Then, find the difference between the observed shut-in pressure and the extrapolated well-flowing pressure, $\Delta P_{\Delta t}$, and plot this as a function of Δt. The data should be adjusted to the following equation [34–37]:

$$\Delta P_{\Delta t} = P_{ws} - P_{wext} = \Delta P_{1hr} + m \log \Delta t \tag{3.29}$$

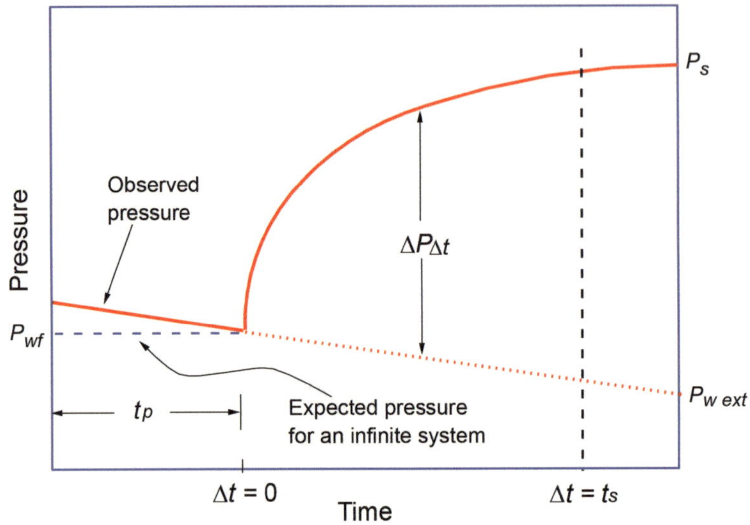

Figure 3.5.
Schematization of pressure buildup in a developed reservoir, after [34–37].

A straight line on this plot gives a slope m given by Eq. (2.33) and intercept:

$$\Delta P_{1hr} = \frac{162.6\, q\mu B}{kh}\left[\log\left(\frac{k}{\phi\mu\, c_t r_w^2}\right) - 3.2275 + 0.86859s\right] \tag{3.30}$$

The permeability is found with Eq. (2.33) and skin factor with a modified version of Eq. (3.6) resulting from changing P_{1hr} by $\Delta P_{1hr}{}^*$.

$$s = 1.1513\left[\frac{\Delta P^*_{1hr}}{m} - \log\left(\frac{k}{\phi\mu\, c_t r_w^2}\right) + 3.2275\right] \tag{3.31}$$

If the pressure drop is linear before shutting-in the well, which normally occurs because of the existence of the pseudo-steady state, Eq. (3.29) becomes:

$$P_{ws} - m * \ \Delta t = \Delta P^*_{1hr} + m\ \log \Delta t \tag{3.32}$$

where m^*, usually has a negative value, is the linear change of pressure drop before shutting-in the well:

$$m* = \frac{dP_{wf}}{dt}\ \text{when}\ t < t_p \tag{3.33}$$

Normally, m^* is negative. The value of $\Delta P^*{}_{1hr}$ in Eq. (3.32) is derived from Eq. (3.30) for the extrapolated linear behavior [37], which is:

$$\Delta P_{1hr}* = P_{wf(\Delta t=0)} + m\left[-\log\left(\frac{k}{\phi\mu\, c_t r_w^2}\right) + 3.2275 - 0.86859s\right] \tag{3.34}$$

So, when the pressure declines linearly before the test, a plot of $(P_{ws} - m^*\Delta t)$ vs. log Δt should give a straight line. The permeability is calculated with Eq. (2.33) and the skin factor with Eq. (3.31) by changing P_{1hr} instead of $\Delta P_{1hr}{}^*$. Usually, production occurs under pseudo-steady-state conditions; therefore, the pressure that the well would have if production were to continue would be given by:

$$P_{ext} = P_{wf}(\Delta t = 0) - m * \Delta t \tag{3.35}$$

And, ΔP^* is calculated as the difference between the observed pressure and the extrapolated pressure:

$$\Delta P^* = P_{ws} - P_{ext} \tag{3.36}$$

3.3.4 *TDS* buildup analysis for developed reservoirs

As demonstrated by Escobar and Montealegre [14], *TDS* technique is also applicable to developed reservoirs deriving D*P** and using the traditional equations of the technique.

Example 3.1

Slider [36, 37] first presented this example and then Escobar and Montealegre [14] reworked by *TDS* technique. A well drilled in a field with a uniform spacing of 40 acres has produced an average flow rate of 280 STB/D for 10 days. The well is shut-in for a pressure buildup study. In the five days prior to shutting-in, the flow pressure at the wellhead drops to around 24 psia/day (1 psia/hr). The oil-gas ratio remained constant during production. The test data are reported in **Table 3.1**. The following information is also available:

B= 1.31 rb/STB, μ = 2 cp, h = 40 ft, r_w = 0.33 ft

Solution by conventional analysis

Estimate P_{ext} by means of Eq. (3.35),

$$P_{ext} = P_{wf}(\Delta t = 0) - m * \Delta t = 1123 - (1)(0) = 1123 \text{ psia}$$

Estimate ΔP^* using the observed pressure minus the extrapolated pressure, Eq. (3.36);

$$\Delta P^* = P_{ws} - P_{ext} = 1123 - 1123 = 0 \text{ psia}$$

The remaining estimated values are given in **Table 3.1**.

Figure 3.6 shows a graph of ΔP^* against the log Δt, from which a slope of 192.92 psia/cycle is obtained, which allows estimating the permeability value with Eq. (2.33):

$$k = \frac{162.2 q\mu B}{hm} = \frac{162.6(280)(2)(1.31)}{40(192.92)} = 15.42 \text{ md}$$

As dP/dt is known, assuming that the drainage area approaches a circle, (r_e = 745 ft) product ϕc_t is solved from Eq. (1.130):

$$\phi c_t = \frac{1.79 qB}{h r_e^2 (dP/dt)} = \frac{1.79(280)(1.31)}{(40)(745^2)(24)} = 1.24 \times 10^{-6} \text{ /psia}$$

It is seen from **Figure 3.6** that the intercept, ΔP^*_{1hr} = 1287.6 psia. The skin factor is calculated from Eq. (3.31):

$$s = 1.1513\left[\frac{1287.6}{192.92} - \log\left(\frac{15.42}{(2)(1.24 \times 10^{-6})(0.33^2)}\right) + 3.2275\right] = 2.47$$

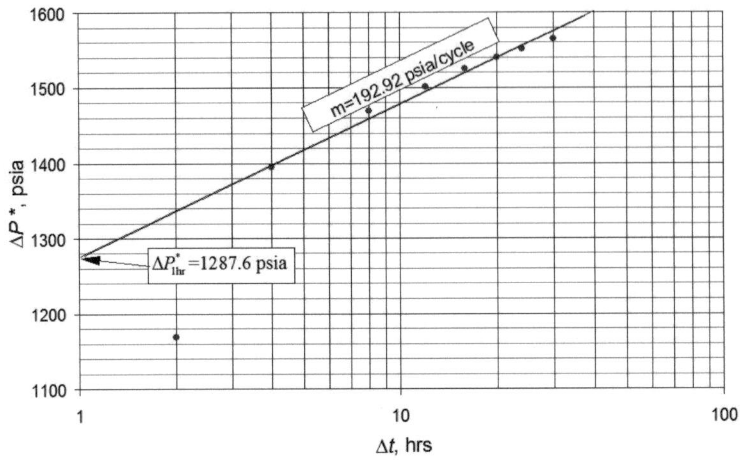

Figure 3.6.
Semi-log plot of DP* *against* Dt.

Solution by *TDS* technique

To apply *TDS* technique to this example, derivative of ΔP^*, see the last column in **Table 3.1**. The pressure and pressure derivative plot is built and given in **Figure 3.8**. Read from this plot the characteristic points, namely, t_r = 24 hr, $[t^*(\Delta P^*)']_r$ = 70.998 psia, and $(\Delta P^*)_r$ = 1551 psia. The permeability and the skin factor are found with Eqs. (2.71) and (2.92), respectively:

$$k = \frac{70.6q\mu B}{h(t*[\Delta P*]')_r} = \frac{70.6(280)(2)(1.31)}{40(84)} = 15.4\text{ md}$$

$$s = 0.5\left[\frac{1551}{70.998} - \ln\left(\frac{(24)(15.4)}{(2)(1.24\times10^{-8})(0.33^2)}\right) + 7.43\right] = 1.81$$

A good approximation to the data estimated by the two methods is observed in Example 3.1.

t, hr	P_{ws}, psia	P_{ext}, psia	ΔP^*, psia	$[t^*(\Delta P^*)']$, psia
0	1123	1123	0	
2	2290	1121	1169	606.577
4	2514	1119	1395	225.854
8	2584	1115	1469	97.372
12	2612	1111	1501	83.543
16	2632	1107	1525	73.916
20	2643	1103	1540	70.157
24	2650	1099	1551	70.988
30	2658	1093	1565	77.176

Table 3.1.
Pressure data for Example 3.1 of developed reservoir, after [14] and [35].

3.4. Average reservoir pressure

The average pressure for a reservoir without water intrusion is the pressure that the reservoir would reach if all the wells shut-in for infinite time. The average pressure is useful for [10, 12, 22, 28, 37, 40] (**Figure 3.7**):

1. For reservoir characterization:

    a. $\Delta P = \overline{P} - P_{wf}$ is small per unit of production, what is known as productivity index, J, indicates that there is an active water influx or a very large reservoir.

    b. If P is large per unit of production, it involves drainage from a small reservoir, sand lens, or faulted reservoir.

2. To calculate in-site oil.

3. For ultimate reservoir recovery.

4. The average pressure is a fundamental parameter that must be understood in processes of primary, secondary, and pressure maintenance projects.

The average reservoir pressure in the drainage region can be obtained by using well pressure test analysis. Most of the methods to estimate this parameter will be presented now.

3.4.3 Matthew-Bronz & Hazebrock (MBH) method

This method is considered the most accurate [12, 28] and was corrected by Odeh [37]. Use a Horner plot. It is applied in most situations where it is desired to find the average pressure in a closed reservoir for any well location within a variety of drain forms. The method assumes that there are no variations in fluid mobilities or fluid compressibilities within the drain region. This limitation can be overcome by using a production time t_p equal t_{pss}. The procedure is outline below:

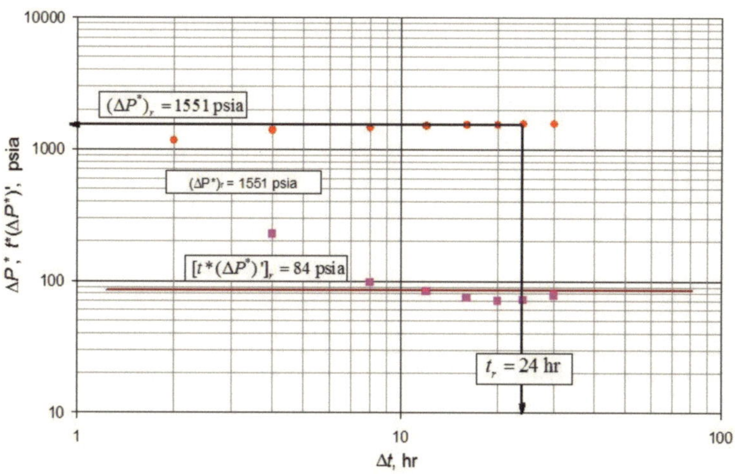

Figure 3.7.
Pressure and pressure derivative of ΔP^* *versus* Δt.

1. If not given, calculate the Horner time, t_p, with Eq. (3.2).

2. The t_p value must be compared with the time required to reach the pseudo-steady-state conditions. Therefore obtain $(t_{DA})_{pss}$ from **Table 2.1**, from the column "Exact for t_{DA} >". For this, the reservoir geometrical shape must be previously known.

3. Calculate the time to reach the pseudo-steady state, t_{pss}:

$$t_{pss} = \frac{\phi \mu c_t A (t_{DA})_{pss}}{0.0002637k} \tag{3.37}$$

4. Estimate the ratio α, $\alpha = t_p / t_{pss}$. If $\alpha > 2.5$, then, set $t = t_{pss}$. If $\alpha < 2.5$ (for very high flow rates, the improvement in the average pressure calculation is significant when α lies between 2.5 and 5), then, set $t = t_p$. Build a plot of P_{ws} versus $(t + \Delta t)/\Delta t$. As seen earlier, the use of t_{pss} in the Horner method can increase the length of the semi-log line, contrary to the MDH plot.

5. With time, t, defined in the previous step, determine t_{pDA}.

$$t_{p_{DA}} = \frac{0.0002637k}{\phi \, \mu \, c_t \, A} t \tag{3.38}$$

6. Extrapolate the semi-log line of the Horner graph and find P^*. See **Figure 3.2**.

7. Using the t_{pDA} calculated in step 5, determine P_{DMBH} from the following equations and tables. Notice that normally, P_{DMBH} is found from charts [12, 28]. However, the appropriate charts provided by [10, 28] were adjusted to polynomials with correlation coefficients greater than 0.999 (**Table 3.2**).

$$P_{DMBH}^2 = a + b\log(t_{pDA}) + c[\log(t_{pDA})]^2 + d[\log(t_{pDA})]^3 + e[\log(t_{pDA})]^4 \tag{3.39}$$

$$\begin{aligned} P_{DMBH}^2 = {} & a + b\log(t_{pDA}) + c[\log(t_{pDA})]^2 + d[\log(t_{pDA})]^3 \\ & + e[\log(t_{pDA})]^4 + f[\log(t_{pDA})]^5 \end{aligned} \tag{3.40}$$

$$P_{DMBH}^2 = a + b\log(t_{pDA}) + c[\log(t_{pDA})]^3 + de^{\log(t_{pDA})} + ee^{-\log(t_{pDA})} \tag{3.41}$$

$$P_{DMBH} = \frac{a + c\log(t_{pDA}) + e[\log(t_{pDA})]^2 + g[\log(t_{pDA})]^3}{1 + b\log(t_{pDA}) + d[\log(t_{pDA})]^2 + f[\log(t_{pDA})]^3} \tag{3.42}$$

Reservoir geometry	*a*	*b*	*c*	*d*	*e*
Hexagon and circle	12.0719262	16.9998709	6.07856232	−0.7618991	−0.5297593
Square	2.06652421	6.99163568	0.08203088	0.75401737	0.52737147
Equilateral triangle	10.1620678	14.9552862	6.63090011	0.42119362	−0.2122641
Rhombus	9.89526391	14.5756539	6.49269093	0.434827	−0.2016777
Right triangle	8.68121352	13.2526116	6.60147238	0.93940158	−0.0660697

Table 3.2.
Parameters for Eqs. (3.39).

$$P_{DMBH} = \frac{a + c\log(t_{pDA}) + e[\log(t_{pDA})]^2}{1 + b\log(t_{pDA}) + d[\log(t_{pDA})]^2 + f[\log(t_{pDA})]^3} \tag{3.43}$$

$$\begin{aligned} P_{DMBH} &= a + b\log(t_{pDA}) + c[\log(t_{pDA})]^2 + d[\log(t_{pDA})]^3 + e[\log(t_{pDA})]^4 \\ &+ f[\log(t_{pDA})]^5 + g[\log(t_{pDA})]^6 + h[\log(t_{pDA})]^7 + i[\log(t_{pDA})]^8 \\ &- 0.082557382[\log(t_{pDA})]^9 - 0.012745849[\log(t_{pDA})]^{10} \end{aligned} \tag{3.44}$$

8. Calculate the average reservoir pressure from:

$$\overline{P} = P^* - \left(\frac{m}{2.3025}\right)P_{DMDH} \tag{3.45}$$

Due to the compensation factors (low values of P^* with corresponding small corrections), any value of t_p used with the MBH method will theoretically give identical results for average reservoir pressure. Practically, a relatively short t_p can eliminate serious numerical problems in the calculation of average pressure. This includes errors caused by long extrapolations and deviations from theoretical assumptions: (1) lack of stabilization of the flow rate prior to closure, (2) migration and change of drainage areas in reservoirs with multiple wells and (3) variations in the compressibility of the system and mobility [12, 28].

3.4.4 Dietz method

This method [12, 28]-assumes that the well flowed sufficiently until it reached the pseudo-steady-state period before shutting-in and that the semi-log straight developed properly. This method is simple and is usually preferred in wells without significant skin factor, $s > -3$ or $r_w' = 0.05\, r_e$. The procedure for this method is:

1. Knowing the reservoir shape and the well location, read C_A from **Table 2.1**.

2. Calculate the Dietz shutting-in time, $(\Delta t)_{\overline{P}}$.

$$\Delta t_{\overline{P}} = \frac{\phi \mu c_t A}{0.0002637 C_A k} \tag{3.46}$$

3. Prepare a MDH plot (optionally find k and s).

4. Enter the MDH plot with the Dietz shutting-in time calculated in step 2 and read the corresponding average reservoir pressure value on the semi-log straight line.

3.4.5 Miller-Dietz-Hutchinson (MDH) method

This was elaborated to estimate the average pressure in circular or square reservoirs. It is applied only in wells that operate under pseudo-steady-state conditions [8, 28]. The procedure is presented as follows:

1. On an MDH graph, choose any point on the semi-log trend and read its coordinates, $(P_{ws})_N$ and Δt_N.

$$\Delta t_{DA}|_N = \frac{0.0002637k}{\phi \mu c_t A} \Delta t_N \tag{3.47}$$

2. Calculate Δt_{DA}.

3. Determine P_{DMDH} corresponding to $(\Delta t_{DA})_N$. This was traditionally done on charts. Again, fitted polynomials are presented here.

$$P_{DMDH} = a + b\Delta t_{DA} + c\Delta t_{DA}^2 \ln(\Delta t_{DA}) + d[\ln(\Delta t_{DA})]^2 + e\ln(\Delta t_{DA}) \tag{3.48}$$

4. Calculate the average reservoir pressure from **Table 3.3**:

$$\overline{P} = P_{ws}|_N + \left(\frac{m}{1.1513}\right) P_{D\,MDH} \tag{3.49}$$

3.4.6 Ramey-Cobb method

They presented a method to extrapolate the average pressure of a Horner plot when $t \geq t_{pss|}$. This method [12, 28, 40] requires information on the shape of the drainage area, the location of the well, and the confirmation that the boundaries are closed. The Ramey-Cobb procedure is (**Table 3.4**):

1. Knowing the reservoir shape and the well location, obtain $(t_{DA})_{pss}$, and calculate t_p and t_{pss}.

$$t_{pss} = \frac{\phi\ \mu\ c_t A (t_{DA})_{pss}}{0.0002637k} \tag{3.50}$$

2. If $t_p < t_{pss}$, then, the method is not reliable. Calculate the Horner time corresponding to the average reservoir pressure.

$$\left(\frac{t_p + \Delta t}{\Delta t}\right)_{\overline{P}} = \frac{0.0002637kC_A}{\phi \mu c_t A} t_P = C_A t_{pDA} \tag{3.51}$$

When $(t_p + \Delta t) = t_p$, Eq. (3.51) reduces to Eq. (3.46).

3. Prepare a Horner plot (optionally find k and s) (**Tables 3.5** and **3.6**).

4. Enter the Horner plot with the result from Eq. (3.51) and read the average reservoir pressure on the straight line trend.

Reservoir geometry	*a*	*b*	*c*	*d*	*e*	*f*	Equation number
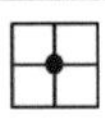	11.3521634	16.1909297	6.18077696	−0.4318785	−0.4463139	0	(3.39)
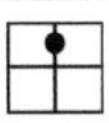	6.51259739	11.7744452	6.73932956	0.04016071	−1.1494588	-0.2753853	(3.40)
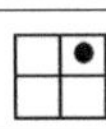	1.94527256	5.8770907	5.91449487	1.72037863	−0.3879657	-0.1920816	
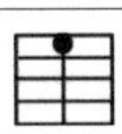	0.03487786	-5.9476006	−1.7074107	4.59461702	−3.5455711	0	(3.41)

Table 3.3.
Parameters for Eqs. (3.39)–(3.41).

Reservoir geometry	*a*	*b*	*c*	*d*	*e*	*f*	*g*	Equation number
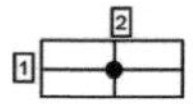	8.65988165	13.5200374	6.80176625	0.89466084	−0.103974	0	0	(3.39)
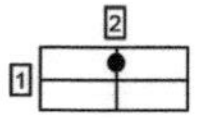	5.40959238	11.624314	7.43959927	−0.1265869	−1.6111914	−0.4024905	0	(3.40)
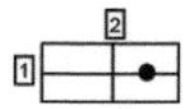	1.57862998	1.36493165	4.38006689	1.68949499	6.42019058	−0.220711	2.40143918	(3.42)
	1.12184958	0.42910105	2.81626528	0.14832439	1.03977832	−0.1368335	0	(3.43)
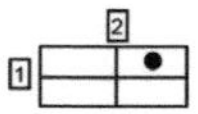	0.75302615	0.53236801	2.38384274	0.7187976	2.48676628	−0.2180852	0.6742364	(3.42)
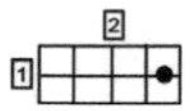	-0.5184336	0.73180568	1.38508346	0.93697751	2.57642261	−0.3389924	0.78194903	(3.42)

Table 3.4.
Parameters for Eqs. (3.39)–(3.42).

Reservoir geometry	*a*	*b*	*c*	*d*	*e*	*f*	*g*	*h*	*i*	Equation number
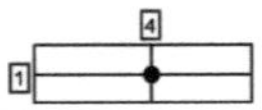	1.70317185	0.70756769	3.57593515	0.50721899	2.93960314	0.02349566	0.76107064	0	0	(3.42)
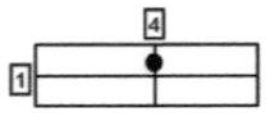	1.063407392	0.304229597	2.668344404	0.427947675	1.16664094	−0.276857388	0	0	0	(3.43)
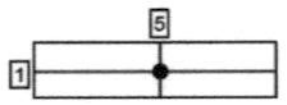	0.916932173	1.030678375	3.271385125	0.785985802	3.353625853	−0.213935954	0.894001955	0	0	(3.42)
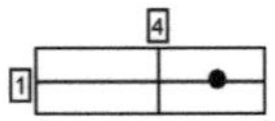	−1.025967698	0.230767114	−0.33785184	1.611531081	2.764160677	−0.02046841	1.191846817	0	0	(3.42)
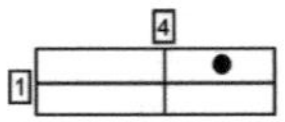	−1.690446615	−0.295741667	3.877346514	1.592302604	−3.675356343	−2.030985974	1.405207073	1.120246808	0.064244461	(3.44)

Table 3.5.
Parameters for Eq. (3.42)–(3.44).

DOI: http://dx.doi.org/10.5772/intechopen.81078

Geometry	*a*	*b*	*c*	*d*	*e*
	−1.8936132	11.06446069	20.39748447	−0.003556014	0.5555625414
	−1.106364859	1.291801492	−0.814213632	−0.002247086	−0.536018398
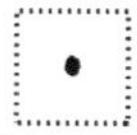	−1.05158653	−0.382513988	−5.024281518	−0.001277198	−0.514385212

Table 3.6.
Parameters for Eq. (3.48).

3.4.7 Arari or direct method

Arari [4] presented in 1987 a simple method to calculate the average reservoir pressure during production or buildup without the help of any graph. This method requires knowing the distance from the well to which the reservoir pressure is the same average pressure. For no-flow boundary reservoirs:

$$\overline{P} = P_{wf} + \frac{162.6q\mu B}{kh}\left(2\log\frac{r_e}{r_w} - 0.5203 + 0.87s\right) \tag{3.52}$$

$$\overline{P} = P_{wf} + \frac{162.6q\mu B}{kh}\left(\log\frac{A}{r_w^2} - 1.1224 + 0.87s\right) \tag{3.53}$$

For constant-pressure boundary reservoirs:

$$\overline{P} = P_{wf} + \frac{162.6q\mu B}{kh}\left(2\log\frac{r_e}{r_w} - 0.4342 + 0.87s\right) \tag{3.54}$$

$$\overline{P} = P_{wf} + \frac{162.6q\mu B}{kh}\left(\log\frac{A}{r_w^2} - 1.036 + 0.87s\right) \tag{3.55}$$

In order to consider different well positions and different reservoir geometries, the flow equations were developed by introducing the Dietz shape geometrical factors in Eqs. (3.539) and (3.55) which are, respectively, transformed into:

$$\overline{P} = P_{wf} + \frac{162.6q\mu B}{kh}\left(\log\frac{A}{C_A r_w^2} + 0.368 + 0.87s\right) \tag{3.56}$$

$$\overline{P} = P_{wf} + \frac{162.6q\mu B}{kh}\left(\log\frac{A}{C_A r_w^2} + 0.454 + 0.87s\right) \tag{3.57}$$

3.4.8 *TDS* technique

3.4.8.1 Circular reservoirs

For a well in the center of a circular reservoir, the average reservoir pressure is obtained from a log-log plot of pressure and pressure derivative versus time according to the following expression [7, 29]:

$$\overline{P} = P_i - \frac{141.2q\mu B}{kh}\left[\left(\frac{(t*\Delta P')_{pss}}{(\Delta P)_{pss} - (t*\Delta P')_{pss}}\right)\ln\left(\frac{r_e}{r_w} - \frac{3}{4}\right)\right] \tag{3.58}$$

where P_i is the initial pressure (in some cases, it can approximate P^*), $(\Delta P)_{pss}$ and $(t^*\Delta P')_{pss}$ are the values of (ΔP) and $(t^*\Delta P')$ in the late straight line of pseudo-steady-state period.

3.4.8.2 Naturally fractured reservoirs

For naturally fractured reservoirs with the dimensionless average pressure and the average reservoir pressure are defined as [29]:

$$\overline{P}_D = \frac{kh}{141.2q\mu B}(\overline{P} - P_{ws}) \tag{3.59}$$

$$\overline{P} = P_{wf} + \Delta P_{pss} + (t * \Delta P')_{pss}\left(1 + \frac{3792.2\phi\mu c_t r_w^2 (1-\omega)^2}{\lambda k t_{pss}}\right) \tag{3.60}$$

Being ω and λ the naturally fractured reservoir parameters which will be discussed about in Chapter 6.

$$\overline{P} = P_{wf} + \Delta P_{pss} + \frac{141.2q\mu B}{kh}\left[\begin{array}{c}\left(\dfrac{(t * \Delta P')_{pss}}{\Delta P_{pss} - (t * \Delta P')_{pss}}\right) \\ \left(\ln\dfrac{r_e}{r_w} - \dfrac{3}{4} + \dfrac{2\pi r_w^2 (1-\omega)^2}{\lambda A}\right)\end{array}\right] \tag{3.61}$$

$$\overline{P} = P_{wf} + \Delta P_{pss} + \frac{141.2q\mu B}{kh}\left[\begin{array}{c}\left(\dfrac{(t * \Delta P')_{pss}}{\Delta P_{pss} - (t * \Delta P')_{pss}}\right) \\ \left(\ln\dfrac{r_e}{r_w} - \dfrac{3}{4} + \dfrac{0.1987 C_A r_w^2 (1-\omega)^2}{\lambda A}\right)\end{array}\right] \tag{3.62}$$

If the dimensionless average pressure is defined as [29]:

$$\overline{P}_D = \frac{kh(P_i - \overline{P})}{141.2q\mu B} \tag{3.63}$$

The average reservoir pressure is given by [29]:

$$\overline{P} = P_i - (t * \Delta P')_{pss}\left(1 + \frac{3792.2\phi\mu c_t r_w^2 (1-\omega)^2}{\lambda k t_{pss}}\right) \tag{3.64}$$

$$\overline{P} = P_i - \frac{141.2q\mu B}{kh}\left[\begin{array}{c}\left(\dfrac{(t * \Delta P')_{pss}}{\Delta P_{pss} - (t * \Delta P')_{pss}}\right) \\ \left(\ln\dfrac{r_e}{r_w} - \dfrac{3}{4} + \dfrac{2\pi r_w^2 (1-\omega)^2}{\lambda A}\right)\end{array}\right] \tag{3.65}$$

$$\overline{P} = P_i - \frac{141.2q\mu B}{kh}\left[\begin{array}{c}\left(\dfrac{(t * \Delta P')_{pss}}{\Delta P_{pss} - (t * \Delta P')_{pss}}\right) \\ \left(\ln\dfrac{r_e}{r_w} - \dfrac{3}{4} + \dfrac{0.1987 C_A r_w^2 (1-\omega)^2}{\lambda A}\right)\end{array}\right] \tag{3.66}$$

The Dietz shape factor can be estimated by [7]:

$$C_A = \frac{2.2458A}{r_w^2}\left\{ e^{\frac{kt_{pss}}{301.77\phi\mu c_t A}\left(\frac{(\Delta P)_{pss}}{(t*\Delta P')_{pss}} - 1\right)} \right\}^{-1} \tag{3.67}$$

For a circular geometry with the well at the center, [26×] arrived to the following expression:

$$\overline{P} = P_i - 1.26(t*\Delta P')_r - 2(t*\Delta P')_r\left[\left(\frac{2\pi r_w^2(1-\omega)^2}{\lambda A}\right)\right] \tag{3.68}$$

They also assumed that:

$$\overline{P}_D = 2\pi t_{DA} \tag{3.69}$$

and,

$$2\pi t_{DA} >> \frac{2\pi r_w^2(1-\omega)^2}{\lambda A} \tag{3.70}$$

After some manipulations, Igbokoyi and Tiab [27] also obtained an expression free of the naturally fractured reservoir parameters:

$$\overline{P} = P_i - (t*\Delta P')_r\left[\left(\frac{(t*\Delta P')_{pss}}{(\Delta P_w)_{pss} - (t*\Delta P')_{pss}}\right)\ln\left(\frac{2.2458A}{C_A r_w^2}\right) + 2s\right] \tag{3.71}$$

3.4.8.3 Bounded elongated systems

The shape factor is given by Eq. (3.67), and the average reservoir pressure equation for these systems is given below [7]:

$$\overline{P} = P_i - 70.6\frac{q\mu B}{kh}\left[\left(\frac{(t*\Delta P')_{pss}}{(\Delta P_w)_{pss} - (t*\Delta P')_{pss}}\right)\ln\left(\frac{2.2458A}{C_A r_w^2}\right)\right] \tag{3.72}$$

3.4.8.4 Hydraulically fractured vertical well in no-flow boundary reservoirs

The shape factor is estimated with Eq. (3.67), and the average reservoir pressure with Eq. (3.69) [7]:

$$\overline{P} = P_i - \frac{q\mu B}{kh}\left\{\frac{0.23373kt_{pss}}{\phi\mu c_t A}\left(\frac{(\Delta P)_{pss}}{(t*\Delta P')_{pss}}\right) - 70.6\ln\left[\left(\frac{x_e}{x_f}\right)^2\left(\frac{2.2458}{C_A}\right)\right]\right\} \tag{3.73}$$

When birradial flow occurs, the area and the average reservoir pressure can be determined from the following equations [7]:

$$A = \frac{k_{BRpssi}}{142.43\phi\mu c_t(x_e/x_f)^{1.123}} \tag{3.74}$$

$$\overline{P} = P_i - 5.64\left[\frac{q\mu B}{kh}\left(\frac{x_e}{x_f}\right)^{0.72}\left(\frac{k}{\phi\mu c_t A}\right)^{0.36}\right]t_{BRpssi}^{0.36} \tag{3.75}$$

For uniform flow fractures and when $x_e/x_f < 8$, birradial flow is difficult to be observed, then the intersection between the linear flow and the pseudo-steady state line, t_{Lpssi}, is used. Then, the area and the average pressure are obtained from:

$$A = 0.0033144\left[\left(\frac{k}{\phi\mu c_t}\right)\left(\frac{x_f}{x_e}\right)^2\right]t_{Lpssi} \tag{3.76}$$

$$\overline{P} = P_i - 4.06\left[\frac{qB\sqrt{\mu}}{h\sqrt{\phi c_t k}}\left(\frac{x_e}{x_f}\right)\right]\sqrt{t_{Lpssi}} \tag{3.77}$$

3.4.8.5 Hydraulically fractured vertical wells in elongated systems

For these systems, the transition between the line of infinite behavior and that of the pseudo-steady-state period is longer compared to the case of square systems in both cases of fracture: infinite conductivity and uniform flux. When the birradial flow line is difficult to be observed, such is the case of xe/xf < 8, the following equation is used to determine permeability [7]:

$$k = \left(\frac{\mu}{\phi c_t A}\right)\left[\frac{8.128qB}{h(t * \Delta P\prime)_{DL1}}\right]^2 \tag{3.78}$$

Since there may be two linear flow regimes, once before radial flow corresponding to flow from the formation to the fracture and the other once radial vanishes, then, $(t^*\Delta P')_{DL1}$ is the value of $(t^*\Delta P')$ at t = 1 hr on the dual-linear flow regime—solving for area it yields [7]:

$$A = \left(\frac{\mu}{\phi c_t k}\right)\left[\frac{8.128qB}{h(t * \Delta P\prime)_{DL1}}\right]^2 \tag{3.79}$$

The point of intersection between the closest parallel-linear line flow: the second linear flow regime, for example, dual-linear and the pseudo-steady-state line, t_{DLpssi}, is unique. With this point, determine the area of the following equation [7]:

$$A = \frac{kt_{DLpssi}}{1207.09\phi\mu c_t} \tag{3.80}$$

This equation should be used for verification purposes of the permeability and area values obtained by Eqs. (3.78) and (3.79). The average reservoir pressure is obtained from [7]:

$$\overline{P} = P_i - \frac{qB}{h}\sqrt{\left(\frac{\mu}{11913.6\phi c_t k A}\right)t_{DLpssi}} \tag{3.81}$$

This equation should be used if k and A can be determined from the nearest parallel boundary; in other words, from the dual-linear flow regime.

3.4.9 Total average reservoir pressure

Golan and Whitson [24] presented a method to estimate the drainage area of wells that produce from a common reservoir. They assumed that the volume drained by a well is proportional to its flow rate. If the properties of the reservoirs are constant and uniform:

$$A_w = A_T\left(\frac{q_w}{q_T}\right) \tag{3.82}$$

All of the above studied methodologies give the value of the average reservoir pressure in the well drainage area. If a number of wells produce from the same reservoir, each well is analyzed separately to give the average reservoir pressure for its own drainage area. The average reservoir pressure can be estimated from the individual average pressures by (possibly from [38] less probably from [9, 24], the author does not remember the exact reference):

$$P = \frac{\sum_i [\overline{P}_i \Delta(F)/\Delta\overline{P}]_i}{\sum_i [\Delta(F)/\Delta\overline{P}]_i} \tag{3.83}$$

$$\Delta(F) = F_{t+\Delta t} + F_t \tag{3.84}$$

$$F_{t+\Delta t} = \int_0^{t+\Delta t} \left[q_o B_o + q_w B_w + (q_g - q_o R_s - q_w R_{sw}) B_g\right] dt \tag{3.85}$$

$$F_t = \int_0^{t} \left[q_o B_o + q_w B_w + (q_g - q_o R_s - q_w R_{sw}) B_g\right] dt \tag{3.86}$$

Bossie-Codreanu [5] suggest that the drainage area can be determined from a Horner or MDH plot by selecting the 3-point coordinates in the straight section of the semi-log graph to determine the slope of the pseudo-steady-state period line, m^*:

- Shutting-in time Δt_1 with corresponding shutting-in pressure P_{ws1}
- Shutting-in time Δt_2 with corresponding shutting-in pressure P_{ws2}
- Shutting-in time Δt_3 with corresponding shutting-in pressure P_{ws3}

The selected shutting-in times satisfy $t_1 < t_2 < t_3$. Then, m^* is approximated by:

$$m* = \frac{(P_{ws2} - P_{ws1})\log(\Delta t_3/\Delta t_1) - (P_{ws3} - P_{ws1})\log(\Delta t_2/\Delta t_1)}{(\Delta t_3 - \Delta t_1)\log(\Delta t_2 \Delta t_1) - (\Delta t_2 - \Delta t_1)\log(\Delta t_3 \Delta t_1)} \tag{3.87}$$

Example 3.2

The data of a pressure buildup test, taken from [40], are reported in **Table 3.7**, along the Horner time and the pressure derivative estimated (using equivalent time, Eq. (3.23)) with a smooth value of 0.1 cycles. The reservoir properties were obtained from a well located in the center of a square shaped reservoir. Given the following data:

$r_w = 4$ in,	h = 44 ft,	ϕ = 12%
μ = 0.76 cp,	B = 1.24 rb/STB,	N_p = 4550 STB
A = 40 acres,	q = 340 BPD,	$c_t = 36\times10^{-6}$ psia^{-1}
P_{wf} = 2980 psia		

Δt, hr	P_{ws}, psia	$(t_p + \Delta t)/\Delta t$	$(t_{pss} + \Delta t)/\Delta t$	ΔP, psia	$t^*\Delta P'$, psia
0	2980			0	0
0.1	3100	3213.00	807.450	120	83.41
0.2	3150	1607.00	404.225	170	100.23
0.3	3200	1071.67	269.817	220	110.92
0.5	3250	643.40	162.290	270	85.89
0.75	3275	429.27	108.527	295	59.48
1	3290	322.20	81.645	310	41.84
2	3315	161.60	41.323	335	34.48
3	3325	108.07	27.882	345	22.35
4	3330	81.30	21.161	350	20.29
5	3335	65.24	17.129	355	21.45
7	3342	46.89	12.521	362	21.96
10	3350	33.12	9.065	370	23.97
15	3360	22.41	6.376	380	21.42
20	3364	17.06	5.032	384	14.87
30	3370	11.71	3.688	390	12.65
40	3372	9.03	3.016	392	9.02
50	3374	7.42	2.613	394	8.47
60	3375	6.35	2.344	395	7.14
70	3376	5.59	2.152	396	8.55
80	3377	5.02	2.008	397	9.76

Table 3.7.
Pressure buildup test data.

It is required to estimate reservoir permeability and skin factor. Then, find the average reservoir pressure using all the studied methods.

Solution

Find t_p with Eq. (3.2);

$$t_p = \frac{24N_p}{q} = \frac{(24)(4550)}{340} = 321.176 \text{ hr}$$

Estimate the Horner time, $(t_p+\Delta t)/\Delta t$, for each pressure value. This is reported in the third column of **Table 3.7** and builds the Horner plot given in **Figure 3.8**. From the Horner plot given in **Figure 3.8**, the slope and intercept are read to be 44 psia/cycle and 3306 psia. They are, respectively, used to find permeability, Eq. (2.33), and skin factor, Eq. (3.6), thus:

$$k = \frac{162.6q\mu B}{mh} = \frac{(162.6)(340)(0.76)(1.24)}{(44)(44)} = 26.91 \text{ md}$$

$$s = 1.1513\left[\frac{3306 - 2980}{(44)} - \log\left(\frac{26.91}{(0.12)(0.76)(36\times10^{-6})(0.333)^2}\right) + 3.2275\right] = 3.18$$

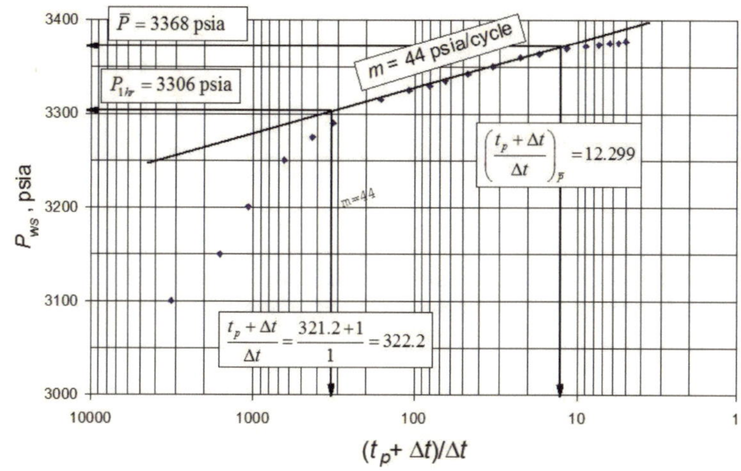

Figure 3.8.
Horner plot for Example 3.2.

Average reservoir pressure by MBH method

Determine $(t_{DA})_{pss}$ from **Table 2.1** for square shaped reservoirs. It is read $(t_{DA})_{pss}$ = 0.1. Calculate t_{pss} with Eq. (3.37):

$$t_{pss} = \frac{\phi\mu c_t A(t_{DA})_{pss}}{0.0002637k} = \frac{(0.12)(0.76)(36 \times 10^{-6})(40)(43560)}{(0.0002637)(26.91)}(0.1) = 80.645 \text{ hr}$$

Estimate the ratio a = t_p / t_{pss} = 312.176/80.645 = 3.982. Since α > 2, then $t = t_{pss}$. Rebuild the Horner plot as P_{ws} vs. $\log(t_{pss} + \Delta t)/\Delta t$, (see **Table 3.7**). In **Figure 3.9**, draw a straight line along the infinite-acting period (radial flow) and extrapolate to a Horner time of 1. Read the false pressure value, P^* = 3398 psia.

Find the dimensionless production time using Eq. (3.38):

$$t_{pDA} = \frac{0.0002637kt}{\phi\mu c_t A} = \frac{(0.0002637)(26.91)(80.645)}{(0.12)(0.76)(36 \times 10^{-6})(40 \cdot 43560)} = 0.0999 \approx 0.1$$

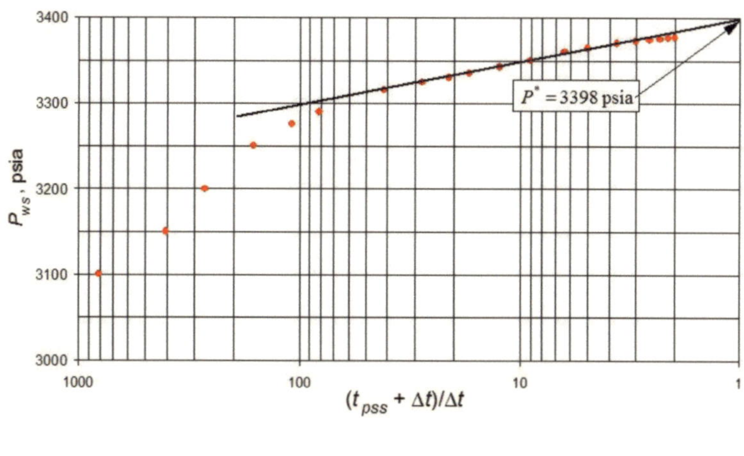

Figure 3.9.
Horner plot with t_{pss} for Example 3.2.

Using Eq. (3.39) with data from **Table 3.3** (first row), the MBH dimensionless pressure is P_{DMBH}=1.152. Then, estimate the average reservoir pressure with Eq. (3.45).

$$\overline{P} = P^* - \frac{m\, P_{DMBH}}{2.303} = 3398 - \frac{(44)(1.152)}{2.303} = 3376 \text{ psia}$$

Average reservoir pressure by Dietz method

Prepare a MDH, P_{ws} vs. log(Δt). See **Figure 3.10**. Determine the shape factor C_A from **Table 2.1** for a well at the center of a square reservoirs. C_A = 30.8828. Find Dietz shutting-in time with Eq. (3.46),

$$\Delta t_{\overline{P}} = \frac{\phi \mu c_t A}{0.0002637 k C_A} = \frac{(0.12)(0.76)(36 \times 10^{-6})(40)(43560)}{(0.0002637)(26.91)(30.8828)} = 26.1136 \text{ hr}$$

Enter with this value in **Figure 3.10** and read an average reservoir pressure of 3368 psia.

Average reservoir pressure by Ramey-Cobb method

Having t_p, t_{pss}, and C_A from previous methods and since $t_p >> t_{pss}$, then estimate Ramey-Cobb shutting-in time from Eq. (3.51);

$$\left(\frac{t + \Delta t}{\Delta t}\right)_{\overline{P}} = \frac{(0.0002637)(26.91)(30.8828)(312.176)}{(0.12)(0.76)(36 \times 10^{-6})(40)(43560)} = 12.299$$

Enter with this value in the Horner plot, **Figure 3.8**, and read an average pressure value of 3368 psia.

Average reservoir pressure by MDH method

Prepare a MDH plot, **Figure 3.10** and choose any convenient point on the semilog straight line. For this case, Δt_N = 10 hr and $(P_{ws})_N$ = 3350 psia were chosen. Calculate the dimensionless shutting-in time using the chosen time in Eq. (3.47):

$$\Delta t_{DA} = \left(\frac{0.0002637k}{\varphi \mu c_t A}\right)\Delta t_N = \frac{(0.0002637)(26.91)}{(0.12)(0.76)(36 \times 10^{-6})(40)(43560)}(10) = 0.0124$$

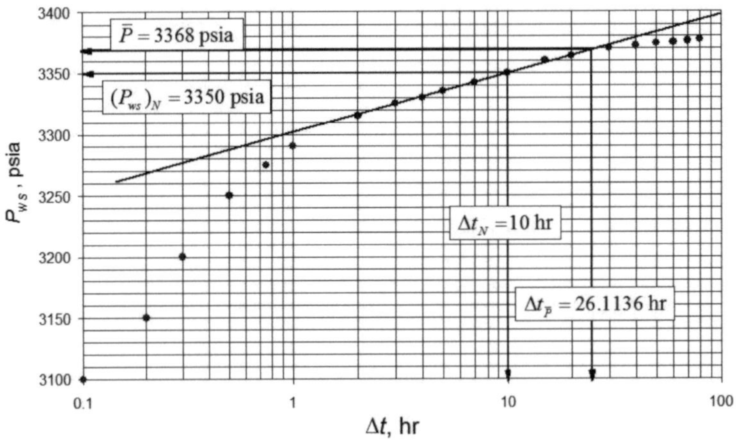

Figure 3.10.
MDH plot for Example 3.2.

Use this value of Δt_{DA} and estimate P_{DMDH} from Eq. (3.48) and **Table 3.6** for a no-flow boundary square reservoir (first row), this gives P_{DMDH} = 0.6. Estimate the average reservoir pressure with Eq. (3.49);

$$\overline{P} = P_{wsN} + \frac{m(P_{DMDH})}{1.1513} = 3350 + \frac{44(0.6)}{1.1513} = 3372.9 \text{ psia}$$

Average reservoir pressure by direct (Arari) method

Using Eq. (3.53), for no-flow boundaries;

$$\overline{P} = 2980 + \left[\frac{\dfrac{162.6(340)(0.76)(1.24)}{(23.52)(44)}}{\left(\log \dfrac{40(43560)}{0.333^2} - 1.1224 + 0.87(1.93)\right)} \right] = 3365.2 \text{ psia}$$

Average reservoir pressure by *TDS* technique

The pressure and pressure derivative versus time log-log plot for Example 3.2 is given in **Figure 3.11**. Notice that after radial flow, the pressure derivative takes a slope of negative one. This may be due to the changes in transmissibility. Anyhow, the pseudo-steady-state period starts at 60 hr. On that line, a point will be chosen for the estimation of the average reservoir pressure. From this plot, the following data are read:

t_r = 7 hr, ΔP_r = 396 psia, $(t^*\Delta P')_{\text{pss}}$ = 8.55 psia
ΔP_{pss} = 384 psia, t_{pss} = 70 h, $(t^*\Delta P')_r$ = 21.86 psia

Estimate permeability and skin factor with Eqs. (2.71) and (2.92),

$$k = \frac{70.6 q \mu B}{h(t * \Delta P')_r} = \frac{(70.6)(340)(0.76)(1.24)}{44(21.86)} = 23.52 \text{ md}$$

$$s = 0.5\left[\frac{362}{21.86} - \ln\left(\frac{23.52(7)}{(0.12)(0.76)(36 \times 10^{-6})(0.3^2)}\right) + 7.43\right] = 1.93$$

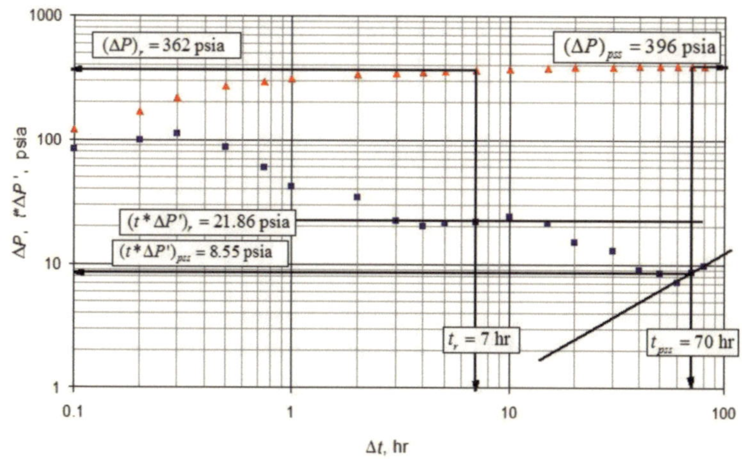

Figure 3.11.
Log-log plot of pressure and pressure derivative against time for Example 3.2.

Taking the case of no-flow boundary rectangular system, Eqs. (3.67) and (3.72), find the shape factor and the average reservoir pressure. Here, the last pressure value, that is, 3377 psia, is taken as P_i.

$$C_A = \frac{2.2458(40)(43560)}{0.333^2}\left\{\exp\left[\frac{0.003314(23.52)(20)}{0.12(0.76)(30\times10^{-6})(40)(43560)}\left(\frac{396}{8.55}-1\right)\right]\right\}^{-1} = 12.943$$

$$\overline{P} = 3377 - 70.6\frac{(340)(0.76)(1.24)}{(23.52)(44)}\left[\begin{array}{c}\left(\dfrac{8.55}{396-8.55}\right)\\ \ln\left(\dfrac{2.2458(40)(43560)}{12.943(0.333^2)}\right)\end{array}\right] = 3369.9\ \text{psia}$$

The results of the estimation of the average reservoir pressure are reported in **Table 3.8**.

3.4.10 Average reservoir pressure in naturally fractured reservoirs from transient-rate analysis

Amin et al. [2] follow the philosophy of the *TDS* technique to determine the average reservoir pressure from TRA (even though, this book does not include such analysis) by means of the following expression (slight simplification is shown here):

$$\overline{P} = P_i - \frac{887.186 q\mu B}{kh}\left[\left(\frac{k_f t_{pss}}{3792.19(\phi c_t)_{m+f}\mu}\right)\left(+\frac{r_w^2(1-\omega)^2}{\lambda A}\right)\right] \tag{3.88}$$

Amin et al. [2] pointed out that from a curve of production rate versus time, a point q_{pss} and t_{pss} that satisfy $t_D > \omega(1-\omega)/\lambda$ (pseudo-steady-state period). In addition, q_{pss} and t_{pss} selected should be those when the flow rate becomes almost constant.

3.4.11 Average reservoir pressure from two-rate tests

Sabet [33] and Dake [10] presented the mathematical development to find the average reservoir pressure from two-rate tests. The final equation uses the value of the first semi-log straight line and the well-flowing pressure after the flow rate has been changed. It is given by:

$$\overline{P} = 2m_1\left\{\log\left(\frac{2.241A}{C_A r_w^2}\right)^2 + 0.435s\right\} + P_{wf@\Delta t'=0} \tag{3.89}$$

Example 3.3

Sabet [34] presented a two-rate test which pressure versus time values are shown in **Table 3.9**. To interpret the test, the following reservoir information, PVT, and flow parameters are given:

r_e = 745 ft	r_w = 0.25 ft	f = 15%	h = 20 ft
m = 1.2 cp	$c_t = 20\times10^{-6}\text{psia}^{-1}$	B = 1.25 bbl/STB	t_p = 300 hr
q_1 = 100 STB/D	q_2 = 50 STB/D	$P_{wf@Dt'=0}$ = 1603.2 psia	

Method	Average reservoir pressure, psia
Ramey & Cobb	3368
MBH	3376
MDH	3372.9
Dietz	3368
TDS	3371.4
Arari	3365.2
Average	**3370.25**

Table 3.8.
Summary of average reservoir pressure results.

Before flow rate change		After flow rate change		
***t*, hr**	**P_{wf}, psia**	***t*, hr**	**P_{wf}, psia**	**Dt'**
288	1607.5	0.33	2475	2.72
289	1607.2	0.42	2482	2.67
290	1606.8	0.5	2487	2.63
291	1606.4	0.58	2492	2.6
292	1606.1	0.67	2497	2.56
293	1605.7	0.75	2500	2.54
294	1605.4	0.83	2502	2.52
295	1605	0.92	2505	2.5
296	1604.6	1	2508	2.48
297	1604.3	1.25	2514	2.43
298	1603.9	1.5	2520	2.39
299	1603.6	1.75	2525	2.36
300	1603.2	2	2531	2.33
		2.5	2542	2.28
		3	2552	2.24
		4	2568	2.18
		5	2582	2.13
		6	2590	2.1
		7	2600	2.06
		7.5	2604	2.05

Table 3.9.
Two-rate test data for Example 3.3, after [34].

Determine average reservoir pressure, skin factor, and demonstrate that the test has not reached the transient period when the flow rate was changed.

Solution

A value of m_1 = 274 psia/cycle and P_{1hr} = 2485 psia were read from the Cartesian graph presented in **Figure 3.12**. Permeability and skin factors are calculated with Eqs. (2.271) and (2.272), respectively, and were found to be 4.5 md and

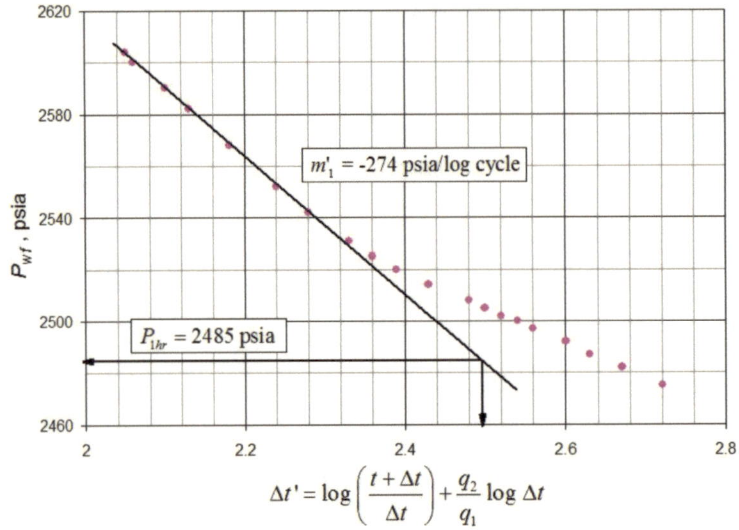

Figure 3.12.
Cartesian plot for two-rate test of Example 3.3.

approximately 3.0. The time to reach the pseudo-steady-state period in the test is estimated with Eq. (2.40) with $r = r_e$, as follows:

$$t_{pss} = \frac{948\ \phi\mu\ c_t r^2}{k} = \frac{948(0.15)(1.2)(20 \times 10^{-6})(745^2)}{4.5} = 420 \text{ hr}$$

Assuming the well is in the center of a rectangular or circular reservoir, it is possible to appreciate the change (t = 300 hr) occurred before reaching the pseudo-steady-state conditions. Eq. (3.89) is used to calculate the average reservoir pressure, thus:

$$\overline{P} = \frac{325.2(100)(1.2)(1.25)}{(84.5)(20)} \log \frac{0.472(745)}{0.33} + 0.87(274)(3) + 1603.2 = 2405.8 \text{ psi}$$

Note that the average pressure value is not correct (it was obtained for explanatory effects) because the well was not producing under pseudo-steady-state period before changing to the second rate.

3.4.12 Average reservoir pressure from multi-rate tests

For this type of tests, it is necessary to construct a Cartesian graph of pressure against the superposition time, X_n, Eq. (2.267), to obtain permeability and damage. Then, an MDH graph is made, and the Dietz method is applied. For this, it is necessary to determine the Dietz shutting-in time by Eq. (3.46). With this value, the average reservoir pressure is read from the MDH plot. Actually, it is possible, then, to apply any of the average pressure methods seen in this chapter. However, for some of the methods, that is, MBH and Ramey-Cobb, the production time, t_p, is required. By definition, it refers to a constant flow rate before shutting-in the well. It does not exist in this case. So, as a recommendation, the equivalent time should be estimated with Eq. (2.269) and this can be used as t_p, and the flow rate is weighted with each period of duration.

However, *TDS* technique plays an important role. Escobar [15] obtained the average pressure equation for circular systems as given below.

$$\overline{P}=P_i-\frac{141.2q_n\mu B}{kh}\left[\left(\frac{(t*\Delta P_q{}')_{pss}}{(\Delta P_q)_{pss}-(t*\Delta P_q{}')_{pss}}\right)\left(\ln\frac{r_e}{r_w}-0.75\right)\right] \tag{3.90}$$

where $P_i$is the initial pressure. $(\Delta P_q)_{pss}$ and $(t^*\Delta P_q{}')_{pss}$ are the corresponding normalized pressure values and their derivative given at an arbitrary time, t_{pss}. For any geometry, the resulting equation is [15]:

$$\overline{P}=P_i-\frac{70.6q_n\mu B}{kh}\left[\left(\frac{(t*\Delta P_q{}')_{pss}}{(\Delta P_q)_{pss}-(t*\Delta P_q{}')_{pss}}\right)\ln\left(\frac{2.2458A}{C_A r_w^2}\right)\right] \tag{3.91}$$

where C_A is found from a slight modification of Eq. (3.67):

$$C_A=\frac{2.2458A}{r_w^2}\left\{e^{\frac{kt_{pss}}{301.77\phi\mu c_t A}\left(\frac{(\Delta P_q)_{pss}}{(t*\Delta P_q{}')_{pss}}-1\right)}\right\}^{-1} \tag{3.92}$$

For multi-rate tests in naturally fractured reservoirs, the respective equations [15] are:

$$\overline{P}=P_i-\frac{141.2q_n\mu B}{kh}\left[\frac{(t*\Delta P_q{}')_{pss}}{(\Delta P_q)_{pss}-(t*\Delta P_q{}')_{pss}}\left(\ln\frac{r_e}{r_w}-0.75+\frac{2r_w^2(1-\omega)^2}{\lambda r_e^2}\right)\right] \tag{3.93}$$

$$\overline{P}=P_i-\frac{141.2q_n\mu B}{kh}\left[\frac{(t*\Delta P_q{}')_{pss}}{(\Delta P_q)_{pss}-(t*\Delta P_q{}')_{pss}}\left(\ln\frac{r_e}{r_w}-0.75+\frac{0.198C_A r_w^2(1-\omega)^2}{\lambda A}\right)\right] \tag{3.94}$$

Example 3.4

This actual field example presented by Escobar [15] comprises two pressure tests performed on an exploratory well in a reservoir which is believed to possess an approximated circular shape. A pressure buildup test, **Table 3.10**, was run so that the average reservoir pressure can be determined by conventional methods. Production was inactive during the following eight months; then, a multi-rate test was performed. Its data are given in **Table 3.11**. The well had produced 190000 STB at a flow rate of 305 BPD before shut-in it for the pressure buildup test. The well data and properties of rock and fluid are as follows:

r_w = 0.33 ft, ϕ = 13%, h = 80 ft
μ = 0.9 cp, $c_t = 1.9 \times 10^{-5}$ psia^{-1} B = 1.3 bbl/STB
A = 62 Ac, P_{wf} $(t = 0)$ = 2143.4 psia, for the pressure buildup test
P_i = 2554 psia, for the multi-rate test.

It is required to estimate the average pressure test from the buildup test using the MBH method and from the multi-rate test using the *TDS* technique.

Solution by MBH method

The production time, t_p, resulted to be 14950.8 hr with Eq. (3.2). From the Horner plot given in **Figure 3.13**, the slope and intercept are read to be 57 psia/cycle and 2427 psia. Permeability is found with Eq. (2.33) and skin factor with Eq. (3.6),

Δt, hr	P_{ws}, psia	ΔP_{ws}, psia	$(\Delta t + t_p)/\Delta t$	$(t_{pss} + \Delta t)/\Delta t$
0	2143.4	0		
0.013	2167.8	24.4	1150062.54	13793.31
0.019	2195.5	52.1	786885.21	9437.84
0.028	2229.8	86.4	533958.14	6404.57
0.037	2265.6	122.2	404076.68	4846.95
0.056	2302.3	158.9	266979.57	3202.79
0.067	2323.5	180.1	223147.27	2677.12
0.079	2341.4	198.0	189251.63	2270.62
0.112	2361.0	217.6	133490.29	1601.89
0.153	2375.6	232.3	97718.65	1172.90
0.214	2389.5	246.1	69864.55	838.85
0.329	2401.7	258.3	45444.16	545.98
0.479	2411.5	268.1	31213.53	375.32
0.608	2416.8	273.4	24591.13	295.90
0.840	2425.4	282.0	17799.57	214.45
1.099	2432.7	289.3	13605.00	164.15
1.486	2437.6	294.2	10062.10	121.66
1.988	2444.9	301.5	7521.52	91.19
2.660	2448.2	304.8	5621.60	68.41
3.482	2455.0	311.6	4294.74	52.49
4.510	2465.3	321.9	3316.03	40.76
5.535	2472.6	329.2	2702.14	33.39
6.506	2479.8	336.4	2299.00	28.56
7.648	2487.3	343.9	1955.86	24.44

Table 3.10.
Pressure buildup test data for Example 3.4.

$$k = \frac{162.6q\mu B}{mh} = \frac{(162.6)(305)(0.9)(1.3)}{(57)(80)} = 26.72 \text{ md}$$

$$s = 1.1513\left[\frac{2427 - 2143.4}{57} - \log\left(\frac{12.7}{(0.13)(0.9)(1.9\times10^{-5})(0.33)^2}\right) + 3.2275\right] = 0.56$$

For the given reservoir, $(t_{DA})_{pss}$ from **Table 2.1** is 0.1. Next, t_{pss} = 179.3 hr from Eq. (3.37) and α ratio is, $\alpha = t_p/t_{pss}$ = 14950.8/179.3= 80.3; then, set $t = t_{pss}$ = 179.3. With this value, a new Horner plot is built and provided in **Figure 3.14**, from which P^* = 3580 psia. Now, determine the dimensionless production time using Eq. (3.38):

$$t_{pDA} = \frac{0.0002637kt}{\phi\mu c_t A} = \frac{(0.0002637)(12.7)(179.3)}{(0.13)(0.9)(1.9\times10^{-5})(62)(43560)} = 0.1$$

Using Eq. (3.39) with data from **Table 3.3** (first row), the MBH dimensionless pressure is P_{DMBH} =1.175. The average reservoir pressure is found with Eq. (3.45).

t, hr	t_{eq}, hr	ΔP_q, psia/(STB/D)	$t^*\Delta P'_q$, psia/(STB/D)	$t_{eq}^*\Delta P'_q$, psia/(STB/D)	q_n,STB/D
0.203	0.214	1.7907	1.2485	1.2969	296
0.253	0.262	2.2985	1.6026	1.6819	295
0.328	0.314	2.9504	2.0571	2.0251	293
0.436	0.456	4.3507	2.2356	2.1813	291
0.602	0.621	5.7416	2.2356	2.0251	292
0.875	0.902	7.1684	1.5587	1.5046	290
1.289	1.327	8.2350	1.1174	1.0771	286
1.924	1.903	9.2014	0.8010	0.7159	285
2.622	2.693	10.0000	0.5742	0.4938	281
3.621	3.762	10.2813	0.4473	0.3808	279
5.335	5.534	10.2813	0.3119	0.2936	278
8.064	8.245	10.2813	0.2001	0.1951	274
13.691	14.336	10.5705	0.1149	0.1000	270
19.910	19.774	10.5705	0.0779	0.0716	266
29.710	30.231	10.5705	0.0920	0.0895	265
44.909	44.468	10.5705	0.0758	0.0743	263
61.223	62.129	10.5705	0.0870	0.0743	258
89.029	89.068	10.8679	0.0823	0.0831	256
124.545	123.500	10.5705	0.1117	0.1136	255
161.241	161.230	10.8679	0.1357	0.1296	253
211.463	211.963	10.8679	0.1648	0.1548	251

Table 3.11.
Multi-rate test data for Example 3.4.

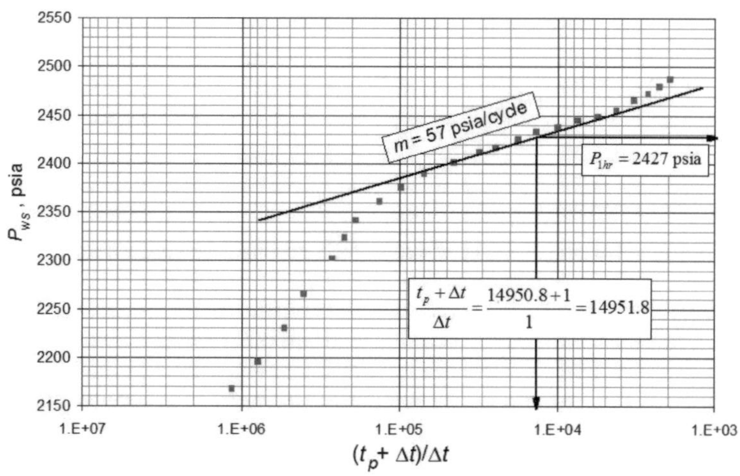

Figure 3.13.
Horner plot for Example 3.4.

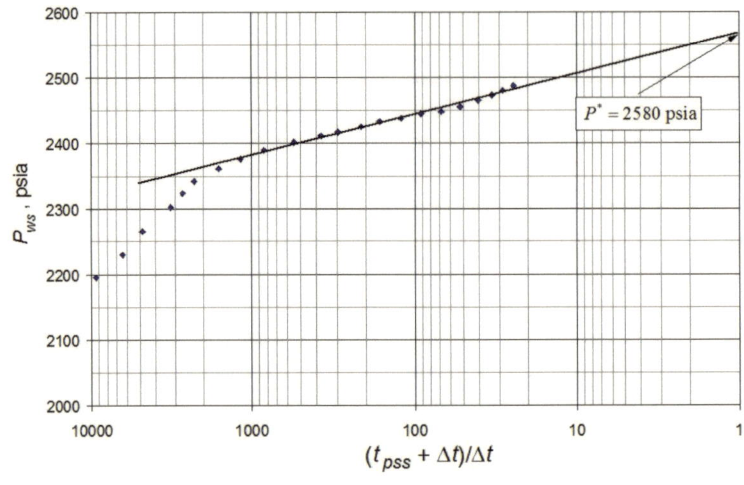

Figure 3.14.
Horner plot with t_{pss} *for Example 3.4.*

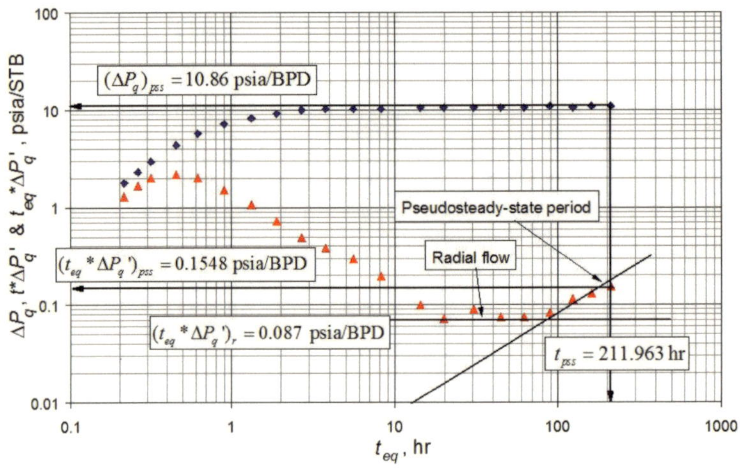

Figure 3.15.
Normalized pressure and pressure derivative versus equivalent time log-log plot for Example 3.4.

$$\overline{P} = P^* - \frac{m\,P_{DMBH}}{2.303} = 2580\frac{(57)(1.175)}{2.303} = 2550.9 \text{ psia}$$

Solution by the *TDS* technique

The following information is read from the pressure derivative plot, **Figure 3.15**,

$(t^*\Delta P'_q)_r = 0.087$ psia/BPD, $(\Delta P_q)_r = 10.86$ psia/BPD, $(t_{eq})_{pss} = 211.963$ hr
$(t^*\Delta P'_q)_{pss} = 0.1548$ psia/BPD, $(\Delta P_q)_{pss} = 10.86$ psia/BPD

Permeability is found from Eq. (2.279), and the average reservoir pressure with Eq. (3.90);

$$k = \frac{70.6\mu B}{h(t * \Delta P'_q)_r} = \frac{70.6(0.9)(1.3)}{(80)(0.087)} = 11.8 \text{ md}$$

Type of test	Method	Average reservoir pressure, psia
Buildup	MBH (Matthews-Brons-Hazebroek)	2550.9
Buildup	Dietz	2542.6*
Buildup	Ramey-Cobb	2538.2*
Buildup	MDH (Miller-Dyes-Hutchinson)	2537.3*
Buildup	Azari	2535.4*
Multi-rate	*TDS*	2549.1

**Reported in [15].*

Table 3.12.
Comparison of results for Example 3.4, after [15].

$$\overline{P} = 2554 - \frac{141.2(251)(0.9)(1.3)}{(11.8)(80)}\left[\frac{\left(\dfrac{0.1648}{10.86 - 0.1648}\right)}{\left(\ln\left(\dfrac{927.18}{0.33}\right) - \dfrac{3}{4}\right)}\right] = 2549.1\text{ psia}$$

Escobar [15] reports the results of this example, **Table 3.12**, along with those from other methods already seen in this chapter. It is observed a very close agreement among the results.

3.4.13. Other methods for estimating the average reservoir pressure

The average pressure can also be estimated using material balance [37],

$$\overline{P} = P_i - \frac{5.615q \cdot t}{c_t V_p};\ V_p \text{ in ft}^3 \tag{3.95}$$

Another formulation to calculate the average reservoir pressure [37] is based upon integrating the reservoir pressure and the volume, thus:

$$\overline{P} = \frac{\sum P_r \Delta V}{V} \tag{3.96}$$

If ΔV is replaced as $2\pi r \Delta r h \phi$ and the pseudo-steady-state solution of the radial flow equation and after several manipulations will give:

$$\overline{P} = P_w + 0.8687m\left(\ln\frac{r_e}{r_w} - 0.75\right) + \Delta P_s \tag{3.97}$$

For wells that are in steady state at the time of shutting-in, constant 0.75 is changed by 0.5.

Slider [37] proposed an equation for the case where there is interference with other wells which includes obtaining the static pressure by extrapolating or correcting the pressure at a time equal to the stabilization time or time to reach the pseudo-steady state, so that:

$$\Delta P_q = p_w - P_{wf} + m * \Delta t \tag{3.98}$$

According to Eqs. (1.130) and (2.40) (with the permeability in Darcies), the above equation becomes:

$$\overline{P} = P_{wf} - \frac{0.04\phi\mu c_t r_e^2}{k}\frac{1.79q}{\phi h c_t r_e^2} + (\Delta P_q)_{t_{pss}} \tag{3.99}$$

Rearranging:

$$\overline{P} = P_{wf} - 0.439m + (\Delta P_q)_{t_{pss}} \tag{3.100}$$

In summary, Slider [37] recommends the following methods for the determination of the average reservoir pressure at the shut-in time:

1. If the well is not acting under either steady or pseudo-steady state, use material balance, Eq. (3.95). This includes wells under infinite behavior or in transition between infinite behavior and steady or pseudo-steady state.

2. If the well is in the center of its drainage area and is in either a steady or pseudo-steady state, use Eq. (3.97), which does not require knowing either the porosity or the compressibility.

3. If the well is located near the center of its drain area and is operating under pseudo-steady state and the change in pressure with respect to time, m^*, is known, Eq. (3.100) can be used.

4. If the well is off-center within the drain area and operates under pseudo-stable state but m^* is unknown, the MBH method is recommended by Slider [37].

5. None of the methods are recommended for a well that is off-center and operates in steady-state conditions at shut-in time.

6. The above recommendations were produced when the pressure derivative did not exist. As could be seen, *TDS* technique is much more practical, and it is not limited to a few shape factors since this parameter is easily found with the technique. Also, *TDS* applies involve equations for fractured wells, horizontal wells (although not given here), and naturally fractured reservoirs. Since drill steam testing, DST consists of some small periods of buildup and drawdown, Chapters 2 and 3 apply to DST.

Nomenclature

A	area, ft^2
A_T	total field drainage area, Acres
A_w	well drainage area, Acres
B_g	gas volume factor, ft^3/STB
B_o	oil volume factor, bbl/STB
B_w	oil volume factor, bbl/STB
b	fraction of penetration/completion
b_{DLF}	intercept of P-vs.-$t^{0.25}$ plot during dual-linear flow, psia$^{0.5}$
b_{LF}	intercept of P-vs.-$t^{0.5}$ plot during hemilinear flow, psia$^{0.5}$
b_{PB}	intercept of P-vs-$1/t^{0.5}$ plot during hemilinear flow, hr^{-1}
b_x	distance from closer lateral boundary to well along the x-direction, ft

b_y	distance from closer lateral boundary to well along the *y*-direction, ft
c	compressibility, 1/psia
C	wellbore storage coefficient, bbl/psia
C_A	reservoir shape factor
c_t	total or system compressibility, 1/psia
DF	damage factor
DR	damage ratio
FE	flow index
$f(t)$	time function
h	formation thickness, ft
h_p	length of perforations, ft
I	intercept
J	productivity index, bbl/psia
k	permeability, md
k_g	gas effective permeability, md
k_o	oil effective permeability, md
k_w	water effective permeability, md
ID_{csg}	internal casing diameter, in
Np	oil produced since last stabilization, bbl
m	slope of *P*-vs.-log *t* plot, psia/hr/cycle
m^*	slope of *P*-vs.-*t* plot, psia/hr
m_{DLF}	slope of *P*-vs.-$t^{0.25}$ plot during dual-linear flow, $\text{psia}^{0.5}$/hr
m_{LF}	slope of *P*-vs.-$t^{0.5}$ plot during hemilinear flow, $\text{psia}^{0.5}$/hr
m_{PB}	slope of *P*-vs.-$1/t^{0.5}$ plot during hemilinear flow, $(\text{psia}^{0.5}/\text{hr})^{-1}$
m'	slope of superposition or equivalent time plot, psia/BPD/cycle
$m'b'$	intercept of superposition or equivalent time plot, psia/BPD/cycle
$m(P)$	pseudopressure, psia/cp
OD_{csg}	external casing diameter, in
P	pressure, psia
P	average reservoir pressure, psia
P_D'	dimensionless pressure derivative
P_D	dimensionless pressure
P_i	initial reservoir pressure, psia
PR	productivity ratio
P_r	reservoir pressure, psia
P_t	shut-in casing pressure, psia
P_{wf}	well flowing pressure, psia
P_{ws}	well shut-in or static pressure, psia
P_{1hr}	intercept of the semi-log plot, psia
P^*	false pressure, psia
ΔP_s	pressure drop due to skin conditions, psia
q	liquid flow rate, BPD
q_T	total field flow rate, BPD
q_w	well flow rate, BPD
q_{sc}	gas flow rate, Mscf/D
r_D	dimensionless radius
r	radius, ft
r_e	drainage or external radius, ft
R_s	gas dissolved in oil, scf/STB
r_w	well radius, ft
s	skin factor
s_c	skin due to partial penetration
s_{cp}	skin due to a change in permeability

s_{DL}	geometrical skin factor converging from radial to dual-linear flow
s_L	geometrical skin factor converging from dual-linear to linear flow
s_p	skin factor due to the restricted flow entry
s_{PB}	geometrical skin factor converging from dual-linear to parabolic flow
s_t	total skin factor
s_θ	skin factor resulting from a well deviation angle
T	reservoir temperature, °R, transmissivity, md-ft/cp
t	time, hr
t_p	production (Horner) time before shutting-in a well, hr
t_D	dimensionless time based on well radius
t_{DA}	dimensionless time based on reservoir area
$t^*\Delta P'$	pressure derivative, psia
V	volume, ft^3
V_u	wellbore volume/unit length, bbl/ft
X	distance in the x-direction
X_E	reservoir length, ft
X_N	superposition time, hr
Y	distance in the y-direction
Y_E	reservoir width, ft
W_D	dimensionless reservoir width
Z	height, ft

Greek

Δt	change, drop, hut-in time, hr
$\Delta t'$	flow time after rate change in two-rate tests
ϕ	porosity, fraction
λ	mobility, md/cp. Also, interporosity flow coefficient
ρ	fluid density, lbm/ft3
θ	deviation angle, °
ψ	measured deviation angle, °
ψ'	corrected deviation angle, °
μ	viscosity, cp
ω	dimensionless storativity ratio

Suffices

0	base conditions
1hr	time of 1 h
a	actual
af	after flow
BRpssi	birradial and pseudo-steady-state lines intersection
d	drainage
D	dimensionless
DA	dimensionless with respect to area
DL	dual linear flow
*DL*1	dual linear flow at 1 hr
DLpssi	intersection of pseudo-steady-state line with dual-linear line
*DLSS*1	intercept between dual-linear line and the −1-slope line (SS1)
*DLSS*2	intercept between dual-linear line and the −1-slope line (SS2)
eq	equivalent

eBL	equivalent bilinear
eBR	equivalent birradial
eL	equivalent linear
eSP	equivalent spherical
F	inflection
g	gas
h	horizontal
hs	hemispherical
i	intersection or initial conditions
ideal	ideal
INT	intercept
inv	investigation
L	linear or hemilinear flow
L1	linear flow at 1 hr
lag	lag
Lpssi	intercept of linear and pseudo-steady state lines
N	an arbitrary point during early pseudo-steady-state period
M	muskat
Mint	intercept of Muskat
max	maximum
o	oil
p	production, porous
pDA	dimensionless based on area and production time
PB	parabolic flow
*PBSS*1	intercept between parabolic line and the −1-slope line (SS1)
*PBSS*2	intercept between parabolic line and the −1-slope line (SS2)
pss	pseudo-steady state
*pss*1	pseudo-steady state at 1 hr
r	radial flow
rDLi	intercept of radial and dual linear lines
r_1	radial flow before spherical/hemispherical flow
rg	relative to gas
rLi	intercept of radial and linear lines
ro	relative to oil
rpssi	intersection of pseudo-steady-state line with radial line
rSSi	intersection between the radial line and the −1−slope line
*rSS*1*i*	intersection between the radial line and the −1-slope line (SS1)
*rSS*2*i*	intersection between the radial line and the −1-slope line (SS2)
rw	relative to water
s	skin
sf	sandface
sp	spherical
SS	steady
SSL	start of semi-log line (radial flow)
*SS*1	−1-slope line formed when the parabolic flow ends and steady-state flow regime starts. Well is near the open boundary, and the far boundary is opened
*SS*2	−1-slope line formed when the parabolic flow ends and steady-state flow regime starts. Well is near the open boundary, and the far boundary is closed
sw	spherical/hemispherical wellbore
w	well, water
wa	apparent wellbore

wb	wellbore
wD	dimensionless emphasizing at wellbore
wf	well flowing
$wf@\Delta t'=0$	well flowing at flow rate change
ws	well shut-in
x	maximum point (peak) during wellbore storage
xc	maximum point for centered wells
$X1$	maximum point between Dual linear and parabolic lines
$X2$	maximum point between parabolic and negative unit slope lines
$X3$	maximum point between hemilinear and negative unit slope lines
z	vertical direction

References

[1] Agarwal RG. A New method to account for producing time effects when drawdown type curves are used to analyze pressure buildup and other test data. Society of Petroleum Engineers. 1980. DOI: 10.2118/9289-MS

[2] Amin D, Iman A, Mohammad F, Riyaz K, Milad J. Application of the transient rate decline analysis for determining average reservoir pressure in naturally fractured reservoirs. Petroleum Exploration and Development. 2015;**42**:254-258. DOI: 10.1016/S1876-3804(15)30013-6

[3] Anderson D, Jordan CL, Mattar L. Why plot the equivalent time derivative on shut-in time coordinates?. Society of Petroleum Engineers. 2002. DOI: 10.2118/75703-MS

[4] Arari M. Non-graphical solutions for average reservoir pressure from production or buildup data. Society of Petroleum Engineers. 1987. DOI: 10.2118/17049-MS

[5] Bossie-Codreanu D. A simple buildup analysis method to determine well drainage area and drawdown pressure for a stabilized well. Society of Petroleum Engineers. 1989. DOI: 10.2118/15977-PA

[6] Bourdet D, Ayoub JA, Pirard YM. Use of pressure derivative in well test interpretation. SPE Formation Evaluation; 1989. pp. 293-302. DOI: 10.2118/12777-PA

[7] Chacon A, Djebrouni A, Tiab D. Determining the average reservoir pressure from vertical and horizontal well test analysis using the Tiab's direct synthesis technique. Society of Petroleum Engineers. 2004. DOI: 10.2118/88619-MS

[8] Cobb WM, Smith JT. An investigation of pressure-buildup tests in bounded reservoirs. Society of Petroleum Engineers. 1975. DOI: 10.2118/5133-PA

[9] Craft BC, Hawkins MF. Applied Reservoir Engineering. New Jersey, USA: Prentice-Hall International. Englewood Cliffs; 1991. p. 431

[10] Dake LP. The Practice of Reservoir Engineering—Revised edition. Amsterdam, Holland: Elsevier Developments in Petroleum Science. 2nd Printing; 2004

[11] Dykstra H, Kazemi H, Raghavan R, Gulati MS. Pressure Transient Testing Methods. SPE Reprint Series No. 14, Dallas, Texas: SPE; 1967

[12] Earlougher RC Jr. Advances in Well Test Analysis, Monograph Series Vol. 5. Dallas, TX: SPE; 1977

[13] Escobar FH, Navarrete JM, Losada HD. Evaluation of pressure derivative algorithms for well-test analysis. Society of Petroleum Engineers. 2004. DOI: 10.2118/86936-MS

[14] Escobar FH, Montealegre MM. Application of TDS technique to developed reservoirs. Journal of Petroleum Science and Engineering. 2007;**55**(3-4):252-258

[15] Escobar FH, Ibagón OE, Montealegre MM. Average reservoir pressure determination for homogeneous and naturally fractured formations from multi-rate testing with the TDS technique. Journal of Petroleum Science and Engineering. 2007;**59**: 204-212

[16] Escobar FH, Hernández YA, Hernandez CM. Pressure transient analysis for long homogeneous reservoirs using TDS technique. Journal of Petroleum Science and Engineering. 2007;**58**(1–2):68-82

[17] Escobar FH, Tiab D, Tovar LV. Determination of areal anisotropy from a single vertical pressure test and geological data in elongated reservoirs. Journal of Engineering and Applied Sciences. 2007;**2**(11):1627-1639

[18] Escobar FH, Hernandez YA, Tiab D. Determination of reservoir drainage area for constant-pressure systems using well test data. CT&F—Ciencia, Tecnología y Futuro. 2010;**4**(1):51-72

[19] Escobar FH, Montealegre M, Carrillo-Moreno D. Straight line methods for estimating permeability or width for two-zone composite channelized reservoirs. CT&F—Ciencia, Tecnología y Futuro. 2009;**3**(5):107-124

[20] Escobar FH, Montealegre MM, Carrillo-Moreno D. Pressure and pressure derivative transient analysis without type-curve matching for elongated reservoirs with changes in permeability or width. CT&F—Ciencia, Tecnología y Futuro. 2010;**4**(1):75-88

[21] Escobar FH, Srivastava P, Wu X. A practical method to determine aquifer leakage factor from well test data in CBM reservoirs. Journal of Engineering and Applied Sciences. 2015;**10**(11):4857-4863

[22] Escobar FH, Ghisays-Ruiz A, Srivastava P. Characterization of the spherical stabilization flow regime by transient pressure analysis. Journal of Engineering and Applied Sciences. 2007;**10**(14):5815-5822. ISSN 1819-6608

[23] Gringarten AC. From straight lines to deconvolution: The evolution of the state of the art in well test analysis. Society of Petroleum Engineers. SPE Formation Evaluation; 2008. pp. 41-62. DOI: 10.2118/102079-PA

[24] Golan M, Whitson CH. Well Performance. Boston, MA, USA: International Human Resources Development Corp.; 1986. p. 617

[25] Horne RN. Modern Well Test Analysis: A Computer-Aided Approach. Palo Alto, CA: Petroway Inc.; 1990. p. 185

[26] Horner DR. Pressure buildup in wells. In: Proceeding Third World Petroleum Congress. Leiden, Netherlands. 1951

[27] Igbokoyi AO, Tiab D. Estimation of average reservoir pressure and drainage area in naturally fractured reservoirs. Society of Petroleum Engineers. 2006. DOI: 10.2118/104060-MS

[28] Mathews CS, Russell DG. Pressure buildup and flow tests in wells. SPE Monograph Series. Vol. 1. Dallas, TX: Society of Petroleum Engineers of AIME; 1967. p. 163

[29] Molina MD, Escobar FH, Montealegre MM, Restrepo DP. Application of the TDS technique for determining the average reservoir pressure for vertical wells in naturally fractured reservoirs. CT&F—Ciencia, Tecnología y Futuro. 2005;**2**(6):45-55

[30] Mongi A, Tiab D. Application of Tiab's technique to multi-rate tests. Society of Petroleum Engineers. 2000. DOI: 10.2118/62607-MS

[31] Moncada K, Tiab D, Escobar FH, Montealegre MM, Chacon A, Zamora RA, Nese SL. Determination of vertical and horizontal permeabilities for vertical oil and gas wells with partial completion and partial penetration using pressure and pressure derivative plots without type-curve matching. CT&F—Ciencia, Tecnología y Futuro. 2005;**2**(6)

[32] Muskat M. Use of data oil the build-up of bottom-hole pressures. Transactions of AIME. 1937;**123:**44-48. DOI: 10.2118/937044-G

[33] Sabet M. Well Testing. Houston, TX, USA: Gulf Publishing Co; 1991. p. 460

[34] Slider HC. Application of pseudo-steady-state flow to pressure-buildup analysis. 1966. DOI: 10.2118/1403-MS

[35] Slider HC. A Simplified method for pressure buildup analysis for a stabilized well. Journal of Petroleum Technology. 1971:1155-1160

[36] Slider HC. A Simplified method of pressure buildup analysis for a stabilized well. Society of Petroleum Engineers. 1967. DOI: 10.2118/1765-MS

[37] Slider HC. Worldwide Practical Petroleum Reservoir Engineering Methods. Revised Edition. Tulsa, Oklahoma, USA: PennWell Publishing Co.; 1983. p. 826

[38] Stewart G. Well Test design & Analysis. Tulsa, Oklahoma, USA: PennWell Corporation; 2011. p. 1940

[39] Ramey HR Jr, Cobb WM. A general pressure buildup theory for a well in a closed drainage area (includes associated paper 6563). Society of Petroleum Engineers. JPT. 1971. pp. 1493-1505. DOI: 10.2118/3012-PA

[40] Tiab D. PE-5553: Well Test analysis. Lecture notes. The University of Oklahoma; 1993

[41] Tiab D. Analysis of pressure and pressure derivative without type-curve matching: 1-skin and wellbore storage. Journal of Petroleum Science and Engineering. 1995;**12**:171-181.

[42] Tiab D. Analysis of pressure and pressure derivatives without type-curve matching: 1-skin and wellbore storage. Society of Petroleum Engineers. 1993. DOI: 10.2118/25426-MS

Chapter 4

Distance to Linear Discontinuities

The available pressure analysis methods are based on assumptions of Darcy's law, for example, a homogeneous and horizontal formation of uniform thickness, with isotropic and constant porosity and permeability distributions. The issue of pressure behavior in heterogeneous reservoirs has received considerable attention in recent years. The main reason for this is the need for greater accuracy in reservoir description, which has a significant effect on the design, operation, and therefore, the economic success of the projects involved. Since these methods can be applied only once to the reservoir, the need for a reliable description of the reservoir is obvious [5, 6, 13, 22].

Two techniques can be used in the fields to describe reservoirs: radioactive tracers and pressure transient tests. Pressure transient tests have been used more (and with better results) than tracers. The determination of the volumetric swept efficiency is a problem that has a better potential to be solved by the tracers. Currently, the description of the heterogeneity of the reservoir by adjusting the tracer behavior is affected by the lack of adequate numerical models, the long time spent to obtain the results, and the dependence of the adjustment to the additional parameters that are introduced by tracers themselves (e.g., dispersion coefficients, tracer retention, etc.). It is quite possible that tracers and future pressure transient tests will be used at the same time for the description of the reservoir [6].

Normally, any type of flow barrier cannot be seen in a DST since the time is too short to affect deep the reservoir. However, in cases where flow periods are so long to observe deviations from the semilog slope or deviation from the flat trend of the pressure derivative, which reflects changes in reservoir transmissibility, faults, discontinuities, boundary conditions, or reservoir geometry as illustrated in **Figure 4.1**. Some of the methods for estimating distance to linear boundaries will be shown later [4, 6, 14, 15].

4.1. Types of reservoir discontinuities

The heterogeneities of the reservoir (see **Figure 4.1**) are variations in rock and fluid properties resulting from deposition, folding, faulting, postdepositional changes in reservoir lithology, and changes in properties or types of fluids. The reservoir heterogeneities of the deposit may be small scale, as in carbonate reservoirs where the rock has two constituents, matrix and fractures, and cavities and caves. These can also be larger scale, such as physical barriers, faults, fluid-fluid contacts, thickness changes, lithology changes, several layers with different properties in each layer, etc. In addition to these natural heterogeneities, man can induce artificial heterogeneities around wellbore during drilling (mud invasion), hydraulic fracturing, or fluid injection [6].

Another related feature is the anisotropy in the permeability, when this property varies with flow direction. Anisotropy can also be caused by sedimentary processes (filled cannel deposits) or by tectonism (fractures orientated parallel). Anisotropy takes place in both homogeneous and heterogeneous reservoirs. Therefore,

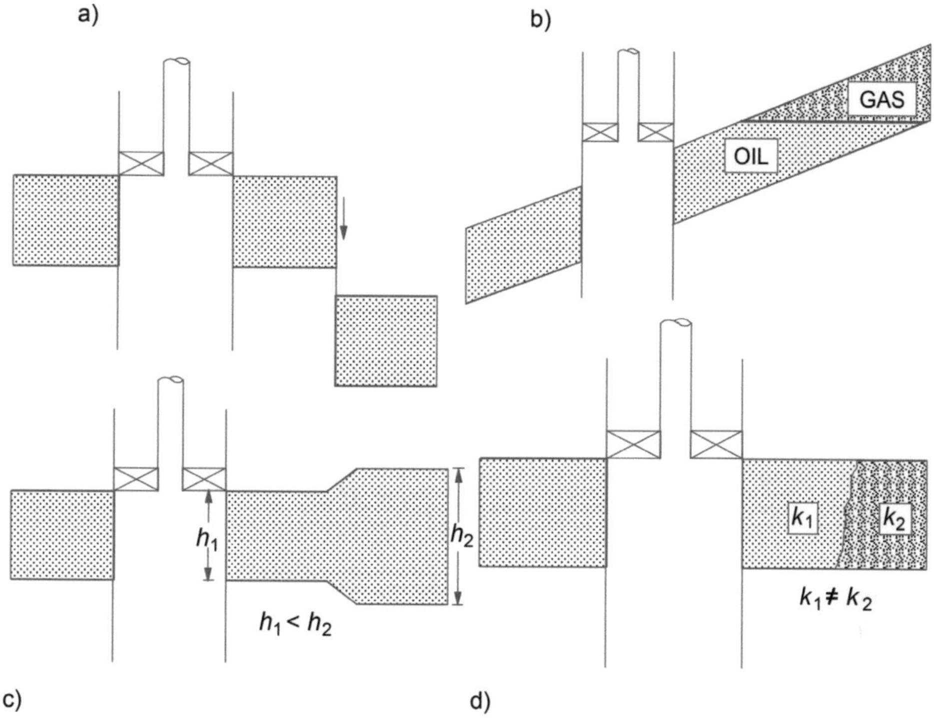

Figure 4.1
Types of discontinuities: (a) no-flow boundary (fault), (b) change of fluid type, (c) change of formation thickness, and (d) permeability change (facies), after [6].

anisotropy does not imply heterogeneity. Most reservoirs have vertical permeability less than horizontal, so there is anisotropy in that sense [5, 6].

4.2. Single-boundary systems

Large-scale heterogeneities can be detected by seismic. However, this technique can be up to one mile or more in error when estimating the well-fault distance. Transient pressure analysis is the cheapest and most accurate form to obtain the distance from a well to a given barrier. In general, to locate faults [23, 24], a test long enough to explore the reservoir in depth is required, at least four times the distance to the fault.

4.2.1 Well-fault distance from pressure buildup tests

Applying the superposition principle, the dimensionless shut-in pressure for a well near a p boundary is given, respectively, by [4–25]:

$$P_{DS} = [P_D(1,\ (t_p+\Delta t)_D)+s] - [P_D(1,\ \Delta t_D)+s] + P_D\left(\frac{2d}{r_w},\ (t_p+\Delta t)_D\right) - P_D\left(\frac{2d}{r_w},\ \Delta t_D\right) \tag{4.1}$$

The $Ei(-s)$ easily applies to find wellbore pressure, practically, at any time. Then, it can be applied to the two first terms of Eq. (4.1). For a production time, t_p, long enough and for testing times very close to the shut-in time, the $Ei(-)$ solution can be fully applied to find the dimensionless pressure drops of Eq. (4.1), so that:

$$P_i - P_{ws} = \frac{70.6q\mu B}{kh}\left[-Ei\left(\frac{-3792\phi\mu c_t d^2}{k\Delta t}\right) + Ei\left(\frac{-3792\phi\mu c_t d^2}{k(t_p + \Delta t)}\right) - Ei\left(\frac{-3792\phi\mu c_t d^2}{kt_p}\right)\right] \tag{4.2}$$

d can be calculated by a trial-and-error procedure using the above equation. This is exact but tedious. However, whenever $t_p >> \Delta t$, the logarithmic approach for $Ei(-x)$ function applies; therefore, Eq. (4.2) becomes:

$$P_{ws} = P_i - \frac{325.2q\mu B}{kh}\log\left(\frac{t_p + \Delta t}{\Delta t}\right) \tag{4.3}$$

Comparison between Eqs. (3.6) and (5.3) shows a doubled slope in the last one. It means that the detection of a fault leads to observe a doubled slope in the Horner plot since Eq. (4.3) can be expressed as:

$$P_{ws} = P_i - 2m\log\left(\frac{t_p + \Delta t}{\Delta t}\right) \tag{4.4}$$

Once the slope is folded, d can be easily calculated by reading the intercept time of the straight line of slope m with the slope straight line $2m$, as illustrated in **Figure 4.2(b)** and **(c)**. This behavior is also presented in a graph of drawdown

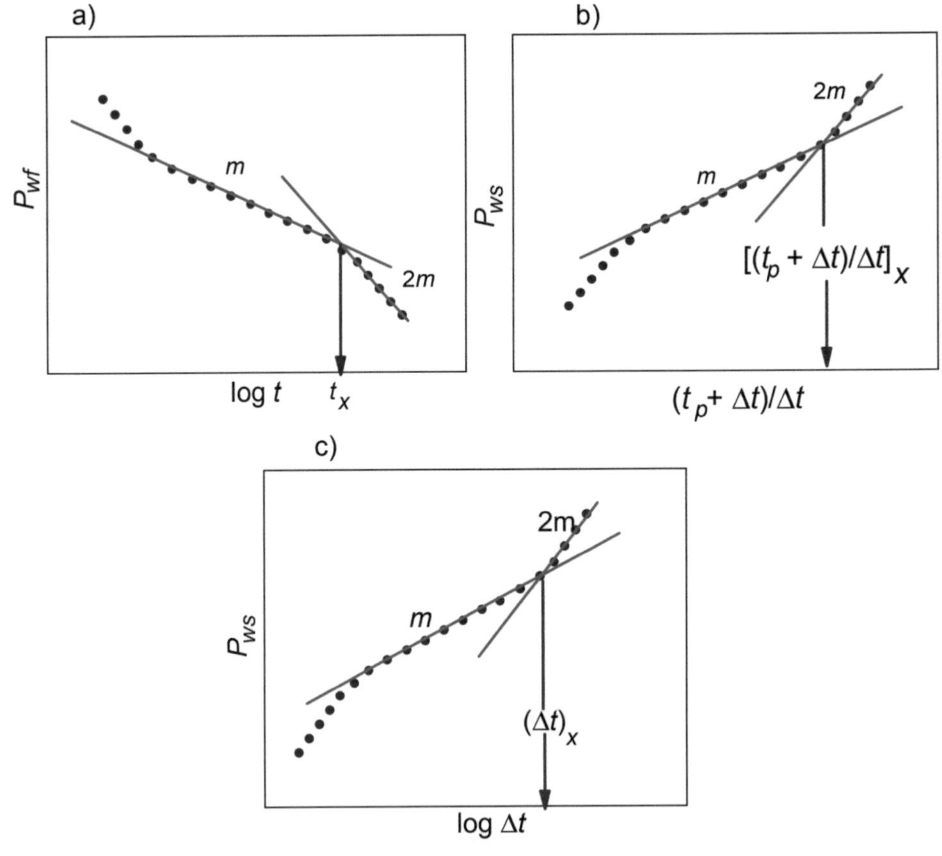

Figure 4.2
Identification of linear boundaries from conventional (semilog) plots: (a) drawdown test—semilog plot, (b) buildup test—Horner plot, (c) buildup test—MDH plot, after [13].

pressure test indicated in **Figure 4.2(a)**. However, the slope of a normal pressure buildup plot will not change at early times. Thus, this early straight line portion with slope m can be used to calculate k, s, and C as discussed in Chapter 3 section 3.2. Extrapolation of the double slope straight line is used for the estimation of the average reservoir pressure as studied in Chapter 3. Care must be taken since a similar behavior, double slope, is presented in multirate tests, injection tests, drawdown tests, etc. Different characteristics of the pressure transient may occur when a well is near multiple barriers. For example, when there are two faults intersecting at right angles to one well (one closer to the other), the slope will double and then doubled again. In general terms, the slopes that are obtained are function of the angles of intersection given by the following equation [20]:

$$\text{New slope} = \frac{360}{\theta} m \tag{4.5}$$

Similarly, the flat lines of radial flow regime will provide another flat pressure derivative line given by:

$$(t^* \Delta P')_{\text{New}} = \frac{360}{2\theta} (t^* \Delta P')_r \tag{4.6}$$

4.2.2 Methods for estimating the distance from a well to a discontinuity

Table 4.1 summarizes the available methods and provides some comments. Notice that developing the double slope takes a long time, actually, more than two log cycles.

Example 4.1

Taken from [25]. The following pressure data were obtained from the Bravo-1 well in West-Texas. This is a limestone reservoir with water influence only in the southern portion. Geological data indicate the presence of a fault (Raven) to the east of the well. See the pressure buildup along with pressure derivative and Horner time data in **Table 4.2**. The properties of rock and fluid are as follows:

r_w = 5 in, h = 18 ft, ϕ = 14%,
$c_t = 22 \times 10^{-5}$/psia, μ = 1.8 cp, B = 1.31 bbl/STB
P_i = 3750 psia, r = 56.8 lbm/ft^3, q = 180 BPD,
N_p = 9000 STB

Find permeability of the reservoir, flow efficiency, and distance to the Raven fault, using the methods of Horner, Earlougher and Kazemi, Kucuk and Kabir, Earlougher, Gray-Martinez and Cinco-Ley, and *TDS* technique [25].

Solution

Reservoir permeability. A Horner plot is given in **Figure 4.2(a)** Horner graph is given (semilog of P_{ws} vs. $(t_p+\Delta t)/\Delta t$), where $t_p = 24N_p/q$ = (24)(9000)/80 = 1200 hr, Eq. (3.2). Take the straight-line portion with slope m = 66 psia/cycle (infinite behavior line). The permeability of the straight line is estimated with Eq. (2.33):

$$k = \frac{162.6 q\mu B}{mh} = \frac{162.6(180)(1.8)(1.31)}{66(18)} = 58.1 \text{ md}$$

Method	Equation	Equation number and comment
Horner [5, 6]	$d = \sqrt{\frac{kt_p}{\phi\mu c_t}\frac{1.48109 \times 10^{-4}}{[(t_p+\Delta t]/\Delta t)]_x}}$ (4.7)	It applies for Δt_D > 25. Less accurate for small t_p values.
	$\Delta t_D = \frac{0.0002637k\Delta t}{\phi\mu c_t d^2}$ (4.8)	
Earlougher and Kazemi, [7, 20, 25]	$d = 0.01217\sqrt{\frac{k\Delta t_x}{\phi\mu\, c_t}}$ (4.9)	It uses either MDH or Horner plot. Strictly valid for $[(t_p+\Delta t)/\Delta t]_x$ $\geq$ 30. It requires long times for the slope to be double. Use Eq. (4.10) to find such time.
	$\Delta t = \frac{380000\phi\mu c_t d^2}{k}$ (4.10)	
Kucuk and Kabir [16]	$d = 0.00431\sqrt{\frac{kt_{s2r}}{\phi\mu c_t}}$ (4.11)	They modified Equation (4.9). t_{s2r} is the beginning of the second semilog line. It is better found on the graph of the derivative.
Earlougher [5, 6, 18]	$d = 0.008119\sqrt{\frac{kt_p}{\phi\mu\, c_t(t_D/r_D^2)}}$ (4.12)	It is accurate for any time. P_D is found with Eq. (4.13). It fails for t_p >> Dt.
	$P_D = \frac{1}{2}\ln\left(\frac{t_p+\Delta t}{\Delta t}\right)_x$ (4.13)	Find (t_D/r_D^2) from **Figure 1.7** or from Eq. (4.14) with x=log(P_D).
	$\frac{t_D}{r_D^2} = 10^{\frac{0.53666069+1.843195406x}{1-0.8502854913x+0.1199676223x^2}}$ (4.14)	
MDH [20, 21]	$d = 0.01217\sqrt{\frac{kt_p}{\phi\mu c_t\Delta t_s}}$ (4.15)	Dt_s is the time found by extrapolating the first slope to the value of P_i.
Sabet [20]	$d = 0.5 \times 10^{-\left\{\frac{P_{2m}-P_{wf}}{\lvert 2m\rvert} - \log t_{2m} - \log\frac{k}{\phi\mu c_t r_w^2} + 3.23 - 0.435s\right\}}$ (4.16)	
Gray [12], Martinez and Cinco-Ley [17]	$d = \sqrt{\frac{t_{dv}\ln(0.86859m_1)}{3792.19\phi\mu c_t}}$ (4.17)	

Method	Equation		Equation number and comment
TDS [14], and [15]	$d = \sqrt{\frac{0.000422kt_{re}}{\phi\mu c_t}}$	(4.18)	Corrected in this book
	$d = 0.01217\sqrt{\frac{kt_F}{\phi\mu c_t}}$	(4.19)	

Table 4.1.
Methods to determine distance well-discontinuity.

Δt, hr	P_{ws}, psia	$(t_p+\Delta t)/\Delta t$	ΔP, psia	$t^*\Delta P'$, psia
0	2900		0	
0.5	3090	2401	190	119.12
0.7	3118	1715.29	218	111.80
1.1	3170	1091.91	270	95.56
1.6	3199	751	299	91.80
2.5	3240	481	340	89.52
3.5	3278	343.86	378	72.10
5	3290	241	390	46.45
7	3302	172.43	402	32.60
9	3310	134.33	410	30.91
13	3320	93.31	420	29.23
20	3333	61	433	27.72
30	3343	41	443	33.68
40	3350	31	450	42.54
50	3363	25	463	48.26
70	3382	18.14	482	57.39
100	3400	13	500	58.58
150	3423	9	550	62.50
250	3450	5.8	550	80.31

Table 4.2.
Pressure buildup test data.

From **Figure 4.2**, P_{1hr} = 3245 psia. Therefore, the mechanical skin factor is obtained from Eq. (3.6), thus:

$$s = 1.1513\left[\frac{3245 - 2900}{66} - \log\left(\frac{58.1}{0.14(1.8)(22 \times 10^{-5})(0.417^2)}\right) + 3.23\right] = 1.7$$

The skin pressure drop is found with Eq. (2.35):

$$\Delta P_s = 0.87(66)(1.7)\text{ psia}$$

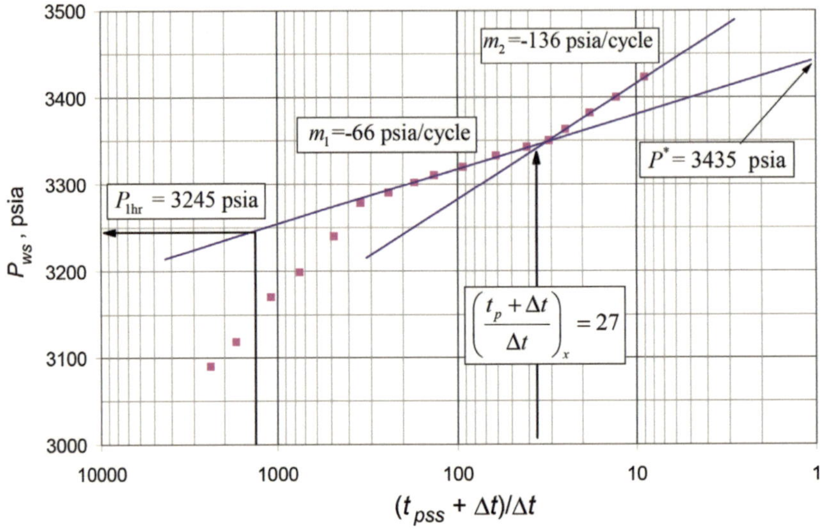

Figure 4.2
Horner plot for Example 4.1.

P^* = 3435 psia from the Horner plot, then *flow efficiency* is estimated with Eq. (2.46):

$$FE = 1 - \frac{97.6}{3435 - 2900} = 81.8\ \%$$

This means that stimulation is necessary. The distance to the linear boundary is found from the following methods:

Horner method

From the Horner plot, a value of $(t_p+\Delta t)/\Delta t)_x$ was found to be 27. Then, using it into Eq. (4.6) will provide:

$$d = 0.01217\sqrt{\frac{k\Delta t_p}{\phi\mu c_t}\frac{1}{\left[t_p + \Delta t/\Delta t\right]_x}} = 0.01217\sqrt{\frac{58.1(1200)}{(0.14)(1.8)(22\times 10^{-5})(27)}} = 83\text{ft}$$

Earlougher and Kazemi method

From the MDH plot, a value of Δt_x of 45 hr is read (**Figure 4.3**). Using it in Eq. (4.9),

$$d = 0.01217\sqrt{\frac{k\Delta t_x}{\phi\mu\, c_t}} = \sqrt{\frac{0.0001481(58.1)(45)}{(0.14)(1.8)(22\times 10^{-5})}} = 83.6\ \text{ft}$$

This method is supposed to work for $[(t_p+\Delta t)/\Delta t]_x \geq 30$ and $[(t_p+\Delta t)/\Delta t]_x = 27$. Use Eq. (4.10) to find the time for the slope to be double:

$$\Delta t = \frac{380000\phi\mu c_t d^2}{k} = \frac{380000(0.14)(1.8)(22\times 10^{-5})(83.6)^2}{58.1} = 253.4\ \text{hr}$$

From the pressure derivative plot (**Figure 4.4**), it is possible to see the second plateau at about 70 hr.

Kucuk and Kabir method

The pressure derivatives indicate that t_{s2r} is about 70 hr. Using Eq. (4.11) gives:

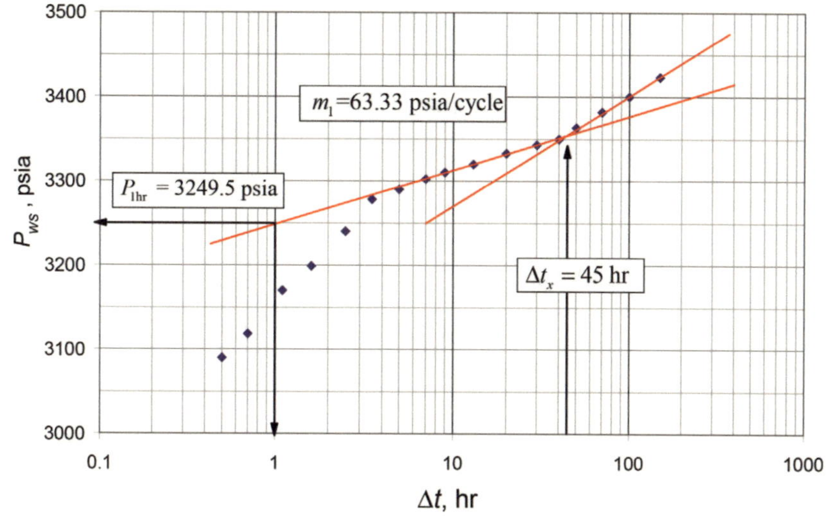

Figure 4.3
MDH plot for Example 4.1.

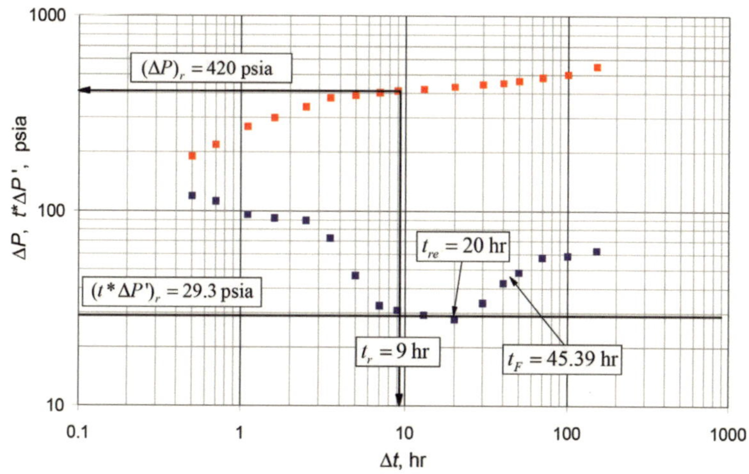

Figure 4.4
Pressure and pressure derivative versus time log-log plot for Example 4.1.

$$d = 0.00431\sqrt{\frac{kt_{s2r}}{\phi\mu c_t}} = \sqrt{\frac{(58.1)(70)}{(0.14)(1.8)(22\times10^{-5})}} = 116.73 \text{ ft}$$

Earlougher method
It is obtained from Eq. (4.13):

$$P_D = \frac{1}{2}\ln\left(\frac{t_p+\Delta t}{\Delta t}\right)_x = \frac{1}{2}\ln(27) = 1.648$$

Then, $x=\log(P_D)=0.217$. Use this value in Eq. (4.14),

$$\frac{t_D}{r_D^2} = 10^{\frac{0.53666069+1.843195406(0.217)}{1-0.8502854913(0.217)+0.1199676223(0.217)^2}} = 13.82$$

This can be verified in **Figure 1.7**. Use the t_D/r_D^2 value in Eq. (4.12):

$$d = 0.008119\sqrt{\frac{kt_p}{\phi\mu\, c_t(t_D/r_D^2)}} = 0.008119\sqrt{\frac{(58.1)(1200)}{(0.14)(1.8)(22\times 10^{-5})\,(13.82)}} = 77.5\ \text{ft}$$

Gray, Martinez, and Cinco-Ley method

$t_{dv} = t_{re} = 20$ hr from the derivative plot (**Figure 4.4**). Use of Eq. (4.17) leads to:

$$d = \sqrt{\frac{t_{dv}\ln(0.86859m_1)}{3792.19\phi\mu c_t}} = \sqrt{\frac{20\ln(0.86859[66])}{3792.19(0.14)(1.8)(22\times 10^{-5})}} = 19.62$$

TDS technique

The buildup pressure derivative with a production time, t_p, of 1200 hr was estimated and reported in **Table 4.2**. The second pressure derivative (not shown in here) was used to better define the inflection point. The pressure derivative was obtained and given in **Figure 4.4** from which the following information was read:

$t_r = 9$ hr, $(t^*\Delta P')_r = 29.3$ psia, $\Delta P_r = 420$ psia,

$t_{re} = 20$ hr, $t_{re} = 45.39$ hr

Find permeability and skin factor with Eqs. (2.71) and (2.92), respectively:

$$k = \frac{70.6q\mu B}{h(t^*\Delta P')_r} = \frac{70.6(180)(1.8)(1.31)}{18(29.3)} = 56.82\ \text{md}$$

$$s = 0.5\left(\frac{420}{29.3} - \ln\left[\frac{(56.82)(9)}{(0.14)(1.8)(22\times 10^{-5})(0.417^2)}\right] + 7.43\right) = 1.98$$

Use Eqs. (4.18) and (4.19) to find the distance from Bravo-1 well to the Raven fault:

$$d = \sqrt{\frac{0.000422kt_{re}}{\phi\mu c_t}} = \sqrt{\frac{0.000422(56.82)(20)}{(0.14)(1.8)(22\times 10^{-5})}} = 92.95\ \text{ft}$$

$$d = 0.01217\sqrt{\frac{kt_F}{\phi\mu c_t}} = 0.01217\sqrt{\frac{(56.82)(45.39)}{(0.14)(1.8)(22\times 10^{-5})}} = 83\ \text{ft}$$

4.2.3 Methods for estimating the distance well-discontinuity from DST

Table 4.3 presents some of the available methods to find linear discontinuity-to-well distance from drill stem tests.

Example 4.2

Earlougher [6] presented DST data from the Red Formation of Major County, Oklahoma. The well was treated for completion work with approximately 480 bbl. The Horner chart is provided below.

$q = 118$ BPD, $c_t = 8.2\times 10^{-6}$ 1/psia, $\mu = 1.3$ cp,

$\phi = 12\%$, $B = 1.1$ bbl/STB, $m = 1321$ psia/cycle,

$t_p = 4$ hr

Method	Equation	Equation number and comment
Horner [4]	$-Ei\left(-\frac{948\phi\mu c_t d^2}{kt_p}\right) = \ln\left(\frac{t_p+\Delta t}{\Delta t}\right)_x$	(4.20) For buildup
Dolan, Einarsen, and Hill [13]	$d = 0.024337\sqrt{\frac{kt_p}{\phi\mu c_t[(t_p+\Delta t]/\Delta t)]_x}}$	(4.21) For buildup
Ishteiwy and Van Poollen [13]	$d = 0.015276\sqrt{\frac{kt_p}{\phi\mu c_t[(t_p+\Delta t]/\Delta t)]_x}}$	(4.22) For buildup
	$\left(\frac{t_p+\Delta t}{\Delta t}\right)_x = \frac{t_{pD}}{1.13}$	(4.23)
	$t_{pD} = \frac{0.0002637}{\phi\mu c_t d^2}$	(4.24)
Bixel, Larkin, and Van Poolen [2] and Bixel and Van Poolen [3]	$d = 0.0307\sqrt{\frac{kt_x}{\phi\mu c_t}}$	(4.25) For drawdown

Table 4.3.
Methods to determine distance well-discontinuity from DST.

Determine the permeability.

Determine the distance to the discontinuity using the methods mentioned.

Since there is no fault near the well, which suggests that it may be discontinuity?

Solution

(a) *Permeability.* Since the slope is given, permeability is found from Eq. (2.33):

$$k = \frac{162.6q\mu B}{mh} = \frac{162.6(118)(1.1)(1.3)}{(15)(1321)} = 1.38 \text{ md}$$

(b) *Distance to the discontinuity*
Horner method
From **Figure 4.5**, $[(t_p+\Delta t)/\Delta t]_x = 1.55$. Using Eq. (4.20),

$$-Ei\left(-\frac{948(0.12)(1.3)(8.2\times10^{-6})d^2}{1.38(4)}\right) = \ln 1.55$$

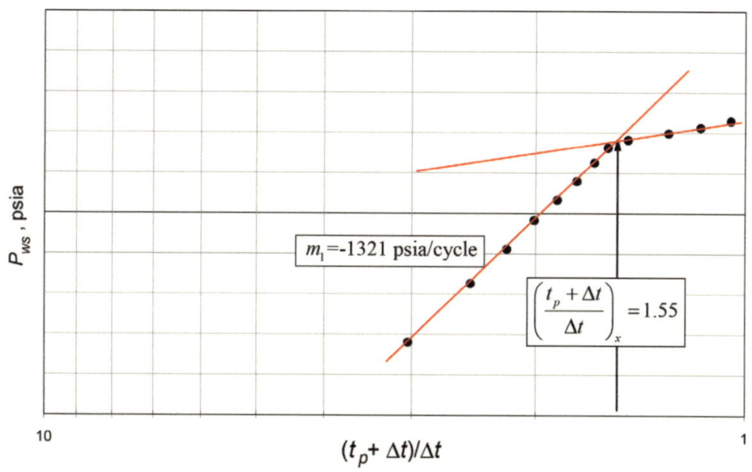

Figure 4.5
Horner plot for Example 4.2, after [13].

$$-Ei\left[(-2.1969 \times 10^{-4})d^2\right] = 0.4383$$

Interpolating from **Table 1.4**,

$$2.1969 \times 10^{-4}d^2 = 0.618$$

Then, $d = 53$ ft

Dolan, Einarsen, and Hill method

Using Eq. (4.21),

$$d = \sqrt{\frac{0.00059229\,kt_p}{\phi\mu c_t[(t_p + \Delta t]/\Delta t)]_x}} = \sqrt{\frac{0.00059229(1.38)(4)}{(0.12)(1.3)(8.2 \times 10^{-6})(1.55)}} = 40.7 \text{ ft}$$

Ishteiwy and van Poollen method

Using Eq. (4.22),

$$d = \sqrt{\frac{0.0002484\,kt_p}{\phi\mu c_t[(t_p + \Delta t]/\Delta t)]_x}} = \sqrt{\frac{0.0002484(1.38)(4)}{(0.12)(1.3)(8.2 \times 10^{-6})(1.55)}} = 25.5 \text{ ft}$$

(c) *Since there is no fault near the well, which suggests that it may be discontinuity?*

The slope decreases; however, there is evidence of a "constant pressure boundary." The improvement in the transmissibility could be due to the treatment that well received before the test.

4.3. Leaky faults

There are cases in which the fault does not fully seal. A nonsealing fault allows the transient wave to cross over the fault and keep traveling. There probably exists a contrast in mobility. Meaning the formation on the other side of the fault may have or not the same properties. There are some other cases where the reservoir has a linear constant-pressure boundary. The aquifer may act either fully or partially.

4.3.1 Nonsealing fault

Escobar et al. [8] used the dimensionless conductivity of the fault/boundary as:

$$F_{CD} = \frac{k_f}{k}\frac{w_f}{L} \tag{4.25}$$

The F_{CD} value typically ranges from 0 to 1.0 or more. A value of zero indicates a sealed boundary or absence of the boundary, and an infinite value indicates a constant pressure or a completely sealed fault.

4.3.1.1 Partially active aquifer

The expected behavior is given in **Figure 4.6**. At late time, the pressure derivative will display a negative slope of 1 (radial stabilization), which reduces its value as τ decreases. A second derivative in this zone will provide a maximum value.

A scalable dimensionless conductivity of the boundary, τ, is defined as [9]:

$$\tau = e^{-F_{CD}} - 1; \quad -1 < \tau < 0 \tag{4.26}$$

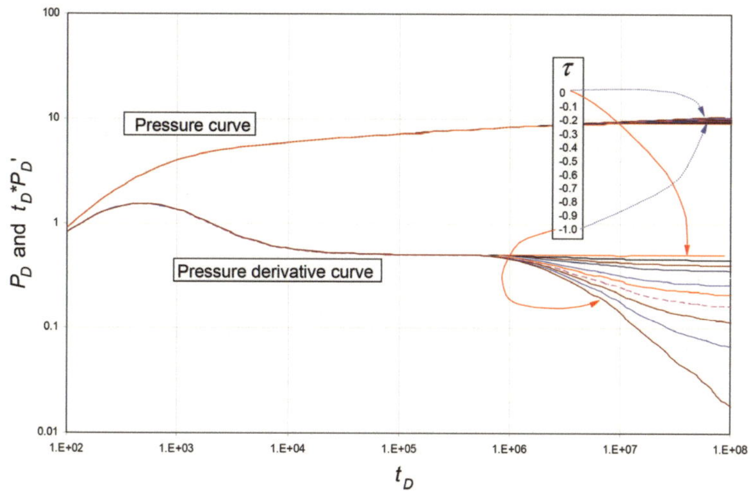

Figure 4.6
Pressure and pressure derivative behavior for a partially active aquifer, after [9].

Negative values of τ indicate the presence of an act aquifer. Note that when τ = 0, F_{CD} = 0 indicating that L = 0, and when t = −1, F_{CD}= ∞, indicating that the boundary conductivity is infinite. Escobar et al. [8] presented the following correlation to find the acting degree of the aquifer.

$$\tau = 0.00375902 - 3.1638126\frac{(t^*\Delta P')'_{2x}}{(t^*\Delta P')_r} \tag{4.27}$$

If exists a second-flat line below the radial flow pressure derivative line, the following expression applies:

$$\tau = 0.983396 - 0.98603107\frac{(t^*\Delta P')'_{2x}}{(t^*\Delta P')_r} \tag{4.28}$$

4.3.1.2 Partially sealing (leaky) fault

This case is given in **Figure 4.7** assuming there is no permeability contrast. This may be an explanation of why sometimes the second slope on the semilog plot is not doubled. A scalable dimensionless conductivity of the boundary (fault transparency), τ, is defined as [9]:

$$\tau = 1 - F_{CD};\ 0 < \tau < 1 \tag{4.29}$$

Positive values of τ indicate the presence of no-flow boundaries. A value of zero indicates that there is no fault/boundary, so the permeability on both sides of the border is the same. Note that when τ = 0, F_{CD}= 1, and when τ = 1, F_{CD} = 0, which indicate that the barrier has a permeability of zero. Escobar et al. [9] presented the following expressions to determine the fault transparency:

$$\tau = 3.173803\frac{(t^*\Delta P')'_{2x}}{(t^*\Delta P')_r} - 0.0015121 \tag{4.30}$$

$$\tau = 1.01338389\frac{(t^*\Delta P')_{r2}}{(t^*\Delta P')_r} - 1.0146535 \tag{4.31}$$

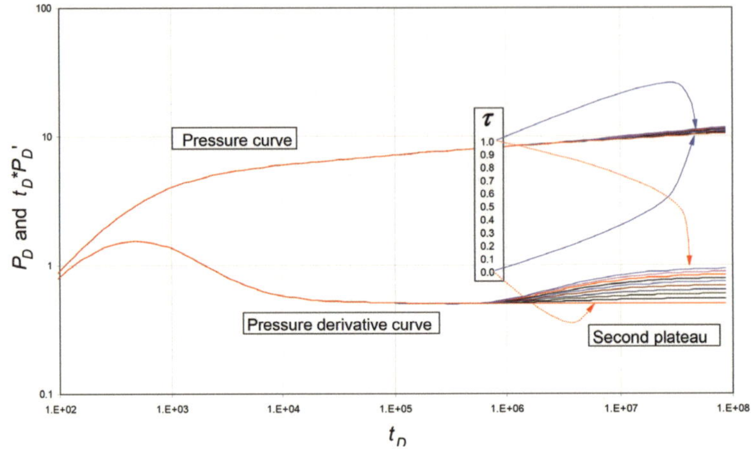

Figure 4.7
Pressure and pressure derivative behavior for a leaky fault, after [9].

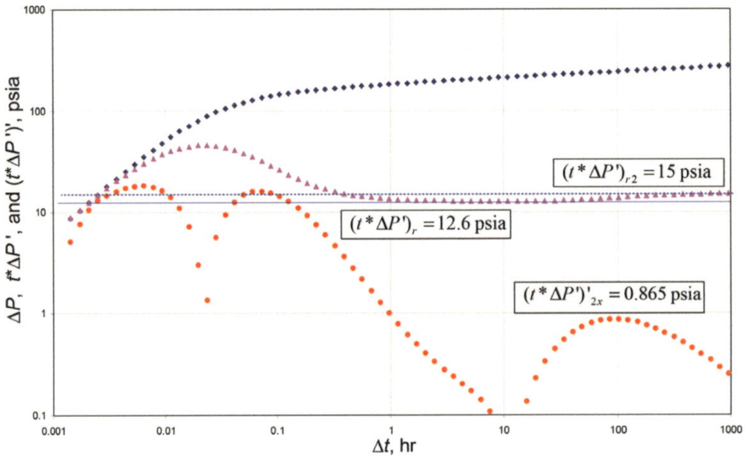

Figure 4.8
Pressure, pressure derivative, and second pressure derivative against time log-log plot for Example 4.3, after [9].

Example 4.3.

Escobar et al. [9] presented an example of a well near a leaky fault and it is required to confirm the transparency fault value. Pressure, pressure derivative, and second-pressure derivative versus time data are given in **Figure 4.8**. Other relevant information is found below:

k = 10 md, h = 50 ft, d = 730 ft, τ = 0.22, ϕ = 23%, $c_t = 1.38 \times 10^{-5}$ 1/psia, μ = 0.3 cp, q = 200 BPD, B = 1.48 rb/STB, r_w = 0.4 ft, C_D = 70, s = 2, P_i = 5200 psia

Solution

The following information was read from **Figure 4.8**.

$(t^*\Delta P')_r$ = 12.6 psia, $(t^*\Delta P')_{2r}$ = 15 psia, $(t^*\Delta P')'_{2x}$ = 0.865 psia

Use of Eqs. (4.30) and (4.31) allows finding the fault transparency:

$$\tau = 3.173803\frac{0.865}{12.6} - 0.0015121 = 0.217$$

$$\tau = 1.01338389 \frac{15}{12.6} - 1.0146535 = 0.192$$

Notice that both values closely match the original transparency fault value of 0.22.

4.3.2 Finite-conductivity faults

Escobar et al. [10, 11] presented *TDS* technique extension for finite-conductivity faults with mobility contrast and without it, respectively. The nonmobility contrast will be shortly discussed here. In these systems in which model was presented by Rahman et al. [19], the reservoir permeability has a lower permeability than that of the fault. Fluid flow takes place both across and along the fault plane (see **Figure 4.9**). The fault enhances the drainage reservoir capacity. It is observed in **Figure 4.9**, a normal flow radial regime is developed at early time around the well. When the transient reaches the fault (the fault may have some damage, s_f), the pressure derivative declines along a negative-unit slope. At this moment, the fault acts as constant pressure linear boundary. Then, as the pressure drops in the fault, a bilinear flow regime results when the flow is established in the fault plane thickness as shown in **Figure 4.10**. Finally, the pressure derivative response comes back to a plateau when radial is restored. This part is not shown in **Figure 4.10**.

New dimensionless quantities are presented here [10]:

$$t_{Df} = \frac{0.0002637kt}{\phi \mu c_t d^2} \tag{4.32}$$

$$h_D = \frac{h}{d} \tag{4.33}$$

$$F_{CD} = \frac{k_f w_f}{k \cdot d} \tag{4.34}$$

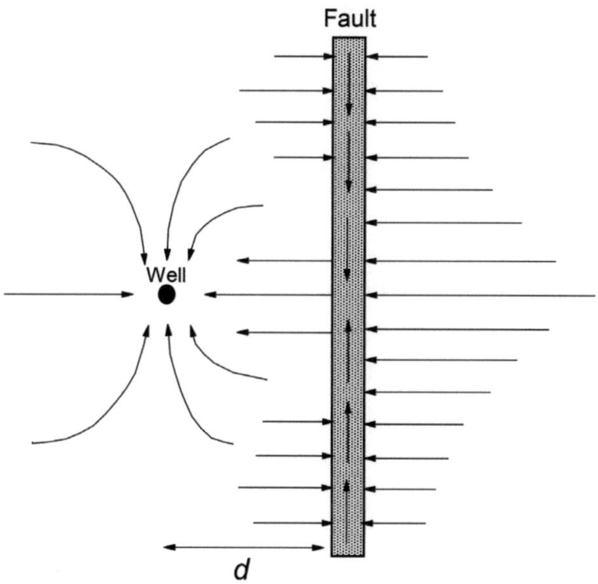

Figure 4.9
Schematic of a typical fault system and flow lines, after [1].

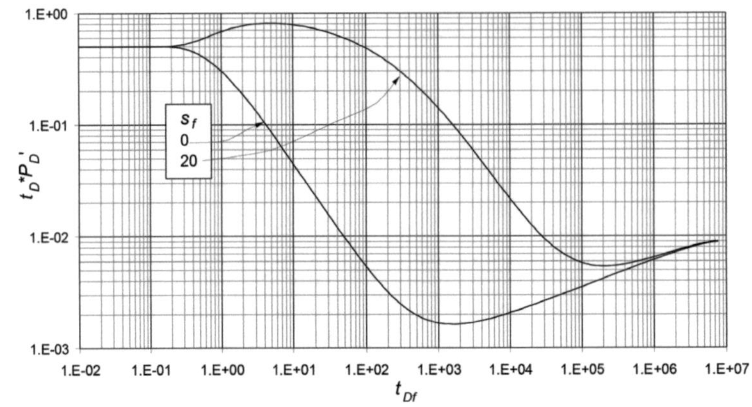

Figure 4.10
Dimensionless pressure derivative for a well near finite-conductivity fault. s_f = 0 *and* 20, *after [10].*

4.3.2.1 TDS technique

For the bilinear flow case, Escobar et al. [10] presented the following expressions for finding the fault conductivity, fault skin factor, and distance from well to fault. Escobar et al. [10] also presented equations for gas flow. Not all the expressions developed by Escobar et al. [10] are reported here.

$$d = 0.0325\sqrt{\frac{kt_{re}}{\phi\mu c_t}} \tag{4.35}$$

$$d = \sqrt{\frac{0.0002637kt_{rssi}}{\phi\mu c_t}} - s_f h \tag{4.36}$$

$$s_f = \frac{d}{h}\left[\sqrt{\left(\frac{3.7351\times10^{-6}k^2ht_{ss}(t^*\Delta P')_{ss}}{q\mu^2B\phi c_t d^2}\right)} - 1\right] \tag{4.37}$$

$$k_f w_f = 121.461\left(\frac{q\mu B}{h(t^*\Delta P')_{BL}}\right)^2\left(\frac{t_{BL}}{k\phi\mu c_t}\right)^{0.5} \tag{4.38}$$

$$k_f w_f = 1.694\times10^{-9}kd\left(\frac{kt_{BLssi}}{\phi\mu c_t L_F^2}\right)^{2.5} 1/\left(1+s_f\frac{h}{L_F}\right)^4 \tag{4.39}$$

If the dimensionless fault conductivity is larger than 2.5×10^8, linear flow will be developed instead of bilinear and the fault has infinite conductivity and the distance from the well to it is found from:

$$d = \frac{6.42\times10^{-6}qB}{h(t^*\Delta P')_L}\sqrt{\frac{\mu t_L}{k\phi c_t}} \tag{4.40}$$

4.3.2.2 Conventional analysis

The bilinear flow model is given by:

$$\Delta P = \frac{44.1q\mu Bt^{1/4}}{h\sqrt{k_f w_f}\sqrt[4]{\phi\mu c_t k}} + \frac{141.2q\mu B}{kh}s_{BL} \tag{4.41}$$

The above equation suggests that a Cartesian plot of ΔP_{wf} versus $t^{0.25}$ gives a linear trend in which slope allows for the estimation of the fault conductivity:

$$k_f w_f = \left[\frac{44.1 q\mu B}{m_{BL} h \sqrt[4]{\phi \mu c_t k}}\right]^2 \tag{4.42}$$

The above equation was also found by Trocchio [28].

The pressure equation for the linear flow is:

$$\Delta P = \frac{1.33 \times 10^{-5} q\mu B t^{1/2}}{h \cdot d(\phi \mu c_t k)^{1/2}} + \frac{141.2 q\mu B}{kh} s_L \tag{4.43}$$

As indicated before, Eq. (4.32) suggests that a Cartesian plot of ΔP_{wf} versus $t^{0.5}$ gives a linear trend in which slope allows the estimation for the distance from the well to the fault:

$$d = \frac{1.33 \times 10^{-5} q\mu B}{m_L h \sqrt{\phi \mu c_t k}} \tag{4.44}$$

The governing pressure equation for the steady state caused by the fault is:

$$\Delta P = -\frac{q\mu^2 B \phi c_t d^2}{3.7351 \times 10^{-6} k^2 h}\left(1 + s_f \frac{h}{d}\right)^2 \frac{1}{t} + \frac{70.6 q\mu B}{kh} \ln\left(\frac{4d^2}{r_w^2} + 8 \times 10^5 \left(s_s \frac{h}{d}\right)^2\right) \tag{4.45}$$

Also a Cartesian plot of ΔP_{wf} versus $1/t$ gives a linear trend in which slope allows the estimation of fault skin factor:

$$s_f = \frac{d}{h}\left[\sqrt{\left(\frac{3.7351 \times 10^{-6} k^2 h m_{ss}}{q\mu^2 B \phi c_t d^2}\right)} - 1\right] \tag{4.46}$$

Trocchio [28] presented minimum fault length, x_{fmin}, the following expression to find the dimensionless end time of the bilinear flow regime and the minimum fracture length:

$$x_{f_{\min}} = \left(\frac{2.5}{4.55\sqrt{\frac{k}{k_f w_f}} \pm \sqrt[4]{\frac{\phi \mu c_t}{0.0002637 k t_{ebf}}}}\right)^2 \tag{4.47}$$

Example 4.4

Escobar et al. [10] presented a pressure test of a well inside finite-conductivity faulted reservoir. Pressure and pressure derivative data are reported in **Figure 4.11**. It is required to estimate distance to fault and fault conductivity using both *TDS* and conventional methodologies.

k = 100 md, h = 100 ft, d = 730 ft, ϕ = 25%, c_t = 1.3792 × 10^{-5} 1/psia,
μ = 0.7747 cp, q = 100 BPD, B = 1.553 rb/STB, r_w = 0.3 ft, d = 250 ft

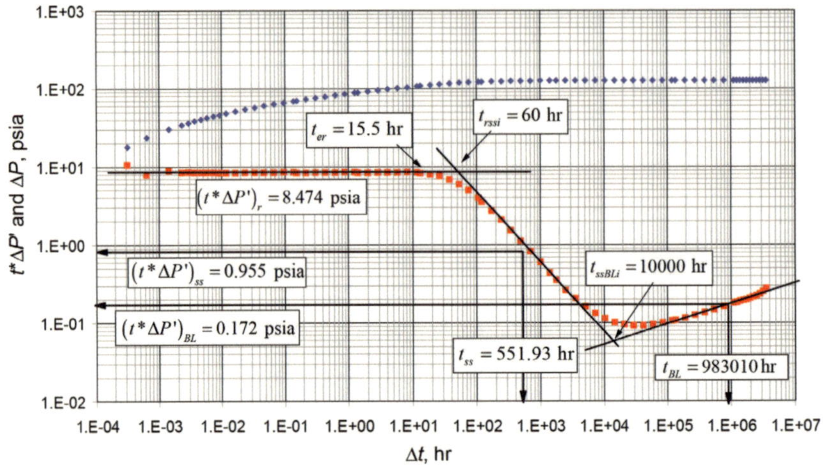

Figure 4.11
Pressure and pressure derivative plot for Example 4.4, after [10].

Parameter	Equation used	Result
d, ft	(4.35)	247.85
d, ft	(4.36)	243.43
s_f	(4.37)	0.00225
$k_f w_f$, md-ft	(4.38)	1.14 × 109
$k_f w_f$, md-ft	(4.3)	1.227 × 109
$k_f w_f$, md-ft	(4.42)*	1.128 × 109
F_{CD}	(4.34)	458900.15

*Conventional analysis.

Table 4.4.
Summary of results for Example 4.4, after [10].

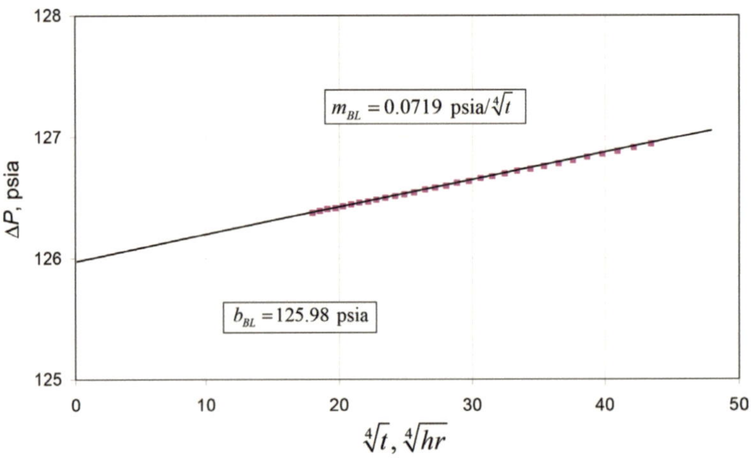

Figure 4.12
Cartesian plot of pressure drop versus the fourth root of time for Example 4.4, after [10].

Solution

The log-log plot of pressure and pressure derivative against production time is given in **Figure 4.11** from which the following information was read:

$(t^*\Delta P')_r$ = 8.474 psia, t_{er} = 15.5 hr, t_{ss} = 551.93 hr, $(t^*\Delta P')_{ss}$ = 0.955 psia, t_{BL} = 983010 hr, $(t^*\Delta P')_{BL}$ = 0.172 psia, t_{rssi} = 60 hr, t_{ssBLi}= 10000 hr, $(t^*\Delta P')_{min}$ = 0.0912 psia

With the read parameters from **Figure 4.11**, the parameters were estimated and reported in **Table 4.4**. For straight-line conventional analysis, only the bilinear flow regime part is plotted in **Figure 4.12** from which a slope value of 0.0719 psia/hr$^{0.25}$ was estimated. Then, Eq. (4.42) to find a finite conductivity of 1.128 $\times$ 10^9 md-ft, which is also reported in **Table 4.4**.

Nomenclature

B	oil volume factor, bbl/STB
c	compressibility, 1/psia
d	distance from well to linear boundary, ft
c_t	total or system compressibility, 1/psia
F_{CD}	fault dimensionless conductivity
FE	flow index
h	formation thickness, ft
k	permeability, md
$k_f w_f$	fault conductivity, md-ft
Np	oil produced since last stabilization, bbl
m	slope of P-vs-log t plot, psia/hr/cycle
m_1	slope of first semilog straight line, psia/hr/cycle
P	pressure, psia
$P_D{}'$	dimensionless pressure derivative
P_D	dimensionless pressure
P_i	initial reservoir pressure, psia
P_{wf}	well flowing pressure, psia
P_{ws}	well shut-in or static pressure, psia
P_{1hr}	intercept of the semilog plot, psia
P^*	false pressure, psia
ΔP_s	pressure drop due to skin conditions, psia
q	liquid flow rate, BPD
r	radius, ft
r_w	well radius, ft
s	skin factor
t	time, hr
$t^*\Delta P'$	pressure derivative, psia
$t^{2*}\Delta P''$	second pressure derivative, psia
t_p	production (Horner) time before shutting-in a well, hr
t_D	dimensionless time based on well radius
t_{dv}	time at which either pressure or derivative deviate from first radial line, hr

Greek

Δ	change, drop
Δt	shut-in time, hr

ϕ	porosity, fraction
ρ	fluid density, lbm/ft^3
τ	scalable dimensionless conductivity
μ	viscosity, cp

Suffices

1hr	time of 1 hr
2m	second semilog straight line
2x	maximum in the second pressure derivative
BL	bilinear flow regime
BLssi	intercept of bilinear and steady-state lines
d	distance
D	dimensionless
DA	dimensionless with respect to area
Df	dimensionless with respect to fault length
f	fault
ebf	end of bilinear flow
F	inflection, better found from second derivative
i	intersection or initial conditions
L	linear flow regime
p	production, porous
r	radial flow
re	end of radial flow regime
r2	second plateau, hemiradial flow
rssi	intercept of radial and steady-state lines
s	skin
ss	a point on the steady-state period
s2r	start of second semilog straight line
w	well
wf	well flowing
ws	well shut-in

References

[1] Abbaszadeh MD, Cinco-Ley H. Pressure-transient behavior in a reservoir with a finite-conductivity fault. SPE Formation Evaluation. 1995; **10**(3):26-32

[2] Bixel HC, Larkin BK, Van Poollen HK. Effect of linear discontinuities on pressure build-up and drawdown behavior. Richardson, TX: Journal of Petroleum Technology. 1963;(Aug):85. DOI: 10.2118/611-PA

[3] Bixel HC, Van Poollen HK. Pressure drawdown and build-up in the presence of radial discontinuities. Journal of Petroleum Engineering. 1967;(Sept):301

[4] Boussalem R, Tiab D, Escobar FH. Effect of mobility ratio on the pressure and pressure derivative of wells in closed composite reservoirs. Society of Petroleum Engineers. 2002. DOI: 10.2118/76781-MS

[5] Dake LP. The Practice of Reservoir Engineering. Revised edition. Amsterdam, Holland: Elsevier Developments in Petroleum Science; 2004. 2nd printing

[6] Earlougher RC Jr. Advances in Well Test Analysis, Monograph Series Vol. 5, SPE, Dallas, TX;1977

[7] Earlougher RC, Kazemi H. Practicalities of detecting faults from buildup testing. Society of Petroleum Engineers. 1980. DOI: 10.2118/8781-PA

[8] Escobar FH, Tiab D, Berumen-Campos S. Well pressure behavior of a finite-conductivity fracture intersecting a finite sealing-fault. Society of Petroleum Engineers. 2003. DOI: 10.2118/80547-MS

[9] Escobar FH, Tiab T, Jokhio SA. Characterization of leaky boundaries using transient pressure analysis. Society of Petroleum Engineers. 2003. DOI: 10.2118/80908-MS

[10] Escobar FH, Martinez JA, Montealegre-Madero M. Pressure transient analysis for a reservoir with a finite-conductivity fault. CT&F. 2013; **5**(2):5-18

[11] Escobar FH, Martinez JA, Montealegre-Madero M. Pressure and pressure derivative analysis in a reservoir with a finite-conductivity fault and contrast of mobilities. "Fuentes" Journal. 2013;**11**(2):17-25. ISSN: 1657-6527

[12] Gray KE. Approximating well-to-fault distance from pressure build-up tests. Journal of Petroleum Technology. Society of Petroleum Engineers. 1965: 761-767. DOI: 10.2118/968-PA

[13] Gibson JA, Campbell AT. Calculating the distance to a discontinuity from D.S.T. Data. Society of Petroleum Engineers. 1970. DOI: 10.2118/3016-MS

[14] Guira B, Tiab D, Escobar FH. Pressure behavior of a well in an anisotropic reservoir. Society of Petroleum Engineers. 2002. DOI: 10.2118/76772-MS

[15] Ispas V, Tiab D. New method of analyzing the pressure behavior of a well near multiple boundary systems. Society of Petroleum Engineers. 1999. DOI: 10.2118/53933-MS

[16] Kucuk FJ, Kabir S. Well test interpretation for reservoirs with a single linear no-flow barrier. Journal of Petroleum Science and Engineering. 1988 August;**1**(3):195-221

[17] Martinez-Romero N, Cinco-Ley H. Detection of linear impermeable barriers by transient pressure analysis.

Society of Petroleum Engineers. 1983. DOI: 10.2118/11833-MS

[18] Odeh AS. Flow test analysis for a well with radial discontinuity. Society of Petroleum Engineers. 1969:328-334. DOI: 10.2118/2157-PA

[19] Rahman NMA, Miller MD, Mattar L. Analytical solution to the transient-flow problems for a well located near a finite-conductivity fault in composite reservoirs. Society of Petroleum Engineers. 2003. DOI: 10.2118/84295-MS

[20] Sabet M. Well Testing. Houston, TX, USA: Gulf Publishing Co.; 1991.460 p

[21] Stanislav JF, Kabir CS. Pressure Transient Analysis. New Jersey: Prentice Hall; 1990. 320 p

[22] Strelsova TD. Well Testing in Heterogeneous Formations. New York: John Wiley and Sons; 1988. 377 p

[23] Tiab D, Crichlow HB. Pressure analysis of multiple-sealing-fault systems and bounded reservoirs by type-curve matching. Society of Petroleum Engineers. 1979:378-392. DOI: 10.2118/6755-PA

[24] Tiab D, Kumar A. Detection and location of two parallel sealing faults around a well. Society of Petroleum Engineers. 1980:1701-1708. DOI: 10.2118/6056-PA

[25] Tiab D. PE-5553: Well Test Analysis. Lecture Notes. The University of Oklahoma, Norman, OK; 1993

[26] Tiab D. Analysis of pressure and pressure derivative without type-curve matching: 1-skin and wellbore storage. Journal of Petroleum Science and Engineering. 1995;**12**:171-181

[27] Tiab D. Analysis of pressure and pressure derivatives without type-curve matching: I-skin and wellbore storage. Society of Petroleum Engineers. 1993. DOI: 10.2118/25426-MS

[28] Trocchio JT. Investigation of fateh Mishrif fluid-conductive faults. Society of Petroleum Engineers. 1990: 1038-1045. DOI: 10.2118/17992-PA

Chapter 5

Multiple Well Testing

The simplest form of interference testing involves two wells: a producer (or injector) and an observation well. The general idea is to produce in one well and observe the pressure drop in another. Multi-interference testing usually involves a producer (or injector) and several observation wells. This is helpful to find horizontal anisotropy as explained by Earlougher and Kazemi [1] by type-curve matching and [2] using *TDS* technique. To perform an interference test, all wells involved shut-in to stabilize their bottom pressures. Then, the pressure recording tools are lowered into the observation well, and the producer (or injector) is opened to production (injection). If there is interference, a pressure drop is recorded in the observation well(s) within a reasonable length of time. Most of the multiple tests are performed in closed reservoirs [1, 3]. Multiple tests are performed for a number of reasons:

- Search for reservoir connectivity and/or continuity of the reservoir [1, 4].
- Detecting directional permeability and other heterogeneities [1].
- Estimate reservoir volume [1].
- Orientation (azimuth) of hydraulic fractures [5, 6].

For a two-well system, the radius of investigation is given by [1]:

$$r_{inv} = 0.029\sqrt{\frac{kt}{\phi\mu c_t}} \tag{5.1}$$

The skin in the active well does not affect the pressure in the observation well. There are two types of tests: interference and pulse.

5.1. Interference testing

These are used to determine [1, 7]:

- Connectivity of the reservoir and transmissibility.
- Direction of flow patterns. This is done by selective opening of wells around the observation well.
- Storage capacity (storage factor) $= S_t = \phi\, c_t\, h$.
- Determination of the nature and magnitude of the anisotropy. The permeability of the reservoir is found in all directions and the direction, θ, of the anisotropy angle [1, 2].

5.1.1 Conventional analysis

5.1.1.1 Earlougher method

Two wells: One active (injector or producer) and the other one of observation preferably shut-in. The pressure in the observation well is [1]:

$$P_{ws} = P_{1hr} + m \log t \tag{5.2}$$

When $t = 1$h, $P_{ws} \approx P_{1hr} \approx P_i$ for new reservoirs. Eq. (5.2) is valid if $t_D/r_D{}^2 > 100$ ($x < 0.0025$), being r the distance between wells. The restriction of $t_D/r_D{}^2 > 100$ is applied with a 1% error [1].

When a plot of P_{ws} versus log t is built, one should obtain a straight line which slope and intercept gives the transmissibility, Eq. (2.33) and porosity, respectively,

$$S_t = \phi c_t \mu = \frac{T}{r^2} e^{\left[2.302\frac{P_i - P_{1hr}}{m} - 7.41316\right]} \tag{5.3}$$

Note that the skin factor does not appear in this equation since there is only fluid flow in the active well and not in the observation well. However, there are exceptions when the well is highly stimulated. Wellbore storage is also minimized in multiple tests but not entirely [1].

Two shut-in wells

The buildup equation is given by [1]:

$$P_{ws} = P_i + m \log \frac{t + \Delta t}{\Delta t} \tag{5.4}$$

t is the total production time (same as t_p in normal buildup) in the active well. Prepare a Horner graph and using the slope value find the transmissibility with Eq. (5.3). Find porosity from:

$$S_t = \frac{T}{r^2} e^{\left[2.302\frac{P_i - P_{wf}(\Delta t = 0)}{m} - \ln\left(1 + \frac{1}{t}\right) - 7.41316\right]} \tag{5.5}$$

5.1.1.2 Ramey method

It includes one active well (producer or injector) and the one of observation preferably shut-in. This method requires type-curve matching with **Figure 1.7**. Once $\Delta P_{ws} = P_i - P_{ws}$ versus at the observation well has been plotted and the best match is obtained:

$(P_D)_M$, $(t_D/r_D{}^2)_M$, ΔP_M, t_M

Use the following equations:

$$T = 162.6 q B \frac{P_{DM}}{\Delta P_M} \tag{5.6}$$

$$S_t = \frac{0.0002637T}{r^2} \frac{t_M}{(t_D/r_D^2)_M} \tag{5.7}$$

Limitations:

$r_D > 20$ (see **Figure 1.6**)

$t_D/r_D{}^2 > 50$ or 100

5.1.1.3 Tiab and Kumar method

As sketched in **Figure 5.1**, it uses some specific points:

P'_m = the maximum value of the pressure derivative in the observation well which is placed at a distance r from the active well [8]. The units are psia/h.

t_m = The time at which P'_m occurs, h

Procedure:

a. Obtain ΔP versus time in the observation well which is preferably shut-in.

b. Calculate $P' = \Delta(\Delta P)/\Delta t$ = change of ΔP/change in test time (later, it was known as the arithmetic derivative).

c. Graph P' versus t in log-log paper, see **Figure 5.2**.

d. Calculate S_t and transmissibility:

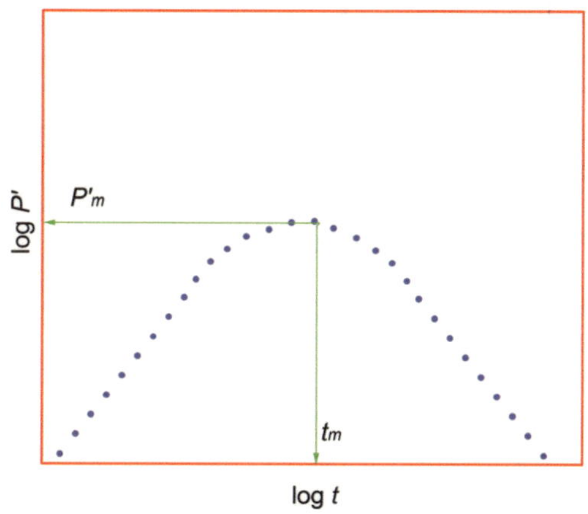

Figure 5.1.
Log-log plot of the arithmetic derivative, after [7, 8].

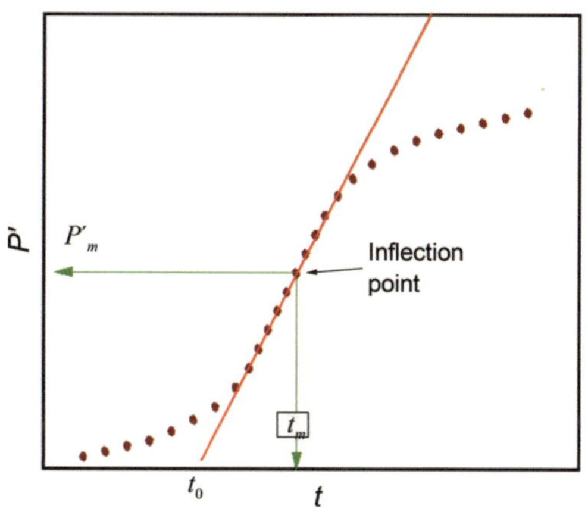

Figure 5.2.
Cartesian plot to find the inflection point, after [7, 8].

$$S_t = 0.0274\frac{qB}{r^2}\left(\frac{1}{P'_m}\right) \tag{5.8}$$

$$T = 948S_t r^2 \frac{1}{t_m} \tag{5.9}$$

From the Cartesian plot (verification purposes):

$$T = 382.2S_t r^2 \frac{1}{t_o} \tag{5.10}$$

It is very difficult to obtain the *P'* due to noisy pressure values. Then, it is recommended to use the Cartesian plot. Select the inflection point there. Extrapolate the line and read the value of t_0 as sketched in **Figure 5.2.**

5.1.2 *TDS* technique

When plotting dimensionless pressure and pressure derivative versus dimensionless time divided by the dimensionless radius squared, a single profile will always be obtained as shown in **Figure 5.3**. This gives two characteristic features: (1) the radial flow regime is similar to that of a wellbore test with a flat derivative with a value of 0.5, Eq. (2.70), which allows finding the permeability from Eq. (2.71) and (2) a unique intersection point between the pressure and the pressure derivative that takes place before the actual flow regime is seen. These features allow obtaining the following observations [9]:

$$(t_D/r_D^2)_{\text{int}} = 0.574952929 \tag{5.11}$$

where suffix *int* denotes intersection. The corresponding values of dimensionless pressure and the dimensionless derivative at this point of intersection are:

$$(P_D)_{t_D/r_D^2 = 0.57495} = 0.32369 \tag{5.12}$$

$$(t_D{}^*P_D{}')_{t_D/r_D^2 = 0.57495} = 0.32369 \tag{5.13}$$

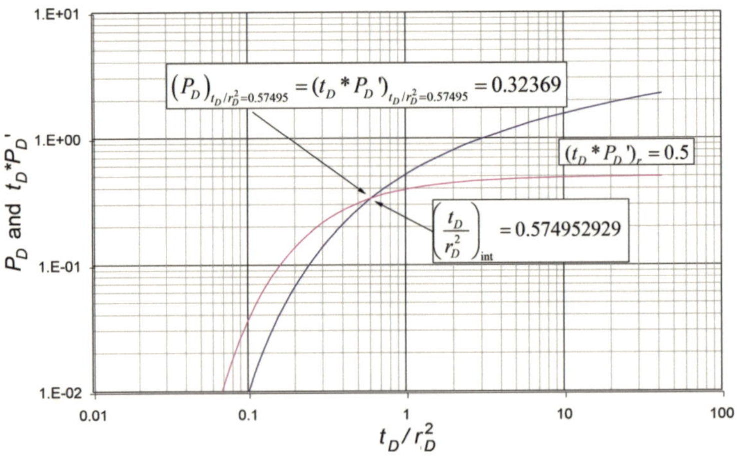

Figure 5.3.
Log-log plot of P_D *and* $t_D{}^*P_D{}'$ *versus* t_D/r_D^2 *for an infinitive reservoir (line source) [2, 9].*

Replacing Eqs. (1.89), (1.94), and (2.57) into the above expressions leads to obtain the following expressions:

$$T = \frac{kh}{\mu} = 45.705\frac{qB}{(\Delta P)_{\text{int}}} \tag{5.14}$$

$$T = \frac{kh}{\mu} = 45.705\frac{qB}{(t^*\Delta P')_{\text{int}}} \tag{5.15}$$

$$S_t = \phi c_t h = 0.000458646\frac{T}{r^2}t_{\text{int}} \tag{5.16}$$

And,

$$T = \frac{kh}{\mu} = \frac{70.6qB}{(t^*\Delta P')_r} \tag{2.76}$$

Ref. [2] extended the application of the *TDS* technique in interference testing to determine areal anisotropy.

Example 5.1

Taken from [7], during an interference test, 3125 STB (stock-tank barrel) of oil was produced by well A. The pressure response was observed at well B, 138 ft away from well A for 300 hr. Test data are reported in **Tables 5.1** and **5.2**. Then, well A was shut-in too, and the pressure response was observed at well B for 100 hr. Additionally, the following data are given:

μ = 1.3 cp,	B = 1.14 bbl/STB,	h = 31 ft
P_i = 2600 psia,	ρ = 55.4 lbm/ft^3,	s = −2.2 (well A)
$c_t = 16 \times 10^{-6}$/psia,	V_u = 0.00697 bbl/ft	

1. Calculate permeability and porosity using: (A) Earlougher's method when well A is active and shut-in, (B) the method of Tiab and Kumar, and (C) *TDS* technique

2. Show that the wellbore storage effects are not important at well A.

Solution

1. Calculate permeability and porosity using A) *The Earlougher Method: Well A is active*.

t, h	*P*, psia	Δ*P*, psia	t*Δ*P*, psia	*t*, h	*P*, psia	Δ*P*, psia	t*Δ*P*, psia
1.1	2595.6	4.4	5.15	10	2575.5	24.5	11.19
1.5	2593.5	6.5	6.29	15	2571	29	11.60
2	2591.4	8.6	8.25	25	2565	35	11.39
2.5	2590	10	8.68	35	2561	39	11.71
3	2587.5	12.5	9.05	60	2555	45	11.50
4	2585	15	9.76	100	2549	51	12.74
5	2583	17	9.46	150	2543.5	56.5	15.61
7.5	2579	21	10.53	300	2530	70	28.14

Table 5.1.
Pressure and pressure derivative data for well B (active).

t, h	P_{ws}, psia	$(t_1+\Delta t)/\Delta t$	t, h	P_{ws}, psia	$(t_1+\Delta t)/\Delta t$
1.0	2541.0	301.00	10.0	2559.0	31.00
2.0	2544.0	151.00	15.0	2563.5	21.00
3.5	2547.0	85.71	25.0	2569.0	13.00
5.0	2551.0	61.00	40.0	2574.0	7.50
7.0	2555.0	43.86	60.0	2577.0	5.00
			100.0	2580.0	4.00

Table 5.2.
Pressure response at well B (shut-in).

It is necessary to construct a graph in semilog of shut-in pressure against time (see **Figure 5.8**). In this graph, a straight line is drawn whose slope, $m = -25.517$ psia/cycle. Since 3125 STB of oil were recovered during 300 hours of production, then flow rate, q, is 250 BPD. The permeability was then calculated using Eq. (2.33):

$$k = -\frac{162.6q\mu B}{mh} = \frac{162.6(250)(1.3)(1.14)}{(25.518)(31)} = 76.15 \text{ md}$$

By linear regression analysis, we find that P_{1hr} = 2600.53 psia. Use Eq (5.2) to find porosity (**Figure 5.4**):

$$\phi = \frac{76.15}{(1.3)(138^2)(16 \times 10^{-6})}e^{\left(\frac{2.302(2600 - 2600.53)}{-25.518} - 7.4316\right)} = 11.94\%$$

Earlougher method: well A is shut-in

Figure 5.5 presents a semilog graph of P_{ws} versus $(t_1+\Delta t)/\Delta t$. From the straight line, we have: $m = -25.749$ psia/cycle and P_{1hr} = 2532.55 psia. Again, permeability is estimated with Eq. (2.33):

$$k = \frac{162.6(250)(1.3)(1.14)}{(26.749)(31)} = 72.65 \text{ md}$$

Find porosity with Eq. (5.5), thus:

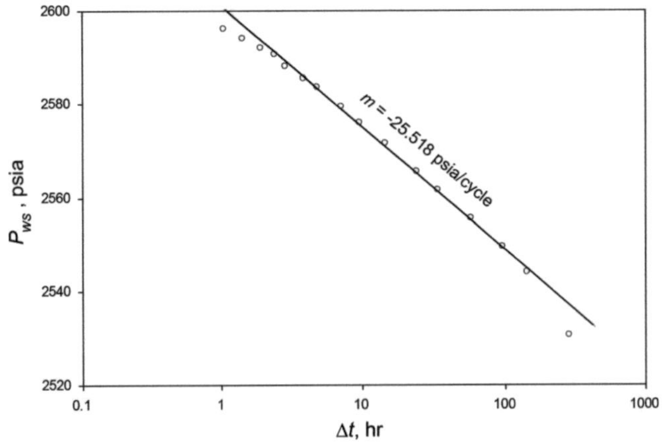

Figure 5.4.
Semilog plot of P_{wf} *versus* Δt, *after [7].*

$$\phi = \frac{72.65}{(1.3)(138^2)(16 \times 10^{-6})} e^{\left(\frac{2.303(2532.55 - 2530)}{-26.749} - \ln\left(1 + \frac{1}{300}\right) - 7.431\right)} = 8.7\%$$

Tiab-Kumar method

The derivative has a smooth of 0.5 cycles; then, it is smoothed possible derivative value. **Figure 5.6** shows $P'_m = 4.19$ psia and $t_m = 1.5$h. Eqs. (5.8) and Eqs. (5.9) are used to find porosity and permeability, respectively:

$$\phi = \frac{0.0274qB}{hr^2c_t}\left(\frac{1}{P'_m}\right) = \frac{0.0274(250)(1.14)}{(31)(138^2)(16 \times 10^{-6})}\left(\frac{1}{4.19}\right) = 19.7\%$$

$$k = 948\phi c_t \mu r^2\left(\frac{1}{t_m}\right) = 948(0.1574)(16 \times 10^{-6})(1.3)(138^2)\left(\frac{1}{1.5}\right) = 39.4 \text{ md}$$

TDS technique

The pressure derivative plot gives a better understanding of the reservoir model. A very clear radial flow regime is seen, and actually, it is possible to observe late pseudosteady-state period meaning that the reservoir boundaries have been felt.

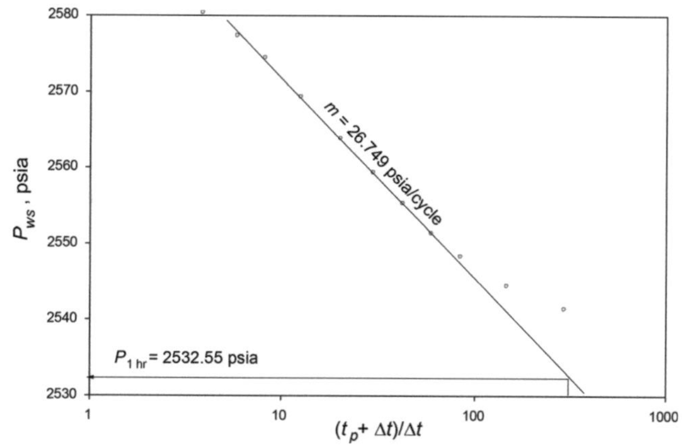

Figure 5.5.
Horner plot of P_{ws} *versus* $(t_1+\Delta t)/\Delta t$*, after [7].*

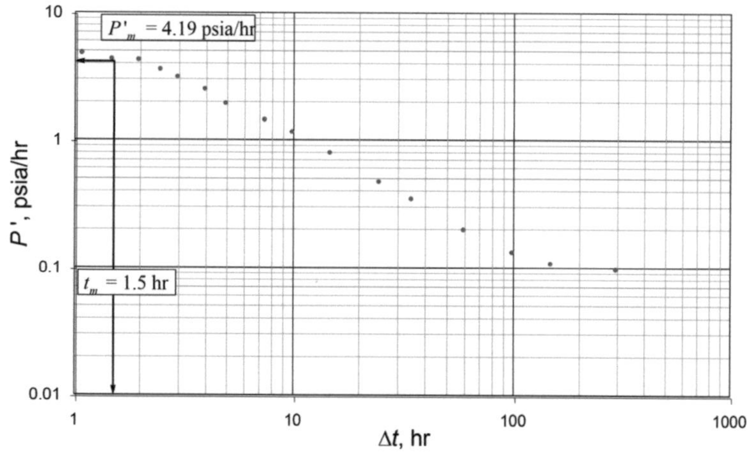

Figure 5.6.
Arithmetic pressure derivative versus time log-log plot for Example 5.1, after [7].

Since actual radial flow regime is observed, the permeability value found from there should be the most accurate one. The following information was read from **Figure 5.7**:

t_{int} = 1.7h, $(t^*\Delta P')_r$ = 11.5psia, $\Delta P_{int} = (t^*\Delta P')_{int}$ = 7.5psia

Find permeability from Eqs. (2.76) and (5.14/5.15) and porosity from Eq. (5.16):

$$k = \frac{70.6q\mu B}{h(t^*\Delta P')_r} = \frac{70.6(250)(1.2)(1.14)}{31(11.6)} = 73.37 \text{ md}$$

$$k = \frac{45.705q\mu B}{h(t^*\Delta P')_{\text{int}}} = \frac{45.705(250)(1.2)(1.14)}{(31)(7.5)} = 72.83 \text{ md}$$

$$\phi = \frac{0.000458646kt_{\text{int}}}{\mu c_t r^2} = \frac{0.000458646(72.83)(1.7)}{(1.3)(12 \times 10^{-6})(138^2)} = 19.1\%$$

2. Show that the wellbore storage effects are not important at well A

As seen in Section 3.2.1, if q_{af}/q < 0.01, it can be concluded that the afterflow or wellbore storage is not affecting the pressure data. To calculate q_{af}, find wellbore storage coefficient with Eq. (2.5) and then q_{af} with Eq. (3.8):

$$C = 144\frac{V_u}{\rho} = 144\frac{0.00697}{56.4} = 0.0178 \text{ bbl/psi}$$

$$q_{af} = \frac{24C}{B}\frac{dP_{ws}}{d\Delta t} = \frac{24(0.0178)}{1.14}4.682 = 1.755 \text{ BPD}$$

The remaining calculations are shown in **Table 5.3**. In this table, it can be seen that the condition q_{af}/q < 0.01 is always fulfilled, so the effects of wellbore storage are not important.

A summary of the results of porosity and permeability is given in **Table 5.4** for comparison purposes. Definitely the radial flow is observed and provided a permeability of 73.37 md from Eq. (2.71). This value closely matches with those from conventional analysis, Eq. (2.33). The intersection point, Eq. (5.14), provided an excellent permeability value which means that that point was properly selected; therefore, the porosity should be about 19% which is well-reported by *TDS* technique and Tiab-Kumar method, but far from conventional analysis. Actually, from the derivative plot, **Figure 5.7**, the intersection point does not coincide with any

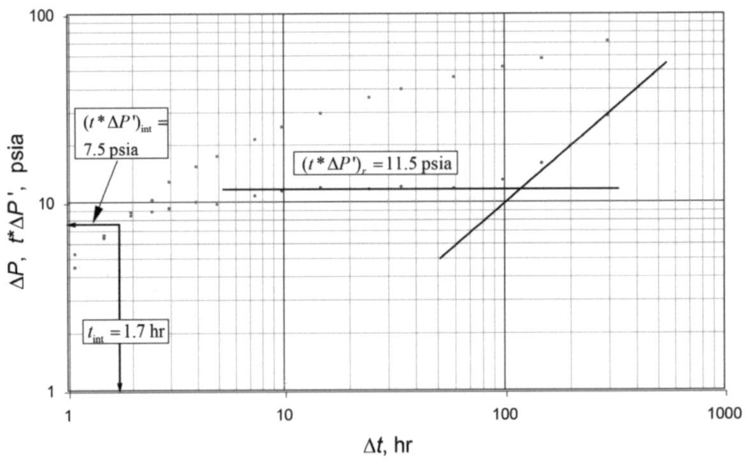

Figure 5.7.
Pressure and pressure versus time log-log plot for Example 5.1, after [7].

t, h	P', psia/h	q_{af}, BPD	q_{af}/q	t, h	P', psia/h	q_{af}, BPD	q_{af}/q
1.1	4.680	1.754	0.00702	10	1.119	0.419	0.00168
1.5	4.190	1.570	0.00628	15	0.773	0.290	0.00116
2	4.125	1.546	0.00618	25	0.456	0.171	0.00068
2.5	3.470	1.300	0.00520	35	0.335	0.125	0.00050
3	3.017	1.130	0.00452	60	0.192	0.072	0.00029
4	2.440	0.914	0.00366	100	0.127	0.048	0.00019
5	1.892	0.709	0.00284	150	0.104	0.039	0.00016
7.5	1.404	0.526	0.00210	300	0.094	0.035	0.00014

Table 5.3.
Arithmetic pressure derivative and afterflow data.

Method	Permeability, md	Porosity, %	Equation number
Earlougher—Active	76.15	11.94	(2.33) and (5.2)
Earlougher—shut-in	72.65	8.7	(2.33) and (5.5)
Tiab-Kumar	39.4	19.7	(5.9) and (5.8)
TDS	73.37		(2.71)
TDS	72.83	19.1	(5.14/5.15) and (5.16)

Table 5.4.
Comparison of results of Example 5.1.

datum in the test, but it was easily eyed interpolated which is not the case for the Tiab-Kumar method which cannot be either interpolated or extrapolated. This old test does not have enough points but recently pressure well tests data have thousands of data points which enabled the use of Tiab-Kumar method.

5.2. Pulse testing

This technique uses a series of short pulses of the flow rate. The pulses are alternating periods of production (or injection) and shut-in with the same flow rate in each production. The pressure response to the pulses is measured in the observation well. The main advantage of pulse testing is the short duration of the pulse. A pulse can last for a few hours or a few days, which disrupts normal operation slightly compared to interference tests [1]. Besides determining conductivity (then, transmissibility and porosity), pulse testing has several applications, that is [6] use them to find the azimuth of a hydraulic fracture ([5] does the same with interference testing) and [10] for estimating permeability distributions.

The nomenclature of a pulse test is given in **Figure 5.8**. The following variables are defined as [1]:

t_L (time lag), is the time between the end of the pulse and the pressure peak caused by the pulse.

$\Delta P/q$ (amplitude). The vertical distance between the tangent to two consecutive peaks and the line parallel to that tangent at the peak of the pulse to be measured, psia

Δt_c, pulse cycle. Time from start to end of a flow period, h.

Δt_p, pulse shut-in period, h.

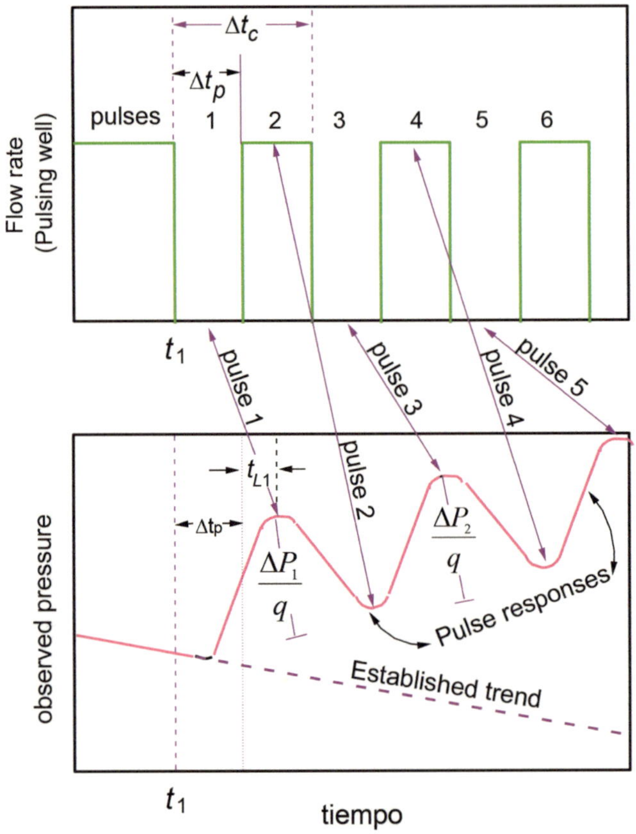

Figure 5.8.
Sketch of a pulse test, after [1].

The sign convention for ΔP is [1, 7]:

$\Delta P > 0$ if $q > 0$ (active producer well), $\Delta P/q > 0$

$\Delta P < 0$ if $q < 0$ (active injector well), $\Delta P/q > 0$

$\Delta P < 0$ for odd peaks

$\Delta P > 0$ for even peaks

5.2.1 Interpretation methods

5.2.1.1 Kamal-Birgham method

Although the methodology was presented by [11], the charts and some equations were corrected later by [12]. The procedure is outline below:

1. Plot $\Delta P/q$ versus t on Cartesian paper

2. From this plot obtain the values of t_L, Δt_c and Δt_p.

3. Calculate the relation $t_L/\Delta t_c$ and $F' = \Delta t_p/\Delta t_c$.

4. Find $[\Delta P_D(t_L/\Delta t_c)^2]$ from **Figures 5.9–5.12**, depending on the pulse, corresponding to F' and $t_L/\Delta t_c$ from step 3 and calculate transmissibility, T, from:

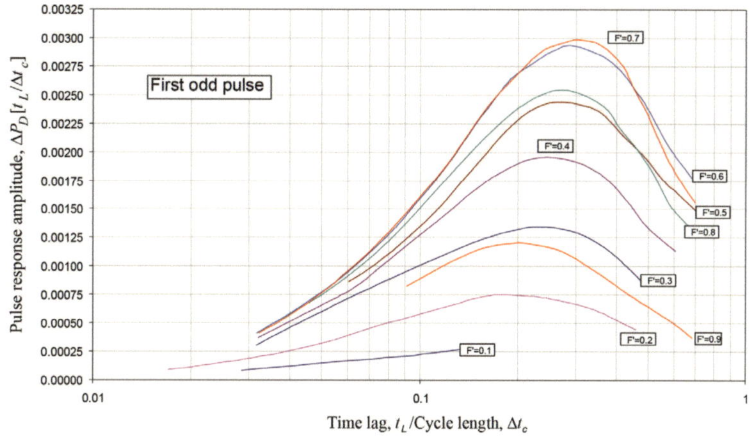

Figure 5.9.
Relationship between transition time and amplitude response for the first odd pulse, after [12].

$$T = \frac{141.2B}{(\Delta P/q)(t_L/\Delta t_c)^2}\left[\Delta P_D\ (t_L/\Delta t_c)^2\right] \quad (5.17)$$

Analyze all pulses since the first one may be affected by wellbore storage.

1. Determine $t_{LD}/r_D{}^2$, dimensionless time lag, from **Figures 5.13–5.16** corresponding to F' y $t_L/\Delta t_c$ obtained in step 3.

2. Calculate S_t:

$$S_t = 0.000263\left(\frac{T}{r^2}\right)\frac{t_L}{(t_{LD}/r_D^2)} \quad (5.18)$$

Wellbore storage effects in the observation well increase with lag time and tend to reduce the amplitude of the first pulses. However, if $r > 32(C/S_t)^{0.54}$ in the response well, storage effects are less than 5% of increase in the transition time and will not affect the amplitude. This is valid if [13]:

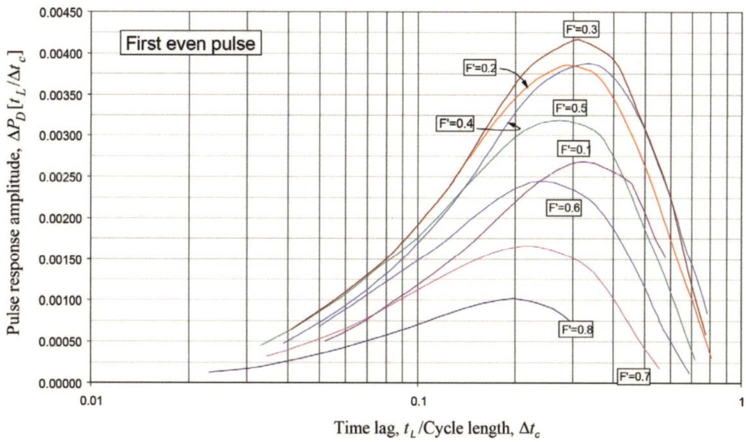

Figure 5.10.
Relationship between transition time and amplitude response for the first even pulse, after [12].

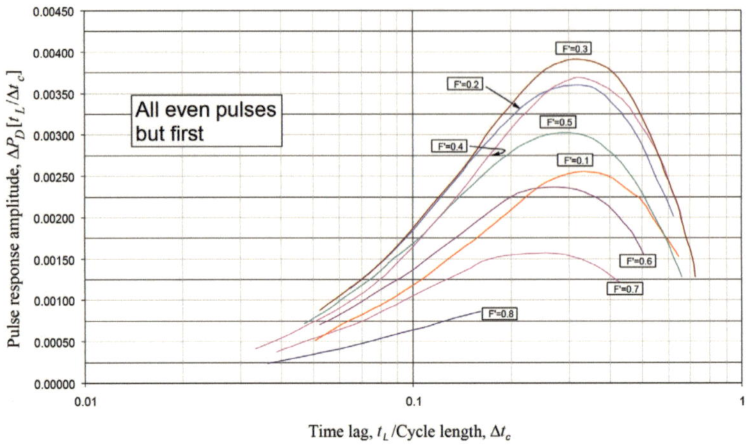

Figure 5.11.
Relationship between transition time and amplitude response for the all even pulses but first, after [12].

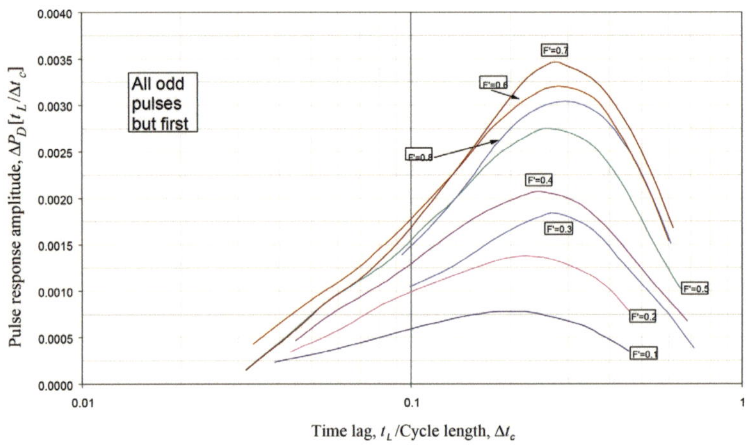

Figure 5.12.
Relationship between transition time and amplitude response for the all odd pulses but first, after [12].

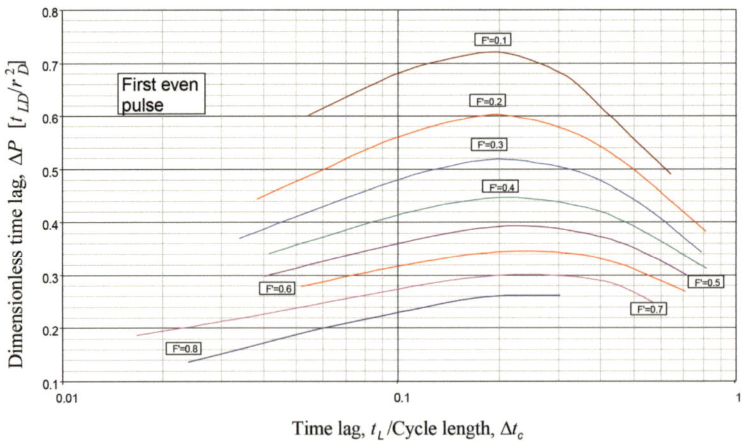

Figure 5.13.
Relationship between transition time and cycle length for the first even pulse, after [12].

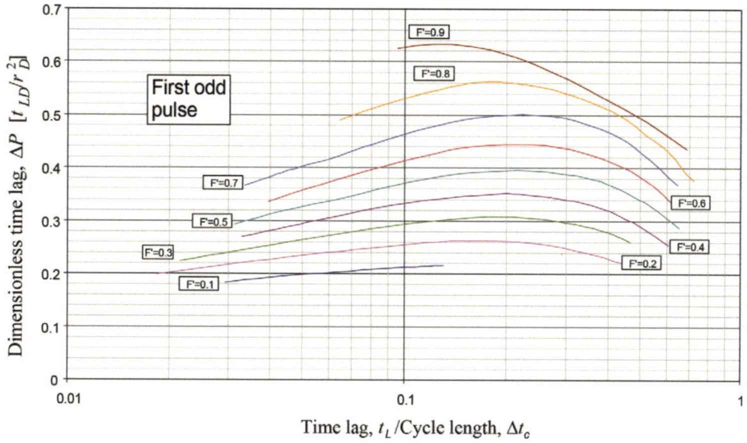

Figure 5.14.
Relationship between transition time and cycle length for the first odd pulse, after [12].

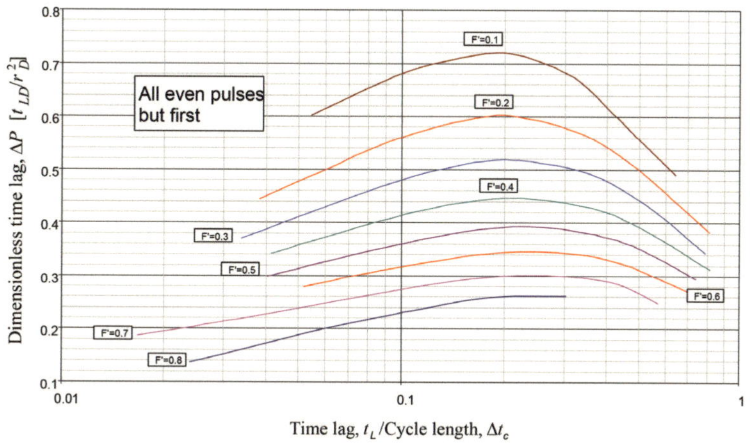

Figure 5.15.
Relationship between transition time and cycle length for all even pulses but first, after [12].

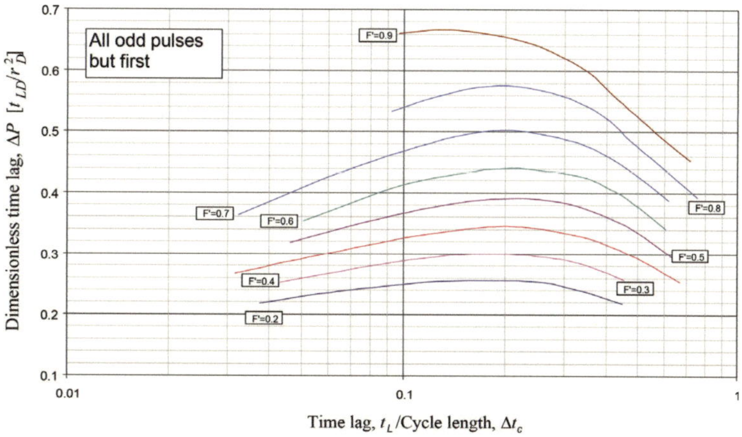

Figure 5.16.
Relationship between transition time and cycle length for all odd pulses but first, after [12].

$$\frac{t_D}{r_D^2} > (230 + 15s)\left(\frac{C_D}{r_D^2}\right)^{0.86} \tag{5.19}$$

Example 5.2

Taken from [7]. The pressure response data given in **Table 5.4** were obtained from a producer well during a multiple test. Additional data concerning this test are shown below (**Table 5.5**):

$\mu = 2.8$ cp, $B = 1.20$ bbl/STB, $h = 30$ ft
$c_t = 12 \times 10^{-6}$/psia, $C = 0.002$ bbl/psia at observation well
Shut-in period = 0.7 h, flow period = 1.63 h
Well distance = 140 ft, $q = 350$ STB/D

1. Calculate the formation permeability and porosity from the third pulse

2. Recalculate k and φ with the other two pulses and compare. Explain the difference.

Solution

The following information was read from **Figure 5.17**:
$\Delta P_1/q$ = 0.007041 psia/BPD, t_{L1} = 0.55 h
$\Delta t_{C1} = \Delta t_{C2} = \Delta t_{C3}$ = 2.33 h, $\Delta t_{p1} = \Delta t_{p2} = \Delta t_{p3}$ = 0.7 h
$\Delta P_2/q$ = 0.0066992 psia/BPD, t_{L2} = −0.0799 h
$\Delta P_3/q$ = 0.007455 psia/BPD, t_{L3} = 0.47 h

1. *Calculate the formation permeability and porosity from the third pulse*

First the ratios $t_L/\Delta t_C$ and F' = $\Delta t_p/\Delta t_C$ are estimated to be $t_{L3}/\Delta t_C$ = 0.47/2.33 = 0.2017 and F' = 0.7/2.33 = 0.3004. With these values, enter **Figure 5.11** and read $[\Delta P_D\,(t_L/\Delta t_C)^2]$ = 0.0033. Calculate permeability with Eq. (5.17),

$$k = \frac{141.2\mu B\left[\Delta P_D (t_L/\Delta t_c)^2\right]}{(\Delta P/q)(t_L/\Delta t_c)^2 h} = \frac{141.2(2.8)(1.2)(0.0033)}{(0.007455)(0.2017^2)(30)} = 172.07 \text{ md}$$

The dimensionless time lag divided by the dimensionless squared radius is found from **Figure 5.15** to be t_{LD}/r_D^2 = 0.52. Estimate porosity with Eq. (5.18):

t, h	ΔP, psia	$\Delta P/q$, psia/BPD	t, h	ΔP, psia	$\Delta P/q$, psia/BPD
0.25	0.175	0.0005	2.75	1.925	0.0055
0.50	0.560	0.0016	3.00	2.975	0.0085
0.75	1.400	0.0040	3.25	3.850	0.0110
1.00	2.625	0.0075	3.50	4.270	0.0122
1.25	3.150	0.0090	3.75	4.060	0.0116
1.50	2.940	0.0084	4.00	3.360	0.0096
1.75	1.890	0.0054	4.25	2.590	0.0074
2.00	1.400	0.0040	4.50	2.100	0.0060
2.25	1.260	0.0036	4.75	2.100	0.0060
2.50	1.505	0.0043	5.00	2.555	0.0073

Table 5.5.
Pulse test data.

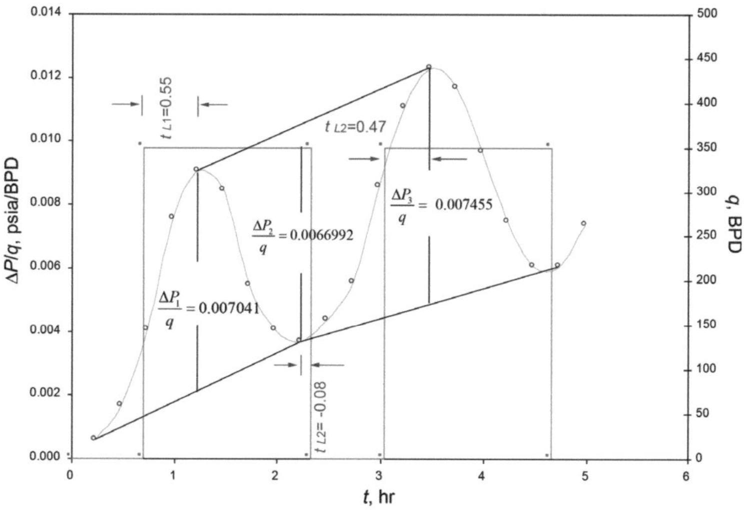

Figure 5.17.
Cartesian plot for Example 5.2.

$$\phi = \frac{0.0002637k}{\mu r^2 c_t}\frac{t_L}{(t_{LD}/r_D^2)} = \frac{0.0002637(172.07)}{(2.8)(140^2)(12\times10^{-6})}\frac{(0.47)}{(0.52)} = 6.22\ \%$$

2. Recalculate k and φ with the other two pulses and compare. Explain the difference.

For pulse 1, $t_L/\Delta t_C$ = 0.55/2.33 = 0.236 and F' = 0.7/2.33 = 0.3004. From **Figure 5.10**, $[\Delta P_D\ (t_L/\Delta t_C)^2]$ = 0.0037. Calculate permeability with Eq. (5.17),

$$k = \frac{141.2\mu B\left[\Delta P_D(t_L/\Delta t_c)^2\right]}{(\Delta P/q)(t_L/\Delta t_c)^2 h} = \frac{141.2(2.8)(1.2)(0.0037)}{(0.007041)(0.236^2)(30)} = 149.21\ \text{md}$$

From **Figure 5.14**, t_{LD}/r_D^2 = 0.25. Porosity is then estimated with Eq. (5.18),

$$\phi = \frac{0.0002637(55.98)}{(2.8)(140^2)(12\times10^{-6})}\frac{(0.55)}{(0.26)} = 12.64\%$$

For Pulse 2: Since the transition time, t_{lag}, is negative (see **Figure 5.17**), which implies that the pressure is beginning to increase after the well is shut-in, as shown in the graph. This behavior is not physically logical and may be caused by some error that occurred during the test.

At the first pulse, the permeability was reduced by 87%, and the porosity was increased by 203%. This was due to an increase in t_L and reduction in the value of pulse amplitude. This can be caused by wellbore storage.

5.2.1.2 TDS technique

Figure 5.18 was built for different distance between wells and different ratios of production-shut-in periods. Comparing to **Figure 5.3**, the same intersection point is given. Also, the radial flow displays the same behavior plus some times this behavior is repeated among the pulses. Based on the above, it is concluded that Eqs. (5.14)—(5.16) and (2.71) also work for pulse testing.

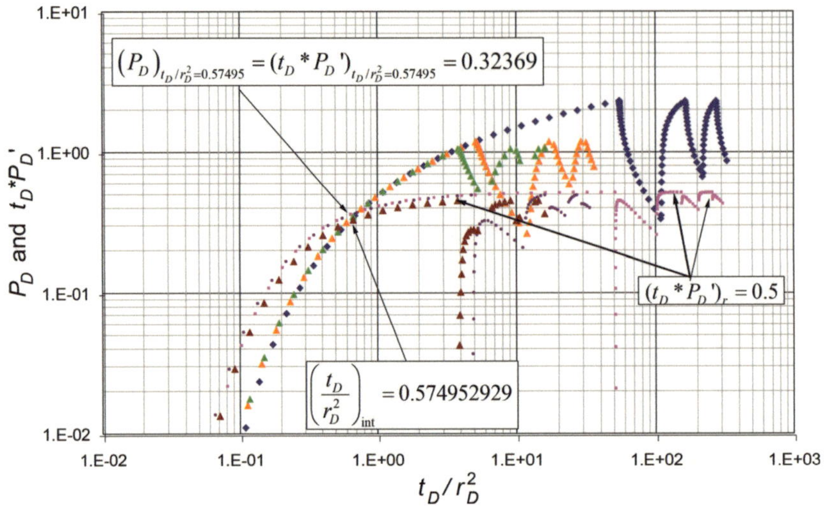

Figure 5.18.
Dimensionless pressure and pressure derivative against dimensionless both time and radius squared, finite-source solution.

Example 5.3

A synthetic pulse test is presented in **Table 5.6** and **Figure 5.19**. The below data were used to generate the test.

μ = 2 cp, B = 1.2 bbl/STB, h = 100 ft

P_i = 3200 psia, r = 100 lbm/ft^3, q = 350 BPD

c_t= 1.2 × 10^{-5}/psia, r_w= 0.3 ft, φ = 10%

k = 400 md

Find permeability and porosity for this example using *TDS* technique.

t, h	Δ*P*, psia	t*Δ*P*, psia	*t*, h	Δ*P*, psia	t*Δ*P*, psia	*t*, h	Δ*P*, psia	t*Δ*P*, psia
	0.001	0.011	6.000	6.065	1.468	10.685	1.315	1.095
0.014	0.004	0.024	6.015	6.062	0.030	10.720	1.509	1.149
0.016	0.009	0.046	6.017	6.056	0.055	10.762	1.713	1.197
0.019	0.020	0.079	6.021	6.044	0.091	10.812	1.924	1.238
0.023	0.038	0.125	6.024	6.024	0.139	10.871	2.142	1.274
0.027	0.065	0.183	6.029	5.995	0.199	10.941	2.366	1.304
0.032	0.103	0.254	6.034	5.955	0.270	11.025	2.596	1.330
0.039	0.156	0.336	6.040	5.902	0.350	11.125	2.829	1.353
0.046	0.223	0.426	6.048	5.835	0.436	11.244	3.066	1.372
0.055	0.305	0.520	6.057	5.754	0.526	11.385	3.307	1.388
0.065	0.405	0.616	6.067	5.657	0.616	11.553	3.550	1.402
0.077	0.520	0.710	6.079	5.546	0.704	11.753	3.796	1.414
0.092	0.651	0.801	6.094	5.421	0.788	11.992	4.043	1.425
0.110	0.798	0.885	6.111	5.283	0.865	12.275	4.292	1.434
0.131	0.958	0.963	6.131	5.132	0.934	12.612	4.543	1.441
0.155	1.131	1.033	6.156	4.970	0.996	13.014	4.794	1.448
0.185	1.315	1.095	6.184	4.799	1.050	13.492	5.047	1.453
0.220	1.509	1.149	6.218	4.619	1.095	14.060	5.300	1.458

t, h	ΔP, psia	$t^*\Delta P$, psia	t, h	ΔP, psia	$t^*\Delta P$, psia	t, h	ΔP, psia	$t^*\Delta P$, psia
0.262	1.713	1.197	6.258	4.433	1.133	14.737	5.555	1.462
0.312	1.924	1.238	6.305	4.240	1.162	15.542	5.809	1.465
0.371	2.142	1.274	6.361	4.043	1.185	16.500	6.065	1.468
0.441	2.366	1.304	6.427	3.842	1.201	16.758	5.885	1.170
0.525	2.596	1.330	6.505	3.640	1.211	16.805	5.685	1.207
0.625	2.829	1.353	6.598	3.435	1.214	16.861	5.480	1.237
0.744	3.066	1.372	6.707	3.231	1.212	16.927	5.270	1.261
0.885	3.307	1.388	6.837	3.028	1.203	17.005	5.056	1.281
1.053	3.550	1.402	6.990	2.826	1.189	17.098	4.839	1.296
1.253	3.796	1.414	7.171	2.627	1.170	17.207	4.620	1.306
1.492	4.043	1.425	7.386	2.432	1.145	17.337	4.400	1.312
1.775	4.292	1.434	7.640	2.241	1.114	17.490	4.178	1.314
2.112	4.543	1.441	7.940	2.057	1.078	17.671	3.957	1.312
2.514	4.794	1.448	8.296	1.878	1.037	17.886	3.736	1.306
2.992	5.047	1.453	8.717	1.707	0.991	18.140	3.517	1.296
3.560	5.300	1.458	9.214	1.544	0.941	18.796	3.085	1.264
4.237	5.555	1.462	9.803	1.390	0.888	19.714	2.667	1.214
5.042	5.809	1.465	10.500	1.246	0.831	21.000	2.269	1.146

Table 5.6.
Pressure and pressure derivative versus time data for Example 5.3.

Solution

The following characteristic features were read from **Figure 5.19**:
t_{int} = 0.131 h, $(t^*\Delta P')_r$ = 1.5 psia, $\Delta P_{int} = (t^*\Delta P')_{int}$ = 0.97 psia
Find permeability from Eq. (2.71) and (5.14/5.15) and porosity from Eq. (5.16):

$$k = \frac{70.6 q\mu B}{h(t^*\Delta P')_r} = \frac{70.6(350)(2)(1.2)}{100(1.5)} = 395.4 \text{ md}$$

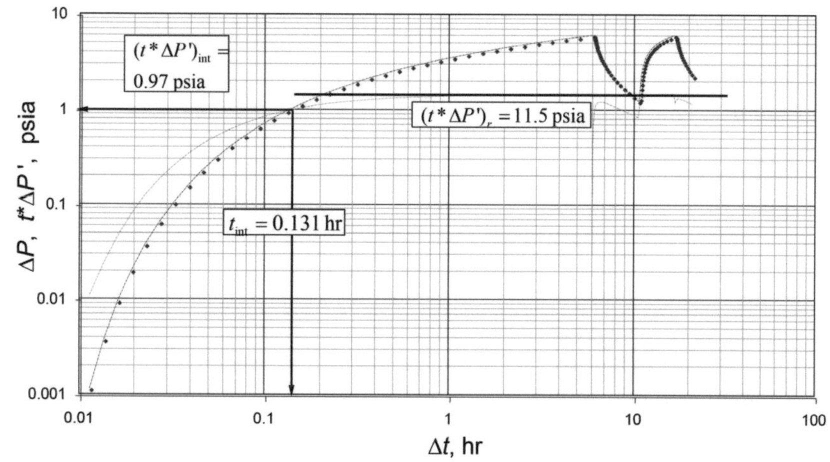

Figure 5.19.
Pressure and pressure derivative versus time for simulated pulse test of Example 5.3.

$$k = \frac{45.705 q \mu B}{h(t^*\Delta P')_{\text{int}}} = \frac{45.705(350)(2)(1.2)}{100(0.97)} = 395.8 \text{ md}$$

$$\phi = \frac{0.000458646 k t_{\text{int}}}{\mu c_t r^2} = \frac{0.000458646(395.4)(0.131)}{(2)(1.2 \times 10^{-5})(100^2)} = 9.9\%$$

The results match quite well with the given porosity and permeability values.

Nomenclature

B	oil volume factor, bbl/STB
c	compressibility, 1/psia
C	wellbore storage coefficient, bbl/psia
d	distance from well to linear boundary, ft
c_t	total or system compressibility, 1/psia
F_{CD}	fault dimensionless conductivity
FE	flow index
h	formation thickness, ft
k	permeability, md
$k_f w_f$	fault conductivity, md-ft
Np	oil produced since last stabilization, bbl
m	slope of P-vs-log t plot, psia/h/cycle
m_1	slope of first semilog straight line, psia/h/cycle
P	pressure, psia
P_D'	dimensionless pressure derivative
P_D	dimensionless pressure
P_i	initial reservoir pressure, psia
P'_m	maximum arithmetic pressure derivative, psia/h
P'	arithmetic pressure derivative, psia/h
P_{wf}	well flowing pressure, psia
P_{ws}	well shut-in or static pressure, psia
P_{1hr}	intercept of the semilog plot, psia
P^*	false pressure, psia
ΔP_s	pressure drop due to skin conditions, psia
q	liquid flow rate, BPD
r	radius, ft
r_w	well radius, ft
s	skin factor
St	reservoir storativity, ft/psia
t	time, h
t_L	lag time, h
t_m	time at which P'_m occurs
t_0	extrapolated time for the inflection point, h
$t^*\Delta P'$	pressure derivative, psia
$t^{2*}\Delta P''$	second pressure derivative, psia
t_p	production (Horner) time before shutting-in a well, h
t_D	dimensionless time based on well radius
t_{dv}	time at which either pressure or derivative deviate from first radial line, hr
T	reservoir transmissibility, md-ft/cp

Greek

$\Delta P/q$	pulse amplitude
Δt	shut-in time, h
Δt_c	pulse cycle, h
Δt_p	pulse shut-in period
Δ	drop, change
φ	porosity, fraction
ρ	fluid density, lbm/ft^3
τ	scalable dimensionless conductivity
μ	viscosity, cp

Suffices

$1hr$	time of 1 h
$2m$	second semilog straight line
$2x$	maximum in the second pressure derivative
BL	bilinear flow regime
$BLssi$	intercept of bilinear and steady-state lines
d	distance
D	dimensionless
DA	dimensionless with respect to area
Df	dimensionless with respect to fault length
f	fault
ebf	end of bilinear flow
F	Inflection. Better found from second derivative
i	intersection or initial conditions
int	intersection between pressure and pressure derivative before radial flow
L	linear flow regime
M	matching point
p	production, porous
r	radial flow
re	end of radial flow regime
$r2$	second plateau, hemi radial flow
$rssi$	intercept of radial and steady-state lines
s	skin
ss	a point on the steady-state period
$s2r$	start of second semilog straight line
w	well
wf	well flowing
ws	well shut-in

References

[1] Earlougher RC, Kazemi H. Practicalities of detecting faults from buildup testing. Society of Petroleum Engineers. 1980. DOI: 10.2118/8781-PA

[2] Escobar FH, Cubillos J, Montealegre-MM. Estimation of horizontal reservoir anisotropy without type-curve matching. Journal of Petroleum Science and Engineering. 2008;**60**:31-38

[3] Sabet M. Well Testing. Houston, TX, USA: Gulf Publishing Co.; 1991. 460 p

[4] Ramey HJ. Interference analysis for anisotropic formations—A case history (includes associated paper 6406). Society of Petroleum Engineers. 1975: 1290-1298. DOI: 10.2118/5319-PA

[5] Cherifi M, Tiab D, Escobar FH. Determination of fracture orientation by multi-well interference testing. Society of Petroleum Engineers. 2002. DOI: 10.2118/77949-MS

[6] Tiab D, Abobise EO. Determining fracture orientation from pulse testing. Society of Petroleum Engineers. 1989. DOI: 10.2118/11027-PA

[7] Tiab D. PE-5553: Well Test Analysis. Lecture notes. The University of Oklahoma, Norman, OK; Fall 1993

[8] Tiab D, Kumar A. Application of the P_D' function to interference analysis. Society of Petroleum Engineers. 1980. DOI: 10.2118/6053-PA

[9] Ouandlous A. Interpretation of interference tests by Tiab's direct synthesis technique [MSc thesis]. University of Oklahoma; 1999

[10] Ahn S, Horne RN. Estimating permeability distributions from pressure pulse testing. Society of Petroleum Engineers. 2010. DOI: 10.2118/134391-MS

[11] Kamal M, Brigham WE. Pulse-testing response for unequal pulse and shut-in periods (includes associated papers 14253, 19365, 20792, 21608, 23476 and 23840). Society of Petroleum Engineers. 1975. DOI: 10.2118/5053-PA

[12] Al-Khalifah AA, Al-Hashim HS, Menouar HK. Revised Pulse Testing Correlation Charts. SPE paper 14253, presented at the 60th Technical Conference and Exhibition of the Society of Petroleum Engineers held in Las Vegas; September 1985

[13] Chu WC, Garcia-Rivera J, Rajagopoal R. Analysis of interference test data influenced by wellbore storage and skin at the flowing well. Society of Petroleum Engineers. Transactions of AIME 237. 1980:171-78. DOI: 10.2118/8029-PA

Chapter 6

Naturally Fractured Reservoirs

In these reservoirs, two different types of porosity are observed. The matrix has less permeability, and its porosity is small compared to that of the fractures, which also has high permeability. However, there are cases where the matrix has zero porosity and permeability, so the flow only occurs from the fractures. This type of behavior occurs in reservoirs with igneous or metamorphic rocks [4].

Naturally, fractured deposits have fractures with permeability, k_f, and porosity, ϕ_f, and a matrix with permeability, k_m, and porosity, ϕ_m. Some reservoirs act as if they were naturally fractured, but they are not. This is the case of dissolved channels, interlayered systems with different permeability (interlayer dolomitic with limestones which have less density or interstratified sandstones with other limolites and fine-grained sandstones). However, naturally fractured models can be applied to these types of reservoirs [29].

In this class of naturally fractured deposits, the two different types of porosity are found as shown on the left side of **Figure 6.1**. A very low porosity is presented in the fine pores, and another high porosity is represented by fissures, cavities, and fractures [46].

When the classification of the naturally fractured reservoirs from the point of view of the flow (engineering) is carried out, the permeability and the porosity of the fracture must be taken into account and a comparison with the permeability and porosity of the matrix must be made. According to the above, Nelson [33] talks about the four types of fracture systems. *Type I* consists of those fractures that provide the storage capacity and permeability of the reservoir. *Type II* is that group of fractures that has a better permeability than that of the matrix. *Type III* is composed of those fractures in which the permeability is negligible, but the storage capacity of hydrocarbons is high. Finally, *Type IV* corresponds to those in which the fractures are filled with minerals, and it is generally not very feasible for the flow to develop [33].

Because of the above, this type of reservoirs is normally known as double porosity reservoirs. Their matrix permeability is negligible compared to that of the bulk fractured systems. Then, it is expected that the well to be fed only by the fractures as sketched in **Figure 6.2.a**. This chapter will devote on this type of systems. There is another kind in which the matrix porosity is not negligible and once depletion caused by fluid withdrawal takes place inside the fracture system, some fluid to the well once comes from the matrix as schematically shown in **Figure 6.2.b**. They are called double-porosity double-permeability reservoirs and *TDS* technique for this type of deposits is provided in [19].

As this point, the reader ought to be aware of one important issue. Most reservoirs, not all of them, are heterogeneous since porous media have chaotic and random distribution. However, as seen in **Figure 6.2**, the fluid comes from one or two media: either matrix or fractures. From the well testing point of view, when a unique system acts, then, the reservoir is recognized as a homogeneous, even though it is really heterogeneous. When two, as depicted in **Figure 6.2**, or more systems act, then the reservoir is meant to be heterogeneous.

Based upon the above, naturally fractured reservoirs are heterogeneous. The idea of a homogeneous channel occurs outside of reality. However, the rock is fractured homogeneously, the percolation of the water causes mineral deposition,

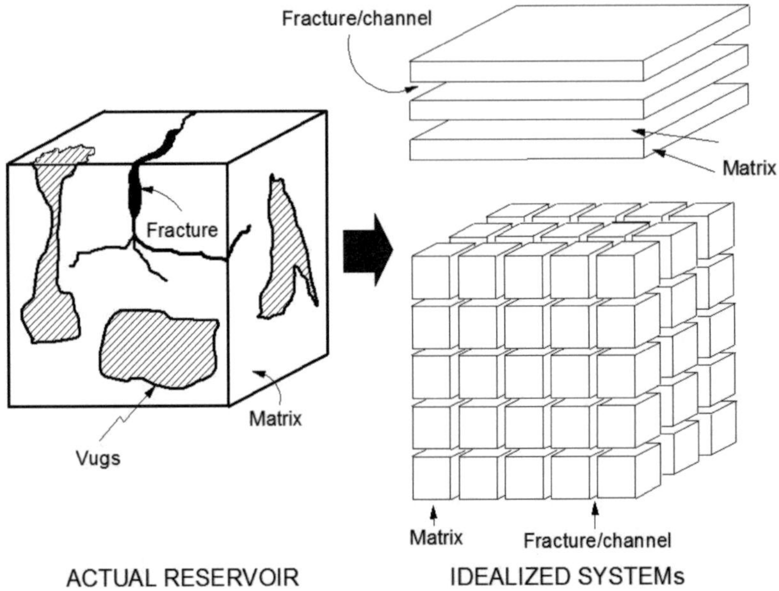

Figure 6.1.
Illustration of a naturally fractured deposit and its ideal representation [46].

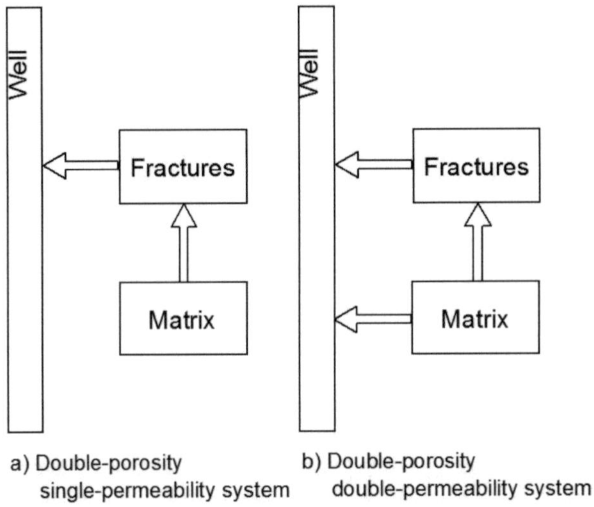

Figure 6.2.
Schematic representation of (a) double-porosity and (b) double-permeability systems, after [27].

which reduces the permeability or completely blocks the channels of the fluid. Therefore, the fractures of homogeneous character change over time, and a heterogeneous rock is obtained. The porosity of the fracture is rarely greater than 1.5 or 2%. Usually, this is less than 1%. The storage capacity of the fracture, $S_f = \phi_f c_f h_f$, is very small, because ϕ_f is small and h_f is extremely low. In contrast, k_f is very high. The storage capacity of the matrix, $S_m = \phi_m c_m h_m$, is greater than the storage capacity of the fractures. Normally, the permeability of the matrix is less than the permeability of the fractures. If these have the same value, the system behaves as homogeneous and without fractures. If the permeability of the matrix is zero and the fractures are randomly distributed, the system has a homogeneous behavior. However, if the permeability of the matrix is zero, but the fractures have a preferential direction, then there is a linear flow. In addition, if the permeability of the matrix is small (usually less than 0.01 md) and the reservoir is widely fractured, the system behaves as homogeneous and without fractures. From the well testing point of view, three conditions must be met to determine if it is actually a naturally fractured deposit [34]:

a. the porosity of the matrix is greater than the porosity of the fractures;

b. the permeability of the matrix is not zero, but its permeability is much smaller than the permeability of the fractures; and

c. the well intercepts the fractures.

6.1. Conventional analysis for characterization of naturally fractured reservoirs

Mavor and Cinco-Ley [31] defined two parameters to characterize naturally fractured formations: the dimensionless storativity ratio, ω, and the interporosity flow parameter or flow capacity ratio, λ. As defined by Eq. (2.1), ω gives what fraction of the total porosity is provided by the fractures, and λ, Eq. (2.2), describes the matrix flow capacity available to the fractures.

$$\omega = \frac{(\phi c_t)_f}{(\phi c_t)_f + (\phi c_t)_m} \tag{6.1}$$

$$\lambda = \frac{4n(n+2)k_m r_w^2}{k_f h_m^2} \tag{6.2}$$

n in Eq. (6.2) depends on the model, n = 1 for strata mode, n = 2 for matchsticks model, and n= 3 for cubic model.

Odeh [34] examined several theoretical models and concluded that fractured deposits (especially with secondary porosity) generally behave as homogenous reservoirs. According to Warren and Root [46], a closure versus log pressure graph $(t_p+\Delta t)/\Delta t$ will yield two portions of parallel straight lines as shown in **Figure 6.3**. The first straight line portion, if seen, can be used to calculate the total product kh by the conventional Horner method. Note that $P_{1\mathrm{hr}}$ is taken from the second straight line. The average reservoir pressure is estimated by extrapolating, also, the second line to $(t_p+\Delta t)/\Delta t$ = 1 to obtain P^* and, then, using conventional techniques. Also, *TDS* technique is recommended [28, 32], as studied in Chapter 3. The vertical distance between the two semilog straight lines, see **Figure 6.3**, identified as ∂P can be used to estimate dimensionless storativity ratio [46]:

$$\omega = e^{-2.303\frac{\partial P}{m}} = 10^{-\frac{\partial P}{m}} \tag{6.3}$$

From the above equation, if $\partial P < 100$, the storage capacity parameter, ω, may be in error. The Warren-and-Root parameter can also be estimated by reading the intersection points among the lines:

$$\omega = t_1/t_2 \tag{6.4}$$

$$\lambda = \frac{(\phi c_t)_f \mu r_w^2}{1.781 k t_1} = \frac{(\phi c_t)_{f+m} \mu r_w^2}{1.781 k t_2} \tag{6.5}$$

when ω approaches zero and $\lambda \le 1 \times 10^{-9}$, all permeability comes from the fractures. **Figure 6.3** can be used to further understand the flow mechanics in naturally fractures formations. At the beginning of the flow, neglecting wellbore storage effects, all the flow comes from the fracture to the well in a radial flow manner; therefore, a flat line will be displayed in the pressure derivative curve. This is labeled as number 1 in **Figure 6.3**. Since the fracture porosity and height are small, it is expected that the fluid depletes inside the fracture system and pressure decline inside the fracture forces the fluid to come from the matrix to the fracture. That transition period is labeled as number 2 and is reflected as a "*v*" shape in the pressure derivative curve. At this point, it is good to know that flow from matrix to fractures can flow under pseudosteady-state (mentioned "*v*") of transient conditions. Once the fractures are filled with fluid, the radial flow regime (horizontal line on the pressure derivative curve) develops and the system behaves as homogeneous.

It is customary to assume that $(c_t)_m = (c_t)_f$ and that is what is going to be treated in this book, just for academic purposes. However, Tiab et al. [43] demonstrated that the fracture compressibility is at least one order of magnitude higher than the matrix compressibility. Neglecting this reality can lead to a huge overestimation of fracture porosity. Actually, the determination of fracture compressibility is a laboratory challenge. The recommended way is using transient pressure analysis as demonstrated by Tiab et al. [43].

The interporosity flow parameter, λ, is a function of the ratio between the matrix permeability and the permeability of the fractures, the shape factor and wellbore radius [38, 46].

$$\lambda = \alpha\left(\frac{k_m}{k_f}\right)r_w^2 \tag{6.6}$$

The α factor is the block shape parameter that depends on the geometry and the shape characteristics of the matrix-fissures system, and is defined by:

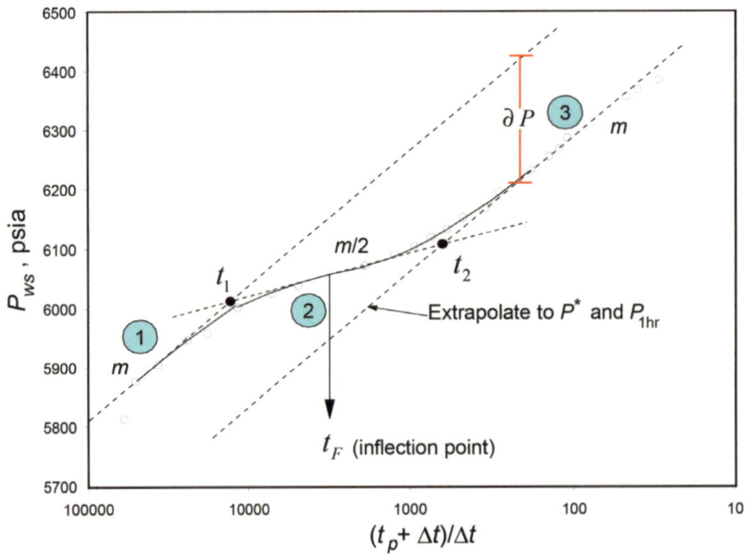

Figure 6.3.
Horner plot for a naturally fractured reservoir [31].

$$\alpha = \frac{A}{V_x} \tag{6.7}$$

where A = surface area of matrix block, ft^2; V = matrix block volume, ft^3; and x = characteristic length of the matrix block.

For the case of cubic block matrix separated by fractures, λ is given by:

$$\lambda = \frac{60}{l_m^2}\left(\frac{k_m}{k_f}\right)r_w^2 \tag{6.8}$$

Being l_m the length of the side of the block. For the case of spherical block matrix separated by fractures, λ is given by:

$$\lambda = \frac{15}{r_m^2}\left(\frac{k_m}{k_f}\right)r_w^2 \tag{6.9}$$

where r_m = radius of the sphere and finally, when the matrix is of blocks of horizontal strata (rectangular slab) separated by fractures, λ is given by (**Figure 6.4**):

$$\lambda = \frac{12}{h_f^2}\left(\frac{k_m}{k_f}\right)r_w^2 \tag{6.10}$$

h_f = thickness of a particular fracture or a high permeability layer.

Another method for estimating the interporosity parameter, λ, was proposed by Uldrich and Ershaghi [45]. This method used the inflection point time described in **Figure 6.3**. However, it is considered here not of practical use since it requires estimation of the $Ei(-x)$ function and chart-information reading.

Eqs. (6.3)–(6.5) has a strong drawback. Since the fractures promote increasing wellbore storage than the expected for a homogeneous system; then, the first or early semilog line is usually masked by wellbore storage. Therefore, conventional analysis cannot be used. To overcome this issue, Tiab and Bettam [42] provided an equation, Eq. (6.11), to find the interporosity flow parameter form the inflection point as Uldrich and Ershaghi [45] did. This equation is applicable to both

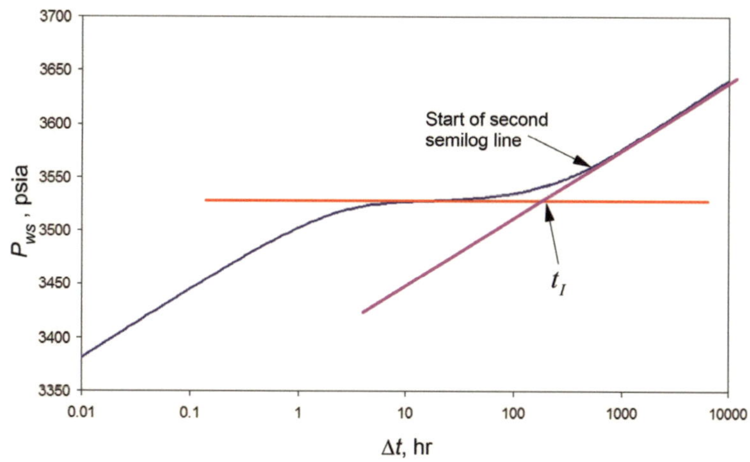

Figure 6.4.
Definition of the intersection point.

drawdown and buildup tests. Once λ is known, t_1 of Eqs. (6.4) and (6.5) can be known. Actually, Tiab et al. [43] demonstrated that wellbore storage affects more the estimation of ω than λ. In other words, the though occurred during the transition is more affected in the pressure scale than in the time scale; therefore, the estimation of the interporosity flow parameter may be acceptable.

$$\lambda = \frac{3792(\phi\, c_t)_{f+m}\mu\, r_w^2}{k_{fb}\Delta t_F}\left[\omega \ln\left(\frac{1}{\omega}\right)\right] \tag{6.11}$$

Another good approximation for finding the interporosity flow parameter is presented by Stewart [38] with the aid of a MDH plot (although a Horner plot can also be taken) as the one given in **Figure 6.5**. A horizontal line passing throughout the transition period is drawn. The intersection of this line with the second semilog line provides t_I which is used in the following expression:

$$\lambda = \frac{1.732(\phi\, c_t)_{m+f}\mu\, r_w^2}{k t_I} \tag{6.12}$$

The beginning of the second semilog straight line, t_{b2}, actual total system behavior response, can also be used. Under this condition, Eq. (6.11) becomes:

$$\lambda = \frac{4(\phi\, c_t)_{m+f}\mu\, r_w^2}{k t_{b2}} \tag{6.13}$$

Bulk-fracture permeability is found from an expression similar to Eq. (2.33);

$$k_{fb} = \frac{162.6 q\mu B}{mh} \tag{6.14}$$

The skin factor can be determined from the first and second semilog line (recommended), respectively:

$$s = 1.1513\left[\frac{P_{1hr} - P_i}{m} - \log_t\left(\frac{k}{\phi_f c_f \mu r_w^2}\frac{1}{\omega}\right) - 3.23\right] \tag{6.15}$$

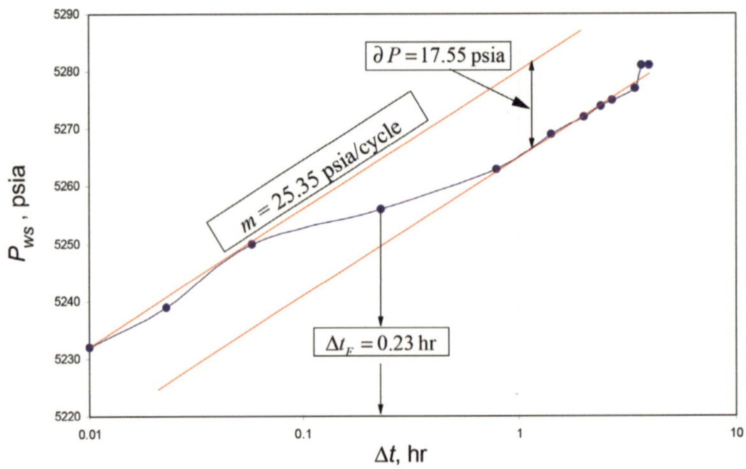

Figure 6.5.
MDH plot well R-6, Example 6.1, after [36].

$$s = 1.1513\left[\frac{P_{1hr} - P_i}{m} - \log_t\left(\frac{k}{(\phi c_t)_{f+m}\mu r_w^2}\right) - 3.23\right] \tag{6.16}$$

For pressure buildup analysis, change P_i by P_{wf} in Eqs. (6.15) and (6.16).

Example 6.1

Determine the bulk fracture permeability and ω and λ from a pressure test run in well R-6 [36], according to the information given below and **Table 6.1**.
h = 1150 ft, r_w= 0.292 ft, q = 17000 STB/D
P_{wf} = 5223 psia, t_p = 408000 h, B = 1.74 rb/STB
μ = 0.47 cp, $(\phi\ c_t)_{m+f}$ = 1.4 × 10^{-06} psia^{-1}, k_m = 0.148 md

Solution

The MDH graph given in **Figure 6.6** confirms the existence of a system with double porosity (also, it can be verified in **Figure 6.7**). The following is read from there:
m = 25.35 psia/cycle, Δt_F = 0.23 h, ∂P = 17.55 psia
The storativity capacity, ω, is estimated from the separation of the parallel lines using Eq. (6.3):

$$\omega = \exp\left(-2.303\frac{\partial P}{m}\right) = \exp\left(-2.303\frac{17.55}{25.35}\right) = 0.2031$$

Use Eq. (6.14) to find permeability,

$$k_{fb} = \frac{141.2q\mu B}{mh} = \frac{162.6(17000)(0.47)(1.74)}{(22)(1150)} = 89.4 \text{ md}$$

Find the interporosity flow parameter from Eq. (6.11):

$$\lambda = \frac{3792(\phi\ c)_{f+m}\mu\ r_w^2}{k_{fb}\Delta t_F}\left[\omega\ln\left(\frac{1}{\omega}\right)\right]$$

$$\lambda = \frac{3792(1.4\times10^{-6})(0.47)(0.292^2)}{(77.54)(0.23)}\left[0.2031\ln\left(\frac{1}{0.2031}\right)\right] = 3.86\times10^{-6}$$

Stewart and Asharsobbi [36] found a value of $\lambda = 3.1\times10^{-6}$.

Δt, hr	P_{ws}, psia	$t^*\Delta P'$, psia	Δt, hr	P_{ws}, psia	$t^*\Delta P'$, psia
0.000	5223		1.4	5269	9.87
0.010	5232	5.61	2.0	5272	9.40
0.023	5239	9.56	2.4	5274	11.72
0.058	5250	8.48	2.7	5275	10.81
0.230	5256	5.06	3.45	5277	8.54
0.780	5263	8.80	3.7	5281	7.78
1.400	5269	9.87	4.0	5281	6.91

Table 6.1.
Pressure and pressure derivative data for well R-6, after [36].

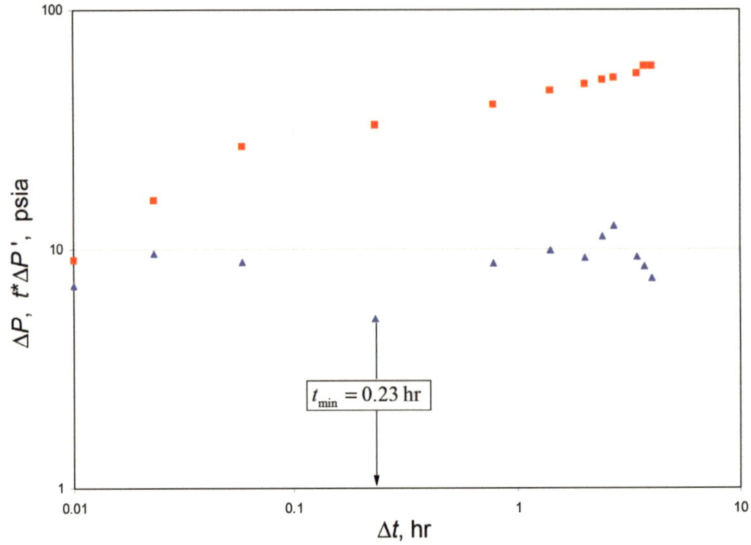

Figure 6.6.
Pressure and pressure derivative plot for well R-6, Example 6.1, after [36].

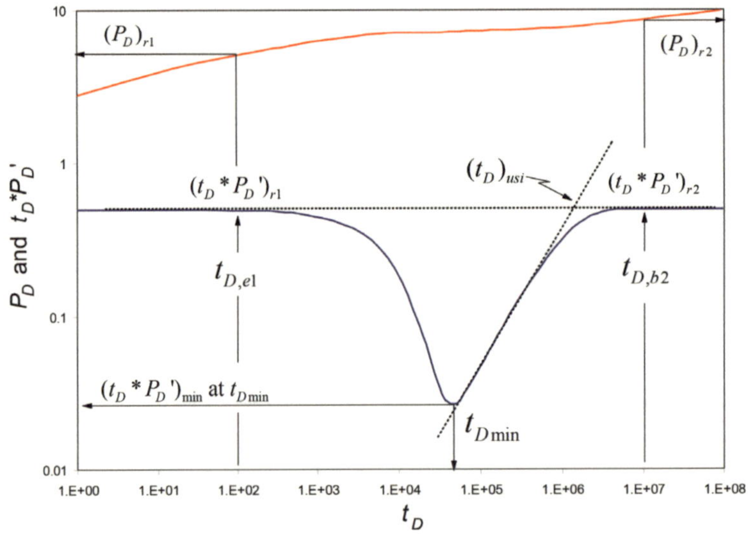

Figure 6.7.
Points and characteristic lines of a naturally fractured reservoir with pseudosteady-state interporosity flow, $\omega = 0.01$, $\lambda = 1 \times 10^{-6}$, *after [6].*

6.2. Type-curve matching for heterogeneous formations

There are several kinds of type curves available for naturally fractured reservoirs.

Some of the pressure type curves are free of wellbore storage effects along with their equation can be found in [29, 34, 39]. Some of them including the pressure derivative function can be found in [31, 35]. However, neither the equations nor the type curves are presented here since *TDS* technique precisely avoids them. Actually, the purpose of this book is to compile *TDS* technique application to several scenarios.

6.3. TDS technique for naturally fractured formations

Warren and Root [46] used this approach to develop an integrated and applicable solution for drawdown and buildup pressure tests in naturally fractured reservoirs with double porosity. From his work can be identified several flow regimes of the semilog analysis. In chronological order, there is a straight line in near time representing only fracture depletion and a straight line in final times, which corresponds to the time when the whole deposit produces as an equivalent homogeneous deposit. At these final times, the semilog straight line is parallel to the first straight line.

New developments by Mavor and Cinco-Ley [31] included wellbore storage and skin effects for the interporosity flow parameter under pseudosteady-state conditions in a naturally fractured reservoir. The solution was given in Laplace space and then was carried out in Laplacian space and inverted numerically using the Stehfest algorithm [37]. As a direct consequence, type curves were developed by Bourdet et al. [3], which included wellbore storage and skin in naturally fractured deposits. Subsequently, reservoir parameters could be estimated when storage would dominate pressure data at early times. An advance in the curves type of naturally fractured deposits occurred with the addition of the derivative curve [3]. Increasing the sensitivity of the derivative curve [9] for naturally fractured deposits results in a better accuracy when applying type-curve matching.

Unfortunately, type-curve matching is a trial-and-error method, which often provides nonunique responses. Besides, it could be really difficult to have all the type curves for all the emerged cases. Therefore, the Tiab's direct synthesis technique [40, 41] extended for naturally fractured formations by Engler and Tiab [6] is presented in this chapter. Actually, a more extensive great work on the subject was performed by Engler [5] who also developed *TDS* technique for horizontal wells in anisotropic formations [7] and naturally fractured deposits [8]. As originally exposed by Tiab [40], this method combines the characteristic points and slopes of a log-log plot of pressure and pressure-derived versus time data with the exact analytical solutions to obtain expressions for reservoir characterization considering that flow from matrix to fractures take place under pseudosteady-state situation.

6.3.1 Mathematical model

An actual naturally fractured formation consists of a heterogeneous system of vugs, fractures, and matrix, which are random in nature. To model this system, it is assumed that the reservoir consists of discrete matrix block elements separated by an orthogonal system of uniform and continuous fractures [46]. These fractures are oriented parallel to the main axes of permeability. Two geometries are commonly assumed, for example, layers and cubes of sugar. The flow between the matrix and the fractures is governed by a pseudosteady-state condition, but only the fluid entering the well comes from the fracture network reach at a constant rate. It is assumed that the fluid is a single phase and slightly compressible. The dimensionless well pressure solution in a reservoir of infinite action along with its dimensionless pressure derivative, with the previous assumptions is given by [46], and also presented by Engler [5] and Engler and Tiab [6]:

$$P_D = \frac{1}{2}\left[\ln t_D + 0.80908 + Ei\left(-\frac{\lambda t_D}{\omega(1-\omega)}\right) - Ei\left(-\frac{\lambda t_D}{1-\omega}\right)\right] + s \tag{6.17}$$

$$t_D{}^*P_D{}' = \frac{1}{2}\left[1 - \exp\left(-\frac{\lambda t_D}{1-\omega}\right) + \exp\left(-\frac{\lambda t_D}{\omega(1-\omega)}\right)\right] \tag{6.18}$$

The dimensionless quantities given by Eqs. (1.94), (1.89), and (2.57) are now rewritten as:

$$t_D = \frac{0.0002637 k_{fb} t}{\mu(\phi c_t)_{m+f} r_w^2} \tag{6.19}$$

$$P_D = \frac{k_{fb} h(P_i - P)}{141.2 q\mu B} \tag{6.20}$$

$$t_D{}^*P_D{}' = \frac{k_{fb} h(t^*\Delta P')}{141.2 q\mu B} \tag{6.21}$$

6.3.2 Characteristic points and lines

Refer to **Figure 6.7** and notice that the radial flow has been interrupted by the transition period during which fractured are fed from the matrix. Each radial flow is labeled $r1$ and $r2$ for distinguishing purposes. As for the case of a homogeneous system, Eq. (2.70) applies (do not consider the dimensionless wellbore storage coefficient), so that Eq. (2.71) applies, now rewritten as:

$$k_{fb} = \frac{70.6 q\mu B}{h(t^*\Delta P')_{r1}} = \frac{70.6 q\mu B}{h(t^*\Delta P')_{r2}} \tag{6.22}$$

Needless to say that in case that both radial flow regimes are seen, bulk fracture permeability can be obtained from any of them using the above expression.

The transition period is affected by the dimensionless storativity coefficient, but independent of the interporosity flow parameter. Engler [5] and Engler and Tiab [6] found an analytical expression for the minimum coordinates by taking the derivative of Eq. (6.18) and equating the result to zero. Subsequently, the dimensionless minimum coordinates are given by:

$$(t_D)_{\min} = \frac{\omega}{\lambda} ln \frac{1}{\omega} \tag{6.23}$$

and,

$$(t_D{}^*P_D{}')_{\min} = 0.5\left(1 + \omega^{1/(1-\omega)} - \omega^{\omega/(1-\omega)}\right) \tag{6.24}$$

Eq. (6.23) was based for [42] to derive Eq. (6.11) which is now rewritten as:

$$\lambda = \frac{3792(\phi\, c_t)_{f+m} \mu r_w^2}{k_{fb} \Delta t_{\min}} \left[\omega \ln\left(\frac{1}{\omega}\right)\right] \tag{6.25}$$

To set Eq. (6.24) in oil-field unit, Engler [5] and Engler and Tiab [6] developed a form to normalize it by division with the radial flow derivative, Eq. (2.70), to yield:

$$(t^*\Delta P')_{\min} / (t^*\Delta P')_{\min} = \left(1 + \omega^{1/(1-\omega)} - \omega^{\omega/(1-\omega)}\right) \tag{6.26}$$

Engler [5] and Engler and Tiab [6] developed the following empirical correlation:

$$\omega = 0.15866\left\{\frac{(t^*\Delta P')_{\min}}{(t^*\Delta P')_r}\right\} + 0.54653\left\{\frac{(t^*\Delta P')_{\min}}{(t^*\Delta P')_r}\right\}^2 \tag{6.27}$$

which is valid from $0 \le \omega \le 0.10$ with an error less than 1.5%. An alternative method for determining ω arises from the defined characteristic times of pressure derivative curve shown in **Figure 6.7**. These include the end of the first horizontal straight line, t_{De1}, the start of the second horizontal straight line, t_{Db2}, and the time corresponding to the minimum derivative, t_{Dmin}. Engler and Tiab [6] developed several empirical correlations by relating such times:

$$\lambda = \frac{\omega(1-\omega)}{50\beta t_{e1}} = \frac{\omega ln(1/\omega)}{\beta t_{min}} = \frac{1}{\beta\, t_{usi}} = \frac{5(1-\omega)}{\beta t_{b2}} \tag{6.28}$$

where

$$\beta = \frac{0.0002637 k_{fb}}{(\phi c_t)_{f+m}\mu r_w^2} \tag{6.29}$$

$$\omega = \exp\left[-\frac{1}{0.9232}\left(\frac{t_{\min}}{50 t_{e1}} - 0.4386\right)\right] \tag{6.30}$$

The correlation for the ratio of the minimum time to the time for the end of the first straight line has an error less than 5% [6].

$$\omega = 0.19211\left\{\frac{5t_{\min}}{t_{b2}}\right\} + 0.80678\left\{\frac{5t_{\min}}{t_{b2}}\right\}^2 \tag{6.31}$$

The correlation using the ratio of the minimum time to the start time of the second straight line, valid for $\omega \le 0.1$, has with an error less than 2%.

For a given dimensionless storativity coefficient, the minimum dimensionless pressure coordinate is independent of the interporosity flow parameter, while the minimum dimensionless time coordinate is a function of λ. Subsequently, Engler and Tiab [6] found that a log plot $(t_D{}^*P'_D)_{min}$ versus log $(\lambda t_D)_{min}$ results in a straight line with unit slope. The corresponding empirical equation is:

$$ln(t_D{}^*P_D{}')_{\min} = ln(\lambda \cdot t_{D\min}) + ln(0.63) \tag{6.32}$$

From which was obtained:

$$\lambda = \left[\frac{42.5h(\phi c_t)_{f+m} r_w{}^2}{qB}\right]\frac{(t^*\Delta P')_{\min}}{t_{\min}} \tag{6.33}$$

An alternative method for determining λ can be carried out by observing a straight line with unit slope characteristic during the last transition period. The smaller dimensionless storativity coefficient (lowest point of the curve) adjusts the data more exactly to the unit slope line. A $\omega < 0.05$ results in a more accurate estimate of λ. For $\omega > 0.05$, λ will be overestimated. The analytical equation for this behavior of the last transition time is [6]:

$$\ln(t_D{}^*P_D{}')_{us} = \ln(\lambda \cdot t_{Dus}/2) \tag{6.34}$$

The intersection of the unit slope line of the transition period with the line of the radial flow regime pressure derivative, Eq. (2.70), shown in **Figure 6.7**, allowed finding a simple expression to determine λ [6]:

$$\lambda = 1/t_{Dusi} \tag{6.35}$$

Replacing Eq. (1.94) in the above expression leads to:

$$\lambda = \left(\frac{(\phi c_t)_{f+m} \mu r_w^2}{0.0002637 k_{fb}} \right) \frac{1}{t_{usi}} \tag{6.36}$$

As for the case of Eq. (2.92), the mechanical skin factor, expected always to be negative for naturally fractured formations, is, found for each one of the radial flow regimes, $r1$ and $r2$, thus, [6]:

$$s = \frac{1}{2} \left[\left(\frac{\Delta P}{t^* \Delta P'} \right)_{r1} - \ln \left(\frac{k_{fb}\, t_{r1}}{(\phi c_t)_f \mu r_w^2 \omega} \right) + 7.43 \right] \tag{6.37}$$

$$s = \frac{1}{2} \left[\left(\frac{\Delta P}{t^* \Delta P'} \right)_{r2} - \ln \left(\frac{k_{fb}\, t_{r2}}{(\phi c_t)_{f+m} \mu r_w^2} \right) + 7.43 \right] \tag{6.38}$$

Eq. (6.37) may be of not practical use since as commented before on the difficulty to obtain a representative value of the fracture compressibility.

6.3.3 Wellbore storage effects

As discussed earlier in this chapter, a direct consequence of wellbore storage is the tendency to mask the early time radial flow period. Therefore, the late or second radial flow line of infinite action is essential for estimating the skin factor and the permeability of the net of fractures. If wellbore storage is presented, it can be obtained, from the early unit slope, by using Eqs. (2.61), (2.69), and (2.80). We must be aware that Eqs. (2.81) and (2.87) were developed for higher skin factors, then, they are not recommended to apply in naturally fractured reservoirs.

The influence of wellbore storage on minimum coordinates is of great importance in the analysis. As **Figure 6.8** shows, the dilemma is whether the minimum observed point is the actual minimum or a "pseudo-minimum" as a direct result of wellbore storage. Engler [5] and Engler and Tiab [6] have shown that the minimum point is not affected by storage for all ω and λ, provided that,

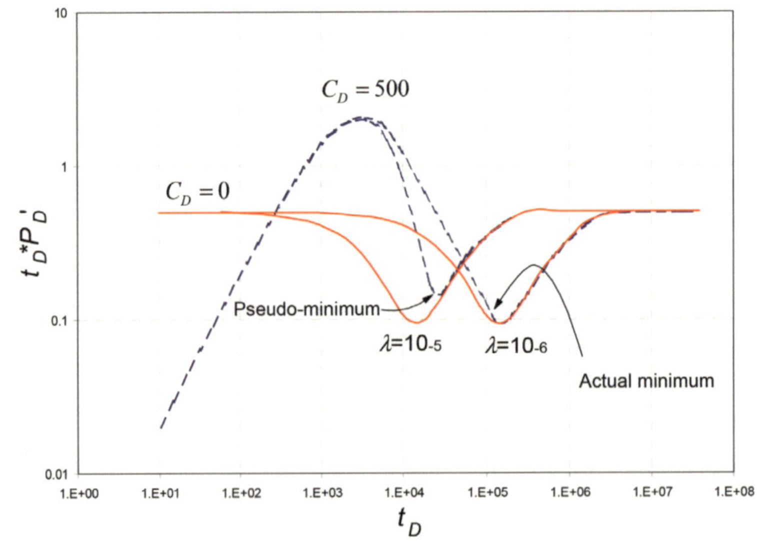

Figure 6.8.
Wellbore storage effect on the minimum value of the pressure derivative, ω = 0.05 and s = 0, after [6].

$$\frac{(t_D)_{\min,o}}{(t_D)_x} \geq 10 \tag{6.39}$$

Accordingly, the procedures described above are valid. When the ratio of the minimum time to the time at the peak is less than the limit defined by Eq. (6.39), a "pseudo-minimum" occurs in the curve of the pressure derivative. An empirical correlation generated during this region provides a method to calculate the interporosity flow parameter [6],

$$[\lambda log(1/\lambda)]_{\min} = \frac{1}{C_D}\left[5.565\frac{t_x}{t_{min,o}}\right]^{10} \tag{6.40}$$

where,

$$\lambda = \left(\frac{[\lambda log(1/\lambda)]_{\min}}{1.924}\right)^{1.0845} \tag{6.41}$$

The corrected ω is found from **Figure 6.9**.

An alternative method for determining λ is based on the ratio of the coordinate of the minimum pressure derivative to the coordinate of the pressure derivative at the peak. This correlation is valid only for $C_D\lambda > 0.001$ [6], and C_D is found from Eq. (2.14);

$$\lambda = \frac{1}{10C_D}\frac{(t^*\Delta P')_{\min}}{(t^*\Delta P')_x} \tag{6.42}$$

Tiab et al. [43] determined that the minimum is not affected by wellbore storage for any value w, provided the conditions given in **Table 6.2** are fulfilled. Then, they proposed a better expression for correcting the minimum:

$$(t^*\Delta P')_{\min} = (t^*\Delta P')_r + \frac{(t^*\Delta P')_{\min O} - (t^*\Delta P')_r[1 + 2D_1D_2]}{1 + D_2\left[\ln\left(\frac{C}{(\phi c_t)_{f+m}hr_w^2}\right) + 2s - 0.8801\right]} \tag{6.43}$$

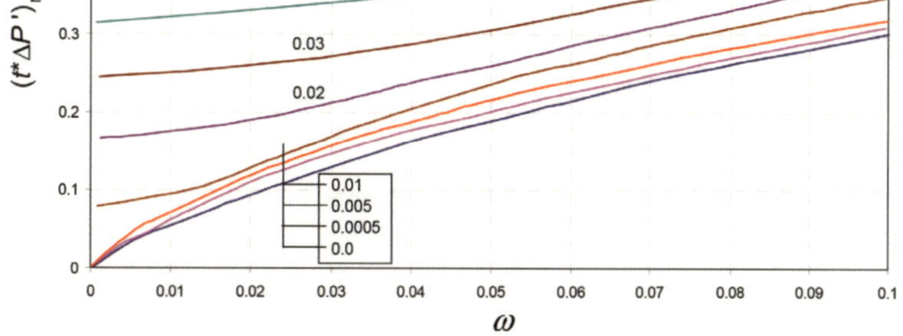

Figure 6.9.
Determination of the dimensionless storativity coefficient using the ratio of the radial with the minimum pressure derivatives, after [6].

λ	C_D
10^{-4}	$C_D > 10$
10^{-5}	$C_D > 100$
10^{-6}	$C_D > 10^3$
10^{-7}	$C_D > 10^4$
10^{-8}	$C_D > 10^5$

Table 6.2.
Conditions for the minimum pressure derivative being affected by wellbore storage, after [43].

where;

$$D_1 = \left[\ln\left(\frac{qBt_{minO}}{(t^*\Delta P')_r(\phi c_t)_{f+m} h r_w^2}\right) + 2s - 4.17 \right] \quad (6.44)$$

and;

$$D_2 = \frac{48.02C}{qB}\left(\frac{(t^*\Delta P')_r}{t_{minO}}\right) \quad (6.45)$$

Being t_{minO} and $(t^*\Delta P')_{minO}$, the value of the coordinates of the minimum point in the derivative without making any correction (observed) when there is wellbore storage effect. Once corrected, the following expression can be applied:

$$\omega = \left(2.9114 + 4.5104\frac{(t^*\Delta P')_r}{(t^*\Delta P')_{min}} - 6.5452e^{0.7912\frac{(t^*\Delta P')_r}{(t^*\Delta P')_{min}}}\right)^{-1} \quad (6.46)$$

Tiab et al. [43] also provided another expression for ω;

$$\omega^{\omega} = e^{-\lambda t_{Dmin}} \quad (6.47)$$

where t_{Dmin} is found using Eq. (1.94) rewritten as;

$$t_{Dmin} = \frac{0.0002637kt_{min}}{(\phi c_t)_{f+m}\mu r_w^2} \quad (6.48)$$

For values of ω less than 0.5 the solution of Eq. (4.47) is:

$$\omega = \left(2.9114 + \frac{3.5688}{\lambda t_{Dmin}} + \frac{6.5452}{\lambda t_{Dmin}}\right)^{-1} \quad (6.49)$$

Although Engler [5] and Engler and Tiab [6] provided step-by-step procedures for the application of *TDS* technique, they are omitted here since it is not mandatory to follow such procedures.

Example 6.2

Tiab e al. [43] presented the derivative plot, **Figure 6.10**, for a pressure test run in a heterogeneous formation. Other relevant information is given below:

q = 960 BPD, B = 1.28 rb/stb, μ =1.01 cp
h = 36 ft, r_w = 0.29 ft, $(\phi c_t)_{m+f} = 0.7 \times 10^{-6}$1/psia

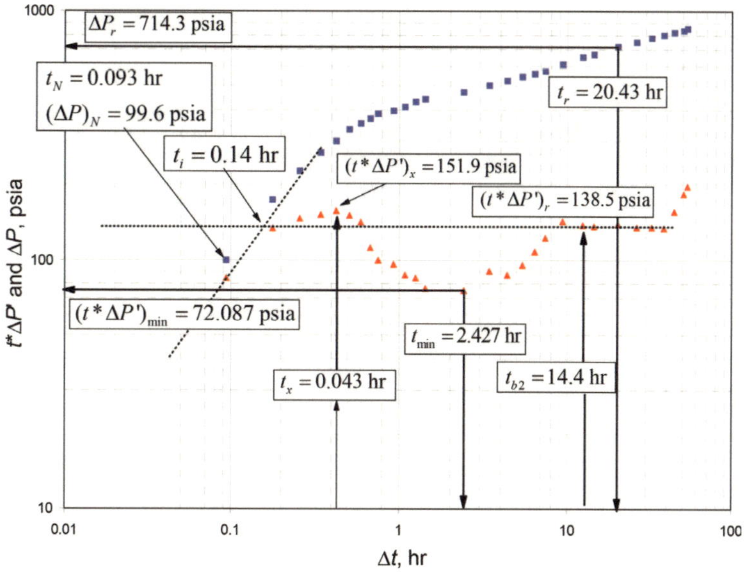

Figure 6.10.
Pressure and pressure derivative against time log-log plot for Example 6.2, after [43].

It is required the interpretation of this test to provide permeability, skin factor, and the Warren-and-Root parameters.

Solution

As observed in the pressure derivative plot, **Figure 6.10**, this pressure test may not be interpretable using conventional analysis since the first radial flow regime is absolutely masked by wellbore storage. The following information is read from such plot:

t_N = 0.00348 h, ΔP_N = 11.095 psia, t_{r2} = 1.8335 h
ΔP_{r2} = 61.5 psia, $(t^*\Delta P')_{r2}$ = 10.13 psia, $t_{min,o}$ = 0.07 h, $(t^*\Delta P')_{minO}$ = 5.32 psia
Permeability is found from Eq. (6.22);

$$k_{fb} = \frac{70.6q\mu B}{h(t^*\Delta P')_{r2}} = \frac{70.6 \times 960 \times 1.01 \times 1.28}{10.13 \times 36} = 238 \text{ md}$$

The wellbore storage coefficient and the dimensionless wellbore storage coefficient are determined with Eqs. (2.61) and (2.14) to be:

$$C = \frac{qB}{24}\left(\frac{t}{\Delta P}\right)_N = \left(\frac{960(1.28)}{24}\right)\frac{0.00348}{11.095} = 0.0161 \text{ bbl/psia}$$

$$C_D = \frac{0.894C}{(\phi c_t)_{f+m} h r_w^2} = \frac{0.894(0.0161)}{0.7 \times 10^{-6}(36)(0.29^2)} = 6792$$

Estimate skin factor from Eq. (6.38):

$$s = \frac{1}{2}\left[\frac{61.5}{10.13} - \ln\left(\frac{(238)(1.8335)}{0.7 \times 10^{-6}(1.01)(0.29^2)}\right) + 7.43\right] = -4.6$$

The interporosity flow parameter is found with Eq. (6.33)

$$\lambda = \left[\frac{42.5h(\phi c_t)_{f+m} r_w^{\,2}}{qB}\right]\frac{(t^*\Delta P')_{min}}{t_{min}} = \left[\frac{42.5(37)0.7\times 10^{-6}(0.29^2)}{960(1.28)}\right]\frac{5.32}{0.07} = 5.57\times 10^{-6}$$

Since $\lambda = 5.57 \times 10^{-6}$ and $C_D = 6792$, by looking at the third row in **Table 6.2**, the coordinates of the minimum point must be corrected before finding ω. Therefore, use Eqs. (6.44), (6.45), and (6.43), so that:

$$D_1 = \left[\ln\left(\frac{960(1.28)(0.07)}{(10.13)}\right) + 2(-4.6) - 4.17\right] = 1.8334$$

$$D_2 = \frac{48.02(0.0161)}{960(1.28)}\left(\frac{10.13}{0.07}\right) = 0.09105$$

$$(t^*\Delta P')\text{min} = 10.13 + \frac{5.32 - (10.13)[1+2(1.8334)(0.09105)]}{1 + 0.09105[\ln\left(\frac{0.0161}{0.7\times 10^{-6}(36)(0.29^2)}\right) + 2(-4.6) - 0.8801]} = 0.9849$$

Then, ω is found from Eq. (6.46), and the correlation is given by Eq. (6.27):

$$\omega = \left(2.9114 + 4.5104\frac{10.13}{0.9849} - 6.5452e^{0.7912\frac{10.13}{0.9849}}\right)^{-1} = 0.024$$

$$\omega = 0.15866\left\{\frac{0.9849}{10.13}\right\} + 0.54653\left\{\frac{0.9849}{10.13}\right\}^2 = 0.0206$$

Without correction, Eq. (6.27) would have given a value of $\omega = 0.234$.

Example 6.3

This was also worked by Tiab et al. [43]. Pressure and pressure derivative against time data are given in **Table 6.3** and plotted in **Figure 6.11** for its interpretation. Other important data are given below:

$q = 3000$ BPD,	$\phi = 0.10$,	$\mu = 1.0$ cp
$c_t = 3.0 \times 10^{-5}$ psia^{-1},	$B = 1.25$ bbl/stb,	$h = 100$ ft
$r_w = 0.40$ ft,	$P_i(t=0) = 4473$ psia	

Calculate the permeability of the fractured system, skin factor, wellbore storage coefficient, interporosity flow parameter, and dimensionless storativity coefficient.

Solution

The following information was read from **Figure 6.11**:

$\Delta P_N = 99.6$ psia,	$(t^*\Delta P')_N = 116.4$ psia,	$t_N = 0.093$ h
$t_{b2} = 14.4$ h,	$t_{r2} = 20.43$ h,	$\Delta P_{r2} = 714.3$ psia
$(t^*\Delta P')_{r2} = 138.5$ psia,	$t_x = 0.43$,	$t_{min,o} = 2.427$ h
$t_{min,o}/t_x = 5 < 10$,	$t_i = 0.14$ h,	$(t^*\Delta P')_{min} = 72.087$ psia

Solution

Use Eq. (6.22) to estimate permeability

$$k_{fb} = \frac{70.6q\mu B}{h(t^*\Delta P')_{r2}} = \frac{70.6(3000)(1.0)(1.25)}{100(138.5)} = 19.1 \text{ md}$$

t, h	P_{wf}, psia	$t^*\Delta P'$, psia	t, h	P_{wf}, psia	$t^*\Delta P'$, psia	t, h	P_{wf}, psia	$t^*\Delta P'$, psia
0.0933	4373.4	84.473	1.0930	4060.3	87.234	12.43	3824.2	137.651
0.1766	4299.1	133.483	1.26	4043.1	84.384	14.43	3804.1	136.857
0.2600	4246.1	146.776	1.427	4032.2	76.719	20.43	3758.7	138.810
0.3433	4203.6	151.595	2.427	4002.8	75.401	26.43	3720.3	135.210
0.4266	4173.8	157.618	3.427	3971.3	90.502	32.43	3695.1	134.790
0.5100	4139.7	150.295	4.427	3948.3	87.168	38.43	3674.6	134.116
0.5933	4118.5	141.355	5.427	3931.6	95.595	44.43	3652.4	156.278
0.6766	4103.5	111.676	6.427	3917.1	108.303	50.43	3636.9	183.611
0.7600	4086.4	99.694	7.427	3898.4	122.336	53.43	3625.2	196.734
0.9266	4075.4	95.720	9.427	3865.3	142.426			

Table 6.3.
Pressure and pressure derivative data against time for Example 6.3, after [43].

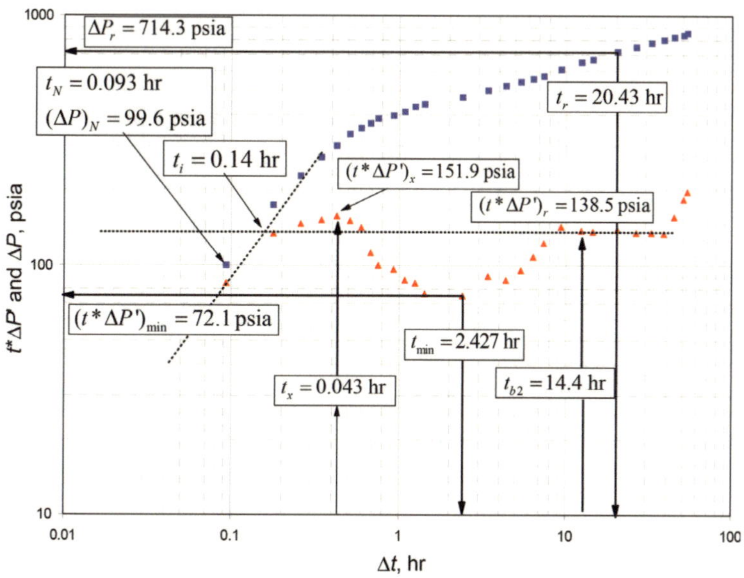

Figure 6.11.
Pressure and pressure derivative against time log-log plot for Example 6.3, after [43].

Find the wellbore storage coefficient using Eqs. (2.61), (2.69), and (2.80).

$$C = \left(\frac{qB}{24}\right)\frac{t_N}{\Delta P_N} = \left(\frac{3000(1.25)}{24}\right)\frac{0.093}{99.6} = 0.146\ \text{bbl/psia}$$

$$C = \left(\frac{qB}{24}\right)\frac{t_N}{(t^*\Delta P')_N} = \left(\frac{3000(1.25)}{24}\right)\frac{0.093}{116.4} = 0.129\ \text{bbl/psia}$$

$$C = \frac{k_{fb}ht_i}{1695\mu} = \frac{19.1(100)(0.14)}{1695\mu} = 0.158\ \text{bbl/psia}$$

Estimate skin factor from Eq. (6.38):

$$s = \frac{1}{2}\left[\frac{714.3}{138.5} - \ln\frac{19.1(20.43)}{3\times10^{-6}(0.1)(1)(0.4^2)} + 7.43\right] = -5.13$$

The ratio between the minimum time and the time in the peak suggests that the minimum coordinates are influenced by wellbore storage. Then, find the dimensionless wellbore storage using Eq. (2.14) with an average C of 0.143 bbl/psia:

$$C_D = \frac{0.8935C}{\phi h c_t r_w^2} = \frac{0.8935(0.143)}{0.1(100)(3\times10^{-6})(0.4^2)} = 26618$$

Eq. (6.40) leads to find:

$$[\lambda \log(1/\lambda)]_{\min} = \frac{1}{C_D}\left[5.565\frac{t_x}{t_{min,o}}\right]^{10} = \frac{1}{26618}\left[5.565\frac{0.43}{2.427}\right]^{10} = 0.000032$$

Then, $\lambda = 5.2\times10^{-7}$. Find again λ with Eq. (6.42):

$$\lambda = \frac{1}{10C_D}\frac{(t^*\Delta P')_{\min}}{(t^*\Delta P')_x} = \frac{1}{10(26618)}\frac{72.087}{151.94} = 1.78\times10^{-6}$$

For the determination of ω, estimate the ratio between the pressure derivatives of minimum point and radial flow regime, $(t^*\Delta P')_{min}/(t^*\Delta P')_r = 72.078/138.5 = 0.52$. Then, find the parameter needed to enter in **Figure 6.9**, thus:

$$C_D\lambda\left[\log\left(\frac{1}{\lambda}\right) + 0.8686s\right] = 26618(1.78\times10^{-6})\left[\log\left(\frac{1}{1.78\times10^{-6}}\right) + 0.8686(-5.13)\right] = 0.061$$

From **Figure 6.9**, it is found that $\omega = 0.07$.

The reader may think that the subject covered by the *TDS* technique in this chapter is the only one as far as naturally fractured reservoirs are concerned. The material exposed in this chapter was the first one ever introduced. *TDS* technique is certainly rich in applications. Regarding naturally fractured systems can be named: double porosity and double permeability for vertical wells [19], already mentioned at the beginning of this chapter, and for horizontal wells [30]. For triple porosity, reservoirs referred to [10]. For horizontal wells in both homogeneous and heterogeneous deposits including the effect of the threshold pressure gradient, the reader is invited to read [21]. The work originally presented for long homogeneous reservoirs [11] was extended to naturally fractured deposits by Escobar et al. [13, 14]. For hydraulically fractured wells, draining heterogeneous formations refer to [13, 20, 26, 44]. Escobar et al. [15] presented *TDS* technique for gas wells in naturally fractured systems. The effect of pseudotime on the Warren-and-Root parameters was observed by [18] for vertical wells and [17] for horizontal wells. TS technique for rate transient analysis of homogeneous and heterogeneous formations was presented by [1] and extended to long reservoirs by [16] and gas bearing long fractured formations by [12]. However, there are publications written by some other researchers and are not reported here.

Lately, shale reservoirs have been the target of many oil and gas companies. Gas reservoirs must be hydraulically fractured for production to occur. Their behavior normally follows those of naturally fractured formations. Some publications regarding shales tested either at constant bottom-hole pressure or constant rate using *TDS* technique are [2, 22–25].

Nomenclature

B	oil volume factor, bbl/STB
b	fraction of penetration/completion
c	compressibility, 1/psia
C	wellbore storage coefficient, bbl/psia
$D1$	minimum point correction parameter
$D2$	minimum point correction parameter
c_t	total or system compressibility, 1/psia
h	formation thickness, ft
k_{ef}	bulk fractured network permeability, md
m	slope of P-vs-log t plot, psia/h/cycle
P	pressure, psia
P_D'	dimensionless pressure derivative
P_D	dimensionless pressure
P_i	initial reservoir pressure, psia
P_{wf}	well flowing pressure, psia
P_{ws}	well shut-in or static pressure, psia
P_{1hr}	intercept of the semilog plot, psia
P^*	false pressure, psia
q	liquid flow rate, BPD
r_D	dimensionless radius
r	radius, ft
r_w	well radius, ft
s	skin factor
t	time, h
t_p	production (Horner) time before shutting-in a well, h
t_D	dimensionless time based on well radius
$t^*\mathrm{D}P'$	pressure derivative, psia
V	volume, ft^3

Greek

α	shape factor
Δ	change, drop
Δt	shut-in time, h
∂P	parallel difference between the two radial flow regime slopes, psia
ϕ	porosity, fraction
λ	interporosity flow coefficient
μ	viscosity, cp
ρ	fluid density, lbm/ft^3
ω	dimensionless storativity coefficient

Suffices

1	intercept between first semilog radial flow and the transition line
1hr	time of 1 h
2, *I*	intercept between second semilog radial flow and the transition line
*b*2	start of second radial flow regime
D	dimensionless
*e*1	end of first radial flow regime
F	inflection
f	fracture network
i	intersection or initial conditions
N	an arbitrary point during early pseudosteady-state period
m	matrix
max	maximum point
min	minimum point
minO	observed minimum point
r	radial flow
r_1	radial flow before transition period
r_2	radial flow after transition period
s	skin
usi	intersect of the pressure derivative lines of the unit-slope line during the transition and second radial flow regime
w	well
wf	well flowing
ws	well shut-in

References

[1] Arab N. Application of Tiab's direct synthesis technique to constant bottom hole pressure test [M.Sc. thesis]. The University of Oklahoma; 2003

[2] Bernal KM, Escobar FH, Ghisays-Ruiz A. Pressure and pressure derivative analysis for hydraulically fractured shale formations using the concept of induced permeability field. Journal of Engineering and Applied Sciences. 2003;**9**(10):1952-1958

[3] Bourdet D, Ayoub JA, Whittle TM, Pirard YM, Kniazeff V. Interpreting well tests in fractured reservoirs. World Oil. 1983;**197**(5):77-87

[4] Earlougher RC Jr. Advances in well test analysis. Monograph Series, Vol. 5. Dallas, TX: Second printing; SPE; 1977

[5] Engler TW. Interpretation of pressure tests in naturally fractured reservoirs by the direct synthesis technique [Ph.D. dissertation]. University of Oklahoma; 1995

[6] Engler T, Tiab D. Analysis of pressure and pressure derivative without type curve matching, 4 Naturally fractured reservoirs. Journal of Petroleum Science and Engineering. 1996;**15:**127-138

[7] Engler TW, Tiab D. Analysis of pressure and pressure derivatives without type-curve matching. 6-Horizontal well tests in anisotropic reservoirs. Journal of Petroleum Science and Engineering. 1996;**15**:153-168

[8] Engler TW, Tiab D. Analysis of pressure and pressure derivatives without type-curve matching. 5-Horizontal well tests in naturally fractured reservoirs. Journal of Petroleum Science and Engineering. 1996;**15**:139-151

[9] Escobar FH, Navarrete JM, Losada HD. Evaluation of Pressure Derivative Algorithms for Well-Test Analysis. Society of Petroleum Engineers. 2004. DOI: 10.2118/86936-MS

[10] Escobar FH, Saavedra NF, Escorcia GD, Polania JH. Pressure and Pressure Derivative Analysis without Type-Curve Matching for Triple Porosity Reservoirs. Society of Petroleum Engineers. 2004. DOI: 10.2118/88556-MS

[11] Escobar FH, Hernandez YA, Hernandez CM. Pressure transient analysis for long homogeneous reservoirs using TDS technique. Journal of Petroleum Science and Engineering. 2007; **58**(1-2):68-82

[12] Escobar FH, Sanchez JA, Cantillo JH. Rate transient analysis for homogeneous and heterogeneous gas reservoirs using the TDS technique. CT&F – Ciencia, Tecnología y Futuro. 2008;**4**(4):45-59

[13] Escobar FH, Martinez J, Cantillo JH. Pressure-Transient Analysis for Naturally Fractured Reservoirs With Transition Period Before and After the Radial-Flow Regime. Society of Petroleum Engineers. 2010. DOI: 10.2118/138490-MS

[14] Escobar FH, Hernandez DP, Saavedra JA. Pressure and pressure derivative analysis for long naturally fractured reservoirs using the TDS technique. Dyna. 2010;**77**(163):102-114

[15] Escobar FH, Muñoz YEM, Cerquera WM. Pressure and pressure derivate analysis vs. pseudotime for a horizontal gas well in a naturally fractured reservoir using the TDS technique. Entornos Journal. 2011;**24**:39-54. ISSN 0124-7905

[16] Escobar FH, Rojas MM, Bonilla LF. Transient-rate analysis for long homogeneous and naturally fractured reservoir by the TDS technique. Journal of Engineering and Applied Sciences. 2012;**7**(3):353-370

[17] Escobar FH, Muñoz YEM, Cerquera WM. Pseudotime function effect on reservoir width determination in homogeneous and naturally fractured gas reservoir drained by horizontal wells. Entornos Journal. 2012;**24**:221-231

[18] Escobar FH, Martinez LY, Méndez LJ, Bonilla LF. Pseudotime application to hydraulically fractured vertical gas wells and heterogeneous gas reservoirs using the TDS technique. Journal of Engineering and Applied Sciences. 2012; **7**(3):260-271

[19] Escobar FH, Vega J, Diaz MR. Pressure and pressure derivative analysis for double-permeability systems without type-curve matching. Journal of Engineering and Applied Sciences. 2012;**7**(10):1314-1320

[20] Escobar FH, Ghisays-Ruiz A, Bonilla LF. New model for elliptical flow regime in hydraulically fractured vertical wells in homogeneous and naturally fractured systems. Journal of Engineering and Applied Sciences. 2014;**9**(9):1629-1636

[21] Escobar FH, Zhao YL, Zhang LH. Interpretation of pressure tests in horizontal wells in homogeneous and heterogeneous reservoirs with threshold pressure gradient. Journal of Engineering and Applied Sciences. 2014;**9**(11):2220-2228

[22] Escobar FH, Montenegro LM, Bernal KM. Transient-rate analysis for hydraulically fractured gas shale wells using the concept of induced permeability field. Journal of Engineering and Applied Sciences. 2014;**9**(8):1244-1254

[23] Escobar FH, Bernal KM, Olaya-Marin G. Pressure and pressure derivative analysis for fractured horizontal wells in unconventional shale reservoirs using dual-porosity models in the stimulated reservoir volume. Journal of Engineering and Applied Sciences. 2014;**9**(12):2650-2669

[24] Escobar FH, Bernal KM, Olaya-Marin G. Pressure and pressure derivative analysis for fractured horizontal wells in unconventional shale reservoirs using dual-porosity models in the stimulated reservoir volume. Journal of Engineering and Applied Sciences. 2014;**9**(12):2650-2669

[25] Escobar FH, Rojas JD, Ghisays-Ruiz A. Transient-rate analysis hydraulically-fractured horizontal wells in naturally-fractured shale gas reservoirs. Journal of Engineering and Applied Sciences. 2014;**10**(1):102-114

[26] Escobar FH, Zhao YL, Fahes M. Characterization of the naturally fractured reservoir parameters in infinite-conductivity hydraulically fractured vertical wells by transient pressure analysis. Journal of Engineering and Applied Sciences. 2015; **10**(12):5352-5362

[27] Escobar FH. Recent Advances in Practical Applied Well Test Analysis. New York: Nova Science Publishers, Inc.; 2015. p. 422

[28] Igbokoyi AO, Tiab D. Estimation of Average Reservoir Pressure and Drainage Area in Naturally Fractured Reservoirs. Society of Petroleum Engineers. 2006. DOI: 10.2118/104060-MS

[29] Kazemi H. Pressure transient analysis of naturally fractured reservoirs

with uniform fracture distribution. Society of Petroleum Engineers. 1969: 451-462. DOI: 10.2118/2156-A

[30] Lu J, Zhu T, Tiab D, Escobar FH. Pressure behavior of horizontal wells in dual-porosity, dual-permeability naturally-fractured reservoirs. Journal of Engineering and Applied Sciences. 2015;**10**(8):3405-3417. ISSN 1819-6608

[31] Mavor MJ, Cinco-Ley H. Transient Pressure Behavior of Naturally Fractured Reservoirs. Society of Petroleum Engineers. 1979. DOI: 10.2118/7977-MS

[32] Molina MD, Escobar FH, Montealegre MM, Restrepo DP. Application of the TDS technique for determining the average reservoir pressure for vertical wells in naturally fractured reservoirs. CT&F—Ciencia, Tecnología y Futuro. 2005;**3**(5):45-55. ISSN 0122-5383

[33] Nelson R. Geologic Analysis of Naturally Fractured Reservoirs. 2nd ed. Houston, TX: Gulf Professional Publishing; 2001. p. 352

[34] Odeh AS. Unsteady-state behavior of naturally fractured reservoirs. SPEJ; Transactions of AIME. 1965;**234**:64-65. Society of Petroleum Engineers. DOI: 10.2118/966-PA

[35] Petak KR, Soliman MY, Anderson MF. Type Curves for Analyzing Naturally Fractured Reservoirs. Society of Petroleum Engineers. 1986. DOI: 10.2118/15638-MS

[36] Stewart G, Asharsobbi F. Well Test Interpretation for Naturally Fractured Reservoirs. Society of Petroleum Engineers. 1988. DOI: 10.2118/18173-MS

[37] Stephest H. Numerical inversion of 39 Laplace transform. Communications of the ACM. 1970;**13**(1):47-49

[38] Stewart G. Well Test Design & Analysis. Tulsa, Oklahoma, USA: PennWell Corporation; 2011. p. 1940

[39] Strobel CJ, Gulati MS, Ramey HJ. Reservoir limit tests in a naturally fractured reservoir—A field case study using type curves. Society of Petroleum Engineers. 1976:1097-1106. DOI: 10.2118/5596-PA

[40] Tiab D. Analysis of pressure and pressure derivative without type-curve matching: 1-skin and wellbore storage. Journal of Petroleum Science and Engineering. 1995;**12**:171-181

[41] Tiab D. Analysis of Pressure and Pressure Derivatives Without Type-Curve Matching: I—Skin and Wellbore Storage. Society of Petroleum Engineers. 1993. DOI: 10.2118/25426-MS

[42] Tiab D, Escobar FH. Determinación del Parámetro de Flujo Interporoso de un Gráfico Semilogarítmico. X Congreso Colombiano del Petróleo. 2003

[43] Tiab D, Igbokoyi AO, Restrepo DP. Fracture porosity from pressure transient data. In: International Petroleum Technology Conference. 2007. DOI: 10.2523/IPTC-11164-MS

[44] Tiab D, Bettam Y. Practical Interpretation of Pressure Tests of Hydraulically Fractured Wells in a Naturally Fractured Reservoir. Society of Petroleum Engineers. 2007. DOI: 10.2118/107013-MS

[45] Uldrich DO, Ershaghi I. A Method for Estimating the Interporosity Flow Parameter in Naturally Fractured Reservoirs. Society of Petroleum Engineers. 1976. DOI: 10.2118/7142-PA

[46] Warren JE, Root PJ. The behavior of naturally fractured reservoirs. Society of Petroleum Engineers. 1963.245-255. DOI: 10.2118/426-PA

Chapter 7

Hydraulically Fractured

Hydraulic fracturing is a technique consisting of cracking the rock with a liquid fluid, normally of non-Newtonian nature, which carries some solid particles (sand or synthetic material) so that reservoir fluids can move easier toward the well. The fluid has three main uses: (1) as pressuring fracturing tool by overpassing the formation fracture gradient, (2) as carrying agent to transport the solid material, called proppant, to the fracture face so, when reducing the pressure, avoids the fracture to close completely and provides flow capability to the fracture, called conductivity, and (3) as lubrication agent. In unconventional fracturing, the fracturing fluid is water and about 4% of the injected mass corresponds to sand.

The orientation of the hydraulic fractures is a function of the distribution of stress in the formation [3, 14]. If the least stress in the formation is horizontal, then a vertical fracture will be obtained. On the other hand, if the least important stress is vertical, then a horizontal fracture will occur [4, 8, 30]. However, there is a general belief that vertical fractures are obtained at depths greater than 3000 ft.

Figure 7.1 is a plane of a bounded circular system, which is a well with a vertical fracture. The fracture length has been exaggerated for explanatory purposes. Generally, the fluid enters the fracture at a uniform flow rate per unit area of the face of the fracture so that there is a pressure drop in the fracture. In this case, the fracture refers to a "uniform flow fracture." However, for some fractures that have infinite permeability (conductivity), the pressures are uniform throughout. Except for fractures with a high content of support material and conductive fractures, it is thought that the uniform flow fracture represents much better reality than the fracture of infinite conductivity [7, 12, 39].

7.1. Well drawdown pressure behavior

The dimensionless pressure in the well for the case of a uniform flow fracture is [4, 5, 7, 8]:

$$P_D = \sqrt{\pi t_{Dxf}}\,erf\left(1/2\sqrt{t_{Dxf}}\right) - 0.5Ei\left(-1/4t_{Dxf}\right) \tag{7.1}$$

And for the case of an infinite conductivity, fracture is:

$$P_D = 0.5\sqrt{\pi t_{Dxf}}\begin{bmatrix} erf\left(0.134/\sqrt{t_{Dxf}}+\right. \\ \left. erf\left(0.866/\sqrt{t_{Dxf}}\right)\right. \end{bmatrix} - \begin{bmatrix} 0.067Ei\left(-0.018/t_{Dxf}\right)- \\ 0.433Ei\left(-0.75/t_{Dxf}\right) \end{bmatrix} \tag{7.2}$$

where

$$t_{Dxf} = t_D\left(r_w/x_f\right)^2 \tag{7.3}$$

If $t_{Dxf} < 0.1$ in Eq. (7.1) and $t_{Dxf} < 0.1$ in Eq. (7.2), these two equations become:

$$P_D = \sqrt{\frac{\pi t_{Dxf}}{\xi}} \tag{7.4}$$

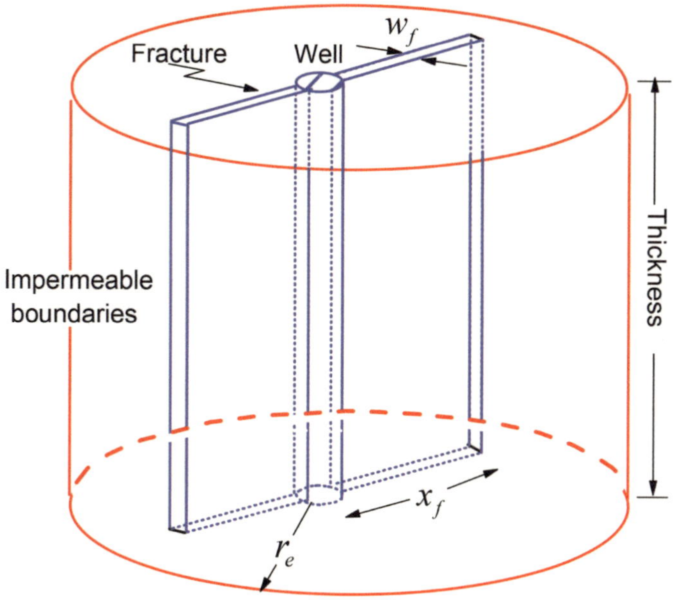

Figure 7.1
Schematic representation of a vertical fracture, after [6].

It is important to note that Eq. (7.4) is a new version proposed by Bettam et al. [2], which considers both homogeneous deposits, $\xi = 1$, and heterogeneous (naturally fractured double porosity) deposits, $\xi = \omega$ (dimensionless storage coefficient). Eq. (7.4) also indicates that at early times the flow within the fracture is linear. In real units, Eq. (7.4) can be written as [8]:

$$P_{wf} = P_i - m_{lf}\sqrt{t} \tag{7.5}$$

where

$$m_{lf} = \frac{-4.064qB}{h}\sqrt{\frac{\mu}{k\xi\phi c_t x_f^2}} \tag{7.6}$$

m_{vf} is the slope of the Cartesian plot of P_{wf} against $t^{1/2}$, which can be used to calculate:

$$kx_f^2 = \left(\frac{-4.064qB}{hm_{lf}}\right)^2 \frac{\mu}{\xi\phi c_t} \tag{7.7}$$

If certainty exists in the determination of the linear flow regime, let us say the pressure derivative is available and reservoir permeability is accurate (it does not need correction), then use Eq. (7.7) directly and find the half-fracture length. The following procedure should be flowed, otherwise. In which the conventional semilog analysis applies to $t_{Dxf} > 10$. Eqs. (7.1) and (7.2) are, respectively, converted to [5, 6, 26, 33]:

$$P_D = \frac{1}{2}\left[\ln\left(t_{Dxf}\right) + 2.80907\right] \tag{7.8}$$

$$P_D = \frac{1}{2}\left[\ln\left(t_{Dxf}\right) + 2.2\right] \tag{7.9}$$

These two equations give the dimensionless pressure for pseudoradial flow, as long as the boundary effects are not found. In a bounded reservoir, the period of infinite action pseudoradial flow only develops completely if $x_e/x_f > 5$.

There is an approximate relationship between the pressure change at the end of the linear flow period, ΔP_{el}, and at the beginning of the semilog straight line, P_{bsl}, [26, 33]:

$$\Delta P_{sbl} \geq 2\Delta P_{el} \tag{7.10}$$

If this relationship is not met, it is because the linear flow period or the radial flow period was incorrectly selected. A couple of pertinent observations:

- A graph of ΔP versus time on a log-log paper will produce a straight line of mean slope during the linear period.
- The above analysis is valid for pressure decline and injection tests.

7.2. Conventional analysis

In vertically fractured wells, pressure buildup and falloff tests are similar to nonfractured wells. As shown in **Figure 7.2**, the semilog slope, m, obtained from traditional analysis of a fractured well is erroneously very small and the value of m decreases progressively as x_f increases [32]. In other words, the fracture presence partially masks the radial flow regime. The pressure derivative certainly allows finding the true start of the semilog line. Because of that fracture effect, permeability, estimated from the Horner or MDH graph, should be corrected as follows:

$$k = k_c F_{cor} \tag{7.11}$$

where

$$F_{cor} = (kh)_{true}/(kh)_{apparent} \tag{7.12}$$

And the uncorrected permeability is found from Eq. (2.33),

$$k_c = 162.6\frac{q\mu B}{mh} \tag{7.13}$$

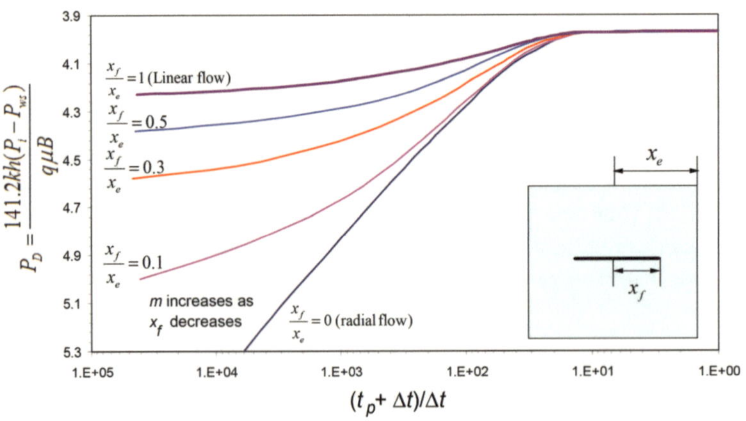

Figure 7.2
Effect of fracture length on the semilog slope [32].

The correction factor, F_{cor}, is read from **Figure 7.3**. The Horner plot is strongly recommended for data analysis of vertically fractured wells. x_e/x_f must be known (normally assumed by a trial-and-error procedure) to use **Figure 7.3**. x_f is simply the half-fracture length, which can be estimated from the slope of the Cartesian plot of pressure versus square root of time, using Eq. (7.7).

F_{cor} can be estimated iteratively as follows:

1. Estimate kx_f^2 from Eq. (7.7).

2. Estimate k_c using Eq. (7.13).

3. Calculate k from Eq. (7.11) using a reasonable assumed value of x_e/x_f (As a first try, assume $x_f = 0.5x_e$) in **Figure 7.3**.

4. Use the value of k from step 3 to estimate x_f with Eq. (7.7).

5. Find x_e with Eq. (7.14).

$$x_e = 0.029\sqrt{\frac{kt_p}{\phi\mu c_t}} \tag{7.14}$$

6. This new value of x_f is used to compute a new value of x_e/x_f to be used in **Figure 7.3**. This would improve the estimation of k.

7. This process continues until two successive x_e/x_f values are equal.

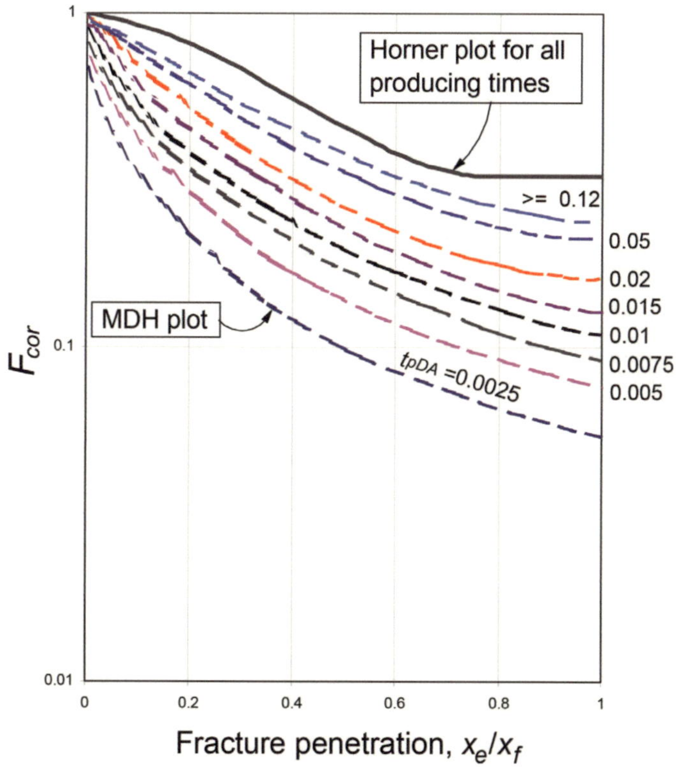

Figure 7.3
Correction factor for kh *estimated from pressure buildup tests in vertically fractured wells, after [31].*

Example

The pressure buildup data obtained after a hydraulic fracturing treatment are shown in **Table 7.1**. The characteristics of the reservoir and well for this test are as follows:

$q = 101$ BPD, $r_w = 0.198$ ft, $h = 42$ ft, $\varphi = 8\%$, $\mu = 0.45$ cp, $B = 1.507$ bbl/STB, $c_t = 17.7 \times 10^{-6}$/psia, $t_p = 364$ hr

Find permeability, skin factor, and half-fracture length using conventional analysis.

Solution

Taken from [34]. In this example, $\xi = 1$ because it is a homogenous deposit. A Horner graph for the data given in **Table 7.1** is presented in **Figure 7.4**. The pressure and pressure derivative plot was just built for verification purposes (**Figure 7.5**). In fact, an early slope of ½ is shown in such plot. An infinite-conductivity or uniform-flux fracture is dealt with since during that period the pressure doubles the pressure derivative. The slope is –510 psia/cycle. The permeability can be estimated from the slope of the semilog straight line, using Eq. (7.13):

$$k_c = \frac{162.6 q\mu B}{mh} = \frac{(162.6)(101)(0.45)(1.507)}{(510)(42)} = 0.52 \text{ md}$$

Figure 7.6 contains a Cartesian plot of pressure versus the square root of time (actually, a tandem time function is recommended to be used instead of a normal square root of time function). To estimate permeability, the trial-and-error process is evoked, so that:

Δt, hr	P_{ws}, psia	$\Delta t^{0.5}$, hr$^{0.5}$	ΔP, psia	$(t_p+\Delta t)/\Delta t$	$t^*\Delta P'$, psia
0.0	1170	0.000			
0.5	1329	0.707	159	729.00	101.02
1.0	1388	1.000	218	365.00	134.47
1.5	1464	1.225	294	243.67	157.31
2	1501	1.414	331	183.00	170.63
3	1570	1.732	400	122.33	204.95
4	1639	2.000	469	92.00	243.36
6	1748	2.449	578	61.67	285.81
10	1899	3.162	729	37.40	300.65
18	2075	4.243	905	21.22	333.95
27	2209	5.196	1039	14.48	357.02
36	2304	6.000	1134	11.11	362.66
45	2375	6.708	1205	9.09	360.86
54	2434	7.348	1264	7.74	375.13
63	2481	7.937	1311	6.78	391.74
71	2516	8.426	1346	6.13	415.29

[*] *Pressure derivative was not given in [34].*

Table 7.1.
Pressure, pressure drop, and pressure derivative versus time data for Example 7.1, after [34][*].

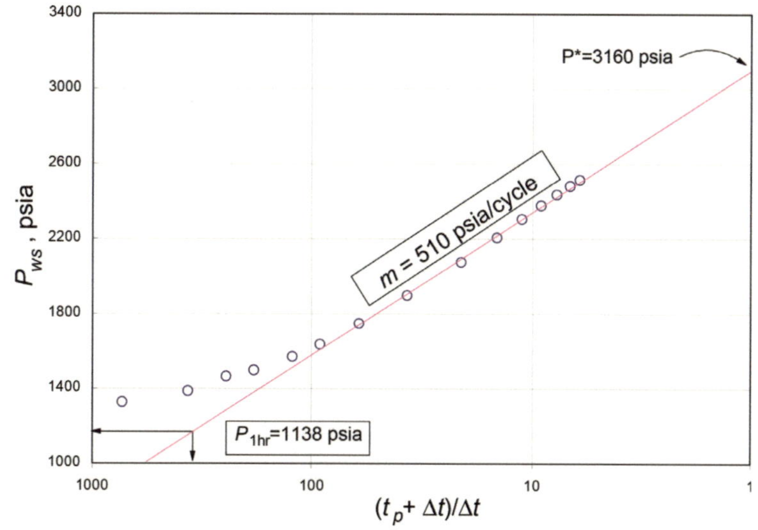

Figure 7.4
Horner plot for Example 7.1, after [34].

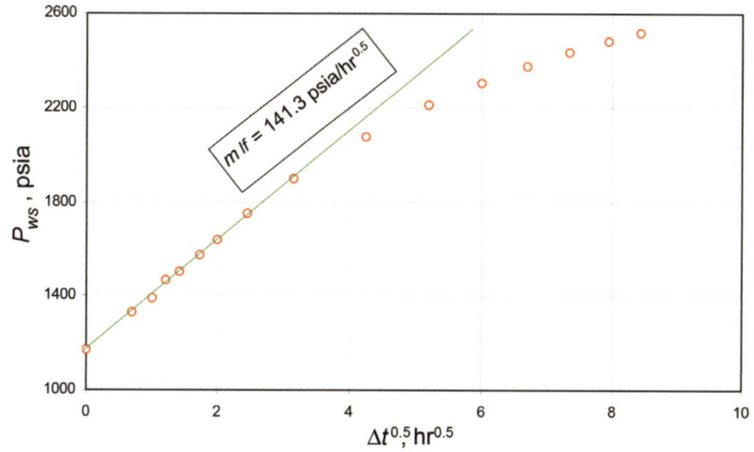

Figure 7.5
Log-log plot of pressure and pressure derivative against time.

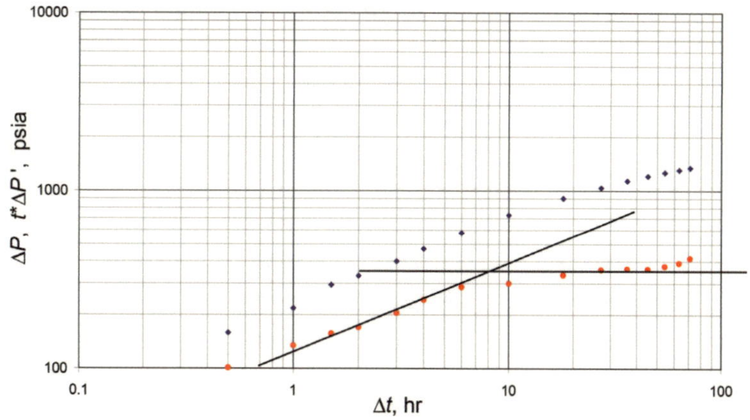

Figure 7.6
Cartesian plot of P_{ws} *against* $t^{0.5}$*, after [34].*

1. Assume $x_f/x_e = 0.5$.

2. With the value of step 1, a correction factor, F_{cor}=0.46, was read from **Figure 7.3** (Horner curve).

3. From **Figure 7.6**, Cartesian plot of P_{ws} versus $t^{0.5}$, the slope m_{lf} is 141.3 psia/h$^{0.5}$.

4. Estimate the kx_f^2 from Eq. (7.7):

$$kx_f^2 = \left(\frac{(4.064)(101)(1.507)}{(42)(141.3)}\right)^2 \frac{(0.45)}{(0.08)(17.7 \times 10^{-6})} = 3452.6 \text{ md} - \text{ft}^2$$

5. Apply the correction factor on Eq. (7.11) to find:

$$k = k_c F_{cor} = (0.52)(0.46) = 0.2392 \text{ md}$$

6. Estimate the half-fracture length:

$$x_f = \sqrt{\frac{kx_f^2}{k}} = \sqrt{\frac{3452.6}{0.2392}} = 120.14 \text{ ft}$$

7. Find x_e from Eq. (7.14):

$$x_e = 0.029\sqrt{\frac{kt_p}{\phi \mu c_t}} = 0.029\sqrt{\frac{(0.2392)(364)}{(0.08)(0.45)(17.7 \times 10^{-6})}} = 339 \text{ ft}$$

8. Calculate the ratio x_f/x_e:

$$\frac{x_f}{x_e} = \frac{120.14}{339} = 0.356$$

Repeat step. 2 through 7 for n iterations until $x_f(i) \approx x_f(i\text{-}1)$. After 12 steps, the permeability is $k = 0.475493$ md. A computer program was written for this purpose. The remaining results are reported in **Table 7.2**. Once the final iteration is achieved, the half-fracture length is found from:

$$x_f = \sqrt{\frac{kx_f^2}{k}} = \sqrt{\frac{3452.6}{0.4754}} = 85.2 \text{ ft}$$

Assumed x_f/x_e	x_e, ft	F_{cor}	k, md	x_f, ft	New x_f/x_e
0.5	337.9897	0.45728	0.237785	120.4982	0.356514
0.356514	385.9966	0.596406	0.310131	105.5117	0.273349
0.273349	449.9846	0.810533	0.421477	90.50785	0.201135
0.201135	471.3004	0.889141	0.462354	86.41439	0.183353
0.183353	476.4818	0.908799	0.472575	85.47472	0.179387
0.179387	477.6316	0.913190	0.474859	85.26894	0.178525

Assumed x_f/x_e	x_e, ft	F_{cor}	k, md	x_f, ft	New x_f/x_e
0.178525	477.8814	0.914146	0.475356	85.22437	0.178338
0.178338	477.9355	0.914352	0.475463	85.21473	0.178298
0.178298	477.9471	0.914397	0.475487	85.21265	0.178289
0.178289	477.9497	0.914407	0.475492	85.2122	0.178287
0.178287	477.9502	0.914409	0.475493	85.2121	0.178287
0.178287	477.9503	0.914409	0.475493	85.21208	0.178287

Table 7.2.
Summary of iterations for Example 7.1.

Since P_{1h} = 1138 psia and using P_{wf} = 1170 psia known from **Table 7.1**, the skin factor is estimated with Eq. (3.6):

$$s = 1.1513\left[\frac{1170 - 1138}{320} - \log\left(\frac{0.671}{(0.08)(0.45)(17.7 \times 10^{-6})(0.28^2)}\right) + 3.23\right] = -4.37$$

As expected, the well is stimulated.

7.3. Type-curve matching

As mentioned before, type-curve matching is attempted to be avoided with *TDS* technique. However, for the interest of the reader, several type-curves have been presented. For this goal, the reader may refer to [5, 7, 29, 31, 32, 39]. The last one included pressure derivative and wellbore storage.

7.4. Fracture conductivity

The product of fracture permeability, k_f, and fracture width, w_f, is known as fracture conductivity, $k_f w_f$,. The conductivity of the dimensionless fracture is expressed mathematically as [6]:

$$C_{fD} = \frac{k_f w_f}{k \cdot x_f} \tag{7.15}$$

The above expression can also be found multiplied by π. However, it is customary, in well test analysis, to be used as given in Eq. (7.15). The uniform flow fracture [25, 26, 35] is one of the concepts introduced in the literature for the interpretation of well test data in fractured wells. This type of conductivity assumes that the flow from the reservoir to the fracture is uniform and a small pressure drop occurs along the fracture [25]. This type of conductivity can be observed in fractures with high damage caused by a zone of low permeability around the fracture. An infinite-conductivity fracture has a conductivity such that the pressure drop along the fracture is considered to be zero. In a log-log plot, this type of fracture is identified by a half slope on the pressure and pressure-derived early data. A fracture is considered to have infinite conductivity and the separation between these two curves should be two times. When its dimensionless fracture conductivity is greater than 300, the fracture has finite conductivity, otherwise [6], which is identified in a log-log plot by a slope of ¼ of the early data on both pressure and pressure derivative. The separation between these two curves should be four times. If this number

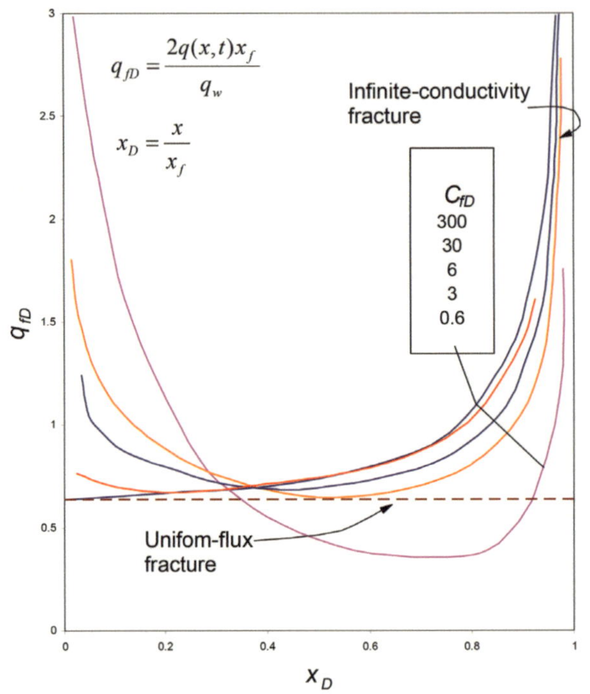

Figure 7.7
Flow distribution along the fracture, after [6].

is higher than four, possible, pseudoskin due to high gas flow rate is presented. A slope ½ may or may not be displayed later. A finite-conductivity fracture involves a pressure drop along the fracture. This pressure drop contributes to the formation of a simultaneous linear flow in the fracture and a linear flow from the formation to the fracture, called bilinear. The long duration of the bilinear flow is a consequence of low fracture conductivity. **Figure 7.7** clearly explains how the flux from formation to the fracture is. There exists a linear flow inside the fracture for finite-conductivity fracture cases.

In *infinite-conductivity* fractures, Tiab [35] showed that the ratio of the length, x_e, with the fracture length, x_f, has some influence on the flow pattern (see **Figure 7.10**). Theoretically, if $x_e = x_f$, only a slope of ½ will be observed indicating the presence of pure linear flow in the formation. However, as the increase of $x_e/x_f \geq 16$ in the straight line of unit slope is short, then only a slop. 0.36 is formed. This is due to the *biradial flow*, Tiab [35] calls it. Other authors have called it elliptical flow. When the relation $x_e/x_f \geq 16$, only the slop. 0.36 is developed and observed.

7.5. Cartesian plot of pressure against one-fourth root of time

The modified bilinear flow equation [2] (originally presented by Cinco et al. [6]), to respond for homogeneous and heterogeneous reservoirs, is presented below:

$$P_D = \frac{2.45}{C_{fD}^{1/2}} \sqrt[4]{\frac{t_{Dxf}}{\xi}} \tag{7.16}$$

After replacing the dimensionless variables (Eqs. 1.89, 7.15, and 7.45), the following expression is obtained:

$$\Delta P = \frac{44.1 q \mu B}{h_f \sqrt{k_f w_f} (\xi \phi \mu c_t k)^{0.25}} t^{0.25} \tag{7.17}$$

The fracture conductivity can be found from the slope, m_{bf}, so that:

$$\sqrt{k_f w_f} = \frac{44.1 q \mu B}{m_{bf} (\xi \phi \mu c_t k)^{0.25} h_f} t^{0.25} \tag{7.18}$$

For this case, x_f can also be expressed using the abovementioned test and error procedure. When the bilinear flow ends, the graph exhibits curvature either concave up or down depending on C_{fD}. If $C_{fD} < 1.6$, there will be concavity downward. If $C_{fD} > 1.6$, the concavity up toward that indicates that the tip of the fracture begins to affect the pressure well behavior.

If the test is not run long enough to terminate the bilinear flow when $C_{fD} > 1.6$, it is not possible to determine the half-fracture length. When $C_{fD} < 1.6$, the flow of fluid in the reservoir has changed from a predominant one-dimensional linear flow and a two-dimensional flow regime. In this case, it is not possible to properly determine x_f even if the bilinear flow ends during the test. These rules can be avoided with *TDS* technique and the pressure derivative curve.

Cinco et al. [6] indicated that C_{fD} can be estimated from a Cartesian graph of P versus $t^{1/4}$ reading the value of ΔP when bilinear flow ends, ΔP_{ebf} by:

$$C_{fD} \approx \frac{194.9 q \mu B}{k h \Delta P_{ebf}} \tag{7.19}$$

Cinco et al. [6] also showed that the end of the bilinear flow line "*ebf*" depends on C_{fD} and can be estimated from:

$$t_{Debf} \approx \frac{0.1}{C_{fD}^2}; \; C_{fD} > 3 \tag{7.20}$$

$$t_{Debf} \approx 0.0205 \left[C_{fD} - 1.5\right]^{-1.53}; \; 1.6 \le C_{fD} \le 3 \tag{7.21}$$

$$t_{Debf} \approx \left[\frac{4.55}{\sqrt{C_{fD}}} - 2.5\right]^{-4}; \; C_{fD} \le 1.6 \tag{7.22}$$

Since C_{fD} and $k_f w_f$ are known, then x_f can be estimated from the definition of C_{fD}.

7.6. Cartesian plot of pressure against the square root of time

Cinco et al. [6] and Cinco-Ley and Samaniego [7] presented the following expressions:

$$\Delta P = \frac{4.064 q B}{h_f x_f} \sqrt{\frac{\mu t}{\xi \phi c_t k}} \tag{7.23}$$

Eq. (7.23) has the modification given by Bettam et al. [2]. The slope m_{lf} is obtained from the Cartesian plot and is useful to find the half-fracture length:

$$x_f = \frac{4.064 q B}{m_{lf} h_f} \sqrt{\frac{\mu}{\xi \phi c_t k}} \tag{7.24}$$

The outer boundary can distort the semilog line if $x_f > x_e/3$. The pressure behavior during the infinite-acting period is very dependent on x_f. For relatively short fractures, the flow is radial but becomes linear as x_f grows and reaches x_e. The m (semilog) obtained from conventional analysis of a fractured well is erroneously very small and the value of m decreases progressively as x_f increases [6, 7, 26, 33], and hence the calculation of the half-fracture length requires trial and error.

The smaller the flow capacity, the longer the curved portion. The beginning of the linear flow in the formation "*blf*" depends on C_{fD} and can be approximated by [6, 7, 26, 33]:

$$t_{Dblf} = \frac{100}{(C_{fD})^2} \tag{7.25}$$

And at the end of the linear flow period, "*elf*" occurs approximately at:

$$t_{Dblf} = 0.016 \tag{7.26}$$

t_{elf} and t_{blf} are the times for the end and beginning of the linear flow regime and serve to determine the dimensionless conductivity of fracture.

$$C_{fD} = 0.0125\sqrt{\frac{t_{elf}}{t_{blf}}} \tag{7.27}$$

The linear flow in the fracture ends as a function of the value of the dimensionless hydraulic diffusivity of the fracture [7] η_{fD}:

$$t_{Dxf} = \frac{0.01(C_{fD})^2}{(\eta_{fD})^2} \tag{7.28}$$

$$\eta_{fD} = \frac{k_f \phi c_{ft}}{k_f \phi c_t} \tag{7.29}$$

The pressure data during the transition period show a curved portion before the line representing the linear flow is obtained. The duration of the curved part represents the transition and depends on the flow capacity of the fracture. For $C_{fD} > 0.5$, the start time of the linear flow regime is governed by:

$$t_{blf} = \frac{227.8\mu c_t x_f^2}{k}(C_{fD}^{-1.39}) \tag{7.30}$$

The linear flow ends at a dimensionless time of approximately 0.016 and the pseudoradial flow starts at a t_D of about 3 and continues until the boundaries have been felt. The pseudoradial flow does not appear if the distance to the border is 10 times smaller than x_f. The equation that approximates this flow regime is [26, 33]:

$$P_{D_f} = 0.5 \ln t_{D_{lf}} + 1.1 \tag{7.31}$$

The uniform flux has less duration than the linear. In the linear, the pseudoradial period is achieved earlier at a $t_D \approx 1$. For uniform flow, the fracture length is:

$$x_f = r_w e^{-s+1} \tag{7.32}$$

$$s = -\ln\frac{x_f}{r_w} - 1 \tag{7.33}$$

In fractured wells, the pseudoradial flow is governed by:

$$P_{Df} = 0.5 \ln t_{Df} + f(C_{fD}) \quad (7.34)$$

The pseudoradial flow period is identical to the radial flow of an unfractured well but with a negative damage factor caused by the influence of the fracture. During this period, the behavior of the pressure is described by:

$$P_{Df} = 0.5 \ln\left(\frac{x_f^2 t_{Dxf}}{r_w^2}\right) + 0.404 + s \quad (7.35)$$

The start of the semilog line is given by [6]:

$$t_{Dssl} = 5 \exp\left[-0.5\left(C_{fD}^{-0.6}\right)\right] \quad (7.36)$$

There is an approximate relationship [6, 26, 33] between ΔP_{elf} and ΔP_{brf}:

$$\Delta P_{brf} = 2\Delta P_{elf} \quad (7.37)$$

This rule is known as the "double *P* rule." For fractured wells, twice the P_{elf} marks the beginning of the pseudoradial flow. Equivalently, a time rule referred to as "rule 10*t*" can be applied at the beginning of the pseudoradial flow, t_{brf}, by:

$$t_{brf} = 10 t_{elf} \quad (7.38)$$

Another approach that can be used to mark the beginning of the radial flow for finite-conductivity fractures is:

$$t_{Dbrf} \approx 5 \exp\left(C_{fD}^{-0.6}\right); \; C_{fD} > 0.1 \quad (7.39)$$

The fracture length can be determined with the following expressions and the aid of **Figure 7.8** for the determination of *n* [6, 7, 26].

$$x_f = n r_w e^{-s} \quad (7.40)$$

$$s = -\ln \frac{x_f}{n r_w} \quad (7.41)$$

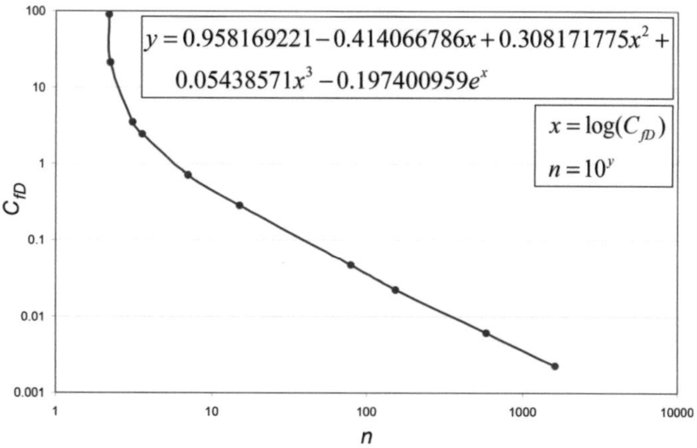

Figure 7.8
Determination of n, *after [6, 7, 26].*

Alternatively, instead of using **Figure 7.8**, the below polynomial fit can be used:

$$y = 0.958169221 - 0.414066786x + 0.308171775x^2 + 0.05438571x^3 - 0.197400959e^x \tag{7.42}$$

$$x = \log(C_{fD}) \tag{7.43}$$

$$n = 10^y \tag{7.44}$$

Theoretically, bilinear flow regime takes place at a dimensionless time given by:

$$t_{Dxf} = \frac{0.1}{C_{fD}^2};\ C_{fD} > 16 \tag{7.45}$$

$$t_{Dxf} = \left(\frac{4.55}{C_{fD}^{1/2}} - 2.5\right) - 4;\ C_{fD} < 16 \tag{7.46}$$

On the other hand, the occurrence of linear flow formation is characterized by a slope of 1/2 in the graph log-log of pressure and pressure derivative. This flow regime will normally be evident and analyzable for fractures with high conductivity ($C_{fD} > 100$). The beginning of the linear flow regime occurs in:

$$t_{Df} C_{fD}^2 = 100 \tag{7.47}$$

To verify that the data used for the analysis actually represent linear flow, Eq. (7.7) was properly applied; the valid range of data occurs during:

$$\frac{100}{C_{fD}^2} < t_{Dxf} < 0.016 \tag{7.48}$$

Based on the time at which the linear flow ends, $t_{Dxf} = 0.016$, it is possible to estimate the permeability of the formation. At the end of the linear flow, the data of P versus $Dt^{1/2}$ deviate from the straight line. Using the time of this deviation with Eqs. (7.15) and (7.50) will yield:

$$k = \frac{101.1 q \mu B}{h \cdot m_{vf} \sqrt{t_{elf}}} \tag{7.49}$$

In Eq. (7.49), t_{elf} represents the end of the linear flow regime.

Further in this chapter, the biradial flow will be characterized (**Figure 7.11**). The pressure equation for such flow was presented by [16]:

$$P_D = 2.14 \left(\frac{x_e}{x_f}\right)^{0.72} \left(\frac{t_{DA}}{\xi}\right)^{0.36} \tag{7.50}$$

After replacing the dimensionless quantities in the above expression:

$$\Delta P = m_{ell} t^{0.36} \tag{7.51}$$

where

$$m_{ell} = 15.53 \left(\frac{qB\mu}{kh}\right) \left(\frac{x_e}{x_f}\right)^{0.72} \left(\frac{k}{\xi \mu \phi c_t A}\right)^{0.36} \tag{7.52}$$

which indicates that a straight line will be obtained from a Cartesian plot of ΔP versus $t^{0.36}$ (for drawdown) or ΔP versus $[(t_p+\Delta t)^{0.36} - \Delta t^{0.36}]$ (for buildup). The slope, m_{ell}, of such line will be useful to find the half-fracture length:

$$x_f = 45.124 x_e \left(\frac{qB\mu}{khm_{ell}}\right)^{25/18} \sqrt{\frac{k}{\xi\mu\phi c_t A}} \tag{7.53}$$

7.7. *TDS* technique for hydraulically fractured vertical wells

This section deals with the analysis of test data from wells that have been fractured hydraulically. Initially, hydraulic fracturing became a good way to increase the productivity of completed wells in low permeability reservoirs. However, lately, it has become a common practice thanks to its impact to increase well productivity and remove damage. The purpose of fracture well tests is to determine fracture and reservoir properties to provide an effective assessment of fracture treatment and to predict long-term productivity for the reservoir. The fracture does not alter the permeability of the reservoir but it alters the average permeability of the system. Basically, fracturing increases the effective radius of the face of the well:

$$r_{wa} = x_f/2 = r_w e^{-s} \tag{7.54}$$

After a well has been fractured, a new group of flow regimes is formed. The main flow regimes are presented in **Figure 7.9** and are as follows [6, 26, 33].

- Linear flow in the fracture
- Bilinear flow (fracture and formation)
- Linear flow in the formation (or elliptical)
- Pseudoradial flow

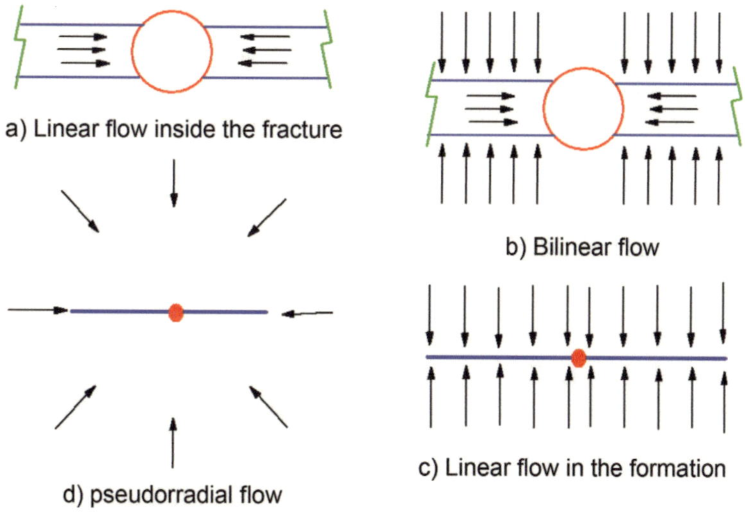

Figure 7.9
Flow regimes governing pressure behavior in a with a finite-conductivity fracture [6, 36].

For infinite conductivity and uniform flow fracture systems, only the third and fourth flow regimes can be seen in the pressure data. Linear flow usually occurs at a very early time, since it is normally masked by wellbore storage effects. The onset of pseudoradial flow can occur at a time that is economically unachievable and therefore cannot occur at any time during a well test. To determine *kh* of the reservoir, it is necessary that the reservoir is in radial flow, unless the interpretation is conducted by *TDS* technique that, in most cases, may be successfully interpreted with having the radial flow regime. A typical case of this is when the fracture is of finite conductivity and the slopes of a half and a quarter are observed; it is possible to obtain the permeability of the point of intersection between these lines. So whenever an analysis of a fractured well test is required, it is important that a prefracture test is involved to determine the *kh* of the reservoir, if conventional methods or type curves are used. If this does not occur, a unique analysis of the data may not be possible, since there are two unknowns: reservoir permeability and fracture length [35, 36].

Wellbore storage may mask the first of the three flow regimes. If this occurs, analysis to determine fracture length is not possible. In this case, the success of the fracture treatment will have to be determined using the calculated skin factor. As a general rule, a fracture is successful if the skin factor is reduced to less than −3. If the effects of storage are short-lived, then bilinear flow or linear flow can be analyzed to determine fracture length and conductivity. For analysis of fractured wells, a new set of dimensionless parameters is used. These are the dimensionless time for a fractured well, t_{Dxf}, (Eq. 7.45) and the dimensionless fracture conductivity, C_{fD} (Eq. 7.15).

$$t_{Dxf} = \frac{0.0002637kt}{\phi\mu\, c_t x_f^2} \tag{7.55}$$

7.7.1 Hydraulic fractured wells in bounded systems

For the case of a uniform-flux fracture, the pressure derivative plots for various x_e/x_f ratios reveal three dominant flow periods. During early times, the flow of fluids is linear and can be identified by a straight line of a slope of 0.5. The linear flow line is used to calculate the average half-fracture length. The infinite-action radial flow regime, which can be identified by a horizontal straight line, is dominated by $x_e/x_f > 8$. This flow regime is used to calculate permeability and skin factor. The third straight line, which corresponds to the pseudosteady-state period, has a unit slope. This line is used to calculate the drainage area and the shape factor. For the case of infinite-conductivity fracture, pressure derivative plots show a fourth dominant flow regime, referred to here as biradial flow. This flow regime, which can be identified by a straight line of slop. 0.36, can also be used to calculate the half-fracture length and permeability [35].

7.7.2 Characteristics of uniform-flux fracture

Figure 7.10 shows a log-log plot of pressure and the pressure derivative group versus dimensionless time for three values of x_e/x_f. These curves have several unique characteristics, which can be used to interpret pressure transient tests in fractured wells without using type-curve matching [35].

(1) For short production times, the flow in the fracture is linear. The duration of this flow regime is a function of the penetration ratio x_e/x_f. The equation corresponding to this straight line at early times is:

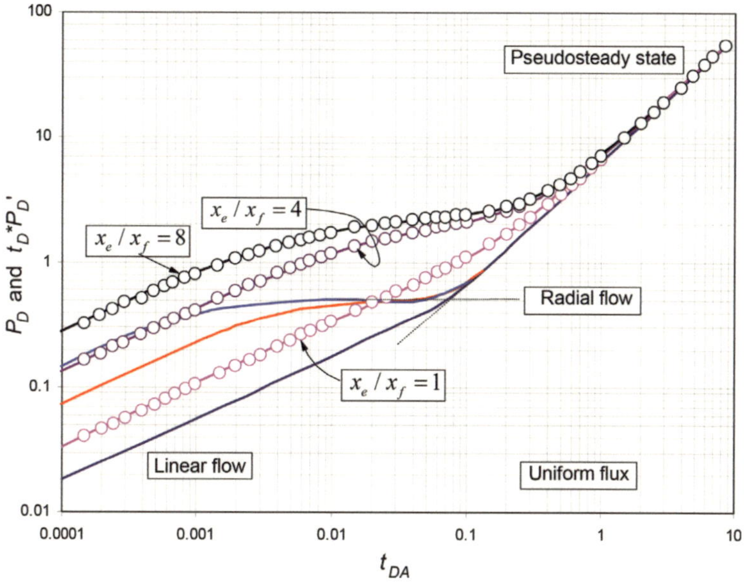

Figure 7.10
Pressure derivative behavior for a uniform-flux fracture inside a square reservoir, after [35].

$$t_{DA}{}^*P_D{}' = 1.772\frac{x_e}{x_f}\sqrt{\frac{t_{DA}}{\xi}} \tag{7.56}$$

Taking logarithm at both sides, it yields:

$$\log(t_{DA}{}^*P_D{}') = 0.5\log\left(\frac{t_{DA}}{\xi}\right) + \log\left(\frac{\sqrt{\pi x_e}}{x_f}\right) \tag{7.57}$$

The slope of this straight line is 0.5, which in itself is a unique feature of the linear flow regime. Substituting the dimensionless quantities in Eq. (7.4) and solving for the well pressure derivative, the following is obtained:

$$t^*\Delta P' = 0.5m_L\sqrt{t} \tag{7.58}$$

where

$$m_L = \frac{4.064qB}{h}\sqrt{\frac{\mu}{\xi\phi c_t k x_f^2}} \tag{7.59}$$

Taking logarithm at both sides of the above expression:

$$\log(t^*\Delta P') = 0.5\log(t) + \log(0.5m_L) \tag{7.60}$$

This expression shows that a graph of $t^*\Delta P'$ versus time in a log-log graph will produce a straight line of slop. 0.5 if the linear flow regime is dominant. Let $(t^*\Delta P')_{L1}$ be the value of $(t^*\Delta P')_{L1}$ at a time $t = 1$ hr in the straight line of the linear flow regime (extrapolated, if necessary). Then, combining Eqs. (7.59) and (7.60) and solving for the half-fracture length, x_f, gives [23] and [35]:

$$x_f = \frac{2.032qB}{h(t^*\Delta P')_{L1}}\sqrt{\frac{\mu}{\xi\phi c_t k}} \tag{7.61}$$

The equation of the linear flow line portion of the pressure curve is:

$$\Delta P = m_L \sqrt{t} \tag{7.62}$$

Let $(\Delta P)_{L1}$ be the value of ΔP in the straight line (extrapolated if necessary) at time $t = 1$ hr. Thus, after substituting for m_L of Eq. (7.62), it results:

$$x_f = \frac{4.064qB}{h(\Delta P')_{L1}} \sqrt{\frac{\mu}{\xi \phi c_t k}} \tag{7.63}$$

(2) After the linear flow regime, radial flow is developed. It is used as seen in Chapter 2. Then, Eqs. (2.71) and (2.92) apply for the estimation of permeability and skin factor.

(3) For long production times, the pressure derivative function will produce a unit-slope straight line. This line corresponds to the pseudosteady-state period, starting at a t_{DA} value of approximately 0.2. The equation of this straight line is given by Eq. (2.96) and it is useful to estimate the drainage area. If the dimensionless quantities are substituted in Eq. (2.96), and solving for ΔP will yield,

$$t^*\Delta P' = \left(\frac{qB}{4.27\phi c_t}\right) t \tag{7.64}$$

This expression leads to find Eqs. (2.98) and (2.99).

(4) The dimensionless pressure during pseudosteady-state period is a linear function of the dimensionless time. The equation corresponding to this period is [35]:

$$P_D = 2\pi t_{DA} + \ln(x_e/x_f) \ln \sqrt{\frac{2.2458}{C_A}} \tag{7.65}$$

Dividing Eq. (7.65) by Eq. (7.56),

$$\frac{P_D}{t_{DA}{}^* P_D{}'} = 1 + \frac{1}{2\pi\, t_D} \ln\left(\frac{x_e}{x_f}\sqrt{\frac{2.2458}{C_A}}\right) \tag{7.66}$$

From which is obtained after replacing the dimensionless variables:

$$C_A = 2.2458\left(\frac{x_e}{x_f}\right)^2 \exp\left[\frac{0.000527kt_{ps}}{\phi\mu\, c_t A}\left(1 - \frac{(\Delta P)_{ps}}{(t^*\Delta P')_{ps}}\right)\right] \tag{7.67}$$

or

$$C_A = 2.2458(x_e/x_f)^2 \tag{7.68}$$

If $(\Delta P)_{ps} = (t^*\Delta P')_{pss}$

(5) The point of intersection of the linear flow line and the infinite-action radial flow line is unique. The coordinates of this point can be obtained by setting Eq. (7.56) to 0.5 and solving for the dimensionless intersection time:

$$t_{DALri} = \frac{1}{4\pi}\left(\frac{x_f}{x_e}\right)^2 \tag{7.69}$$

Substituting Eq. (1.100), setting $A = 4x_e^2$, and solving for x_f^2/k, it yields:

$$\frac{x_f^2}{k} = \frac{t_{Lri}}{1207\xi\phi\mu c_t} \tag{7.70}$$

(6) The linear flow line and the pseudosteady-state line intercept at:

$$t_{DALpssi} = \frac{1}{4\pi}\sqrt{\frac{x_e}{x_f}} \tag{7.71}$$

Substituting Eq. (1.100),

$$kx_f^2 = \frac{7544\xi\phi\mu c_t A^2}{t_{Lpssi}} \tag{7.72}$$

This equation can be used for verification purpose or to calculate k given that x_f is known.

(7) Combining Eqs. (7.69), (7.71), and the time of intercept of the pseudosteady-state with the radial lines provides:

$$\frac{t_{Lri}}{t_{rppsi}} = \frac{t_{rppsi}}{t_{Lpssi}} = \sqrt{\frac{t_{Lri}}{t_{Lpssi}}} = (x_e/x_f)^2 \tag{7.73}$$

This expression can be used for verification purposes. It is also used when designing a pressure test in a well intercepted by a vertical fracture.

7.7.3 Characteristics of infinite-conductivity fractures

Figure 7.11 is a graph of pressure dimensionless and pressure derivative versus dimensionless time based on area for a vertical fracture of infinite conductivity within a square system. This figure shows the existence of four straight lines: (*a*) half-slope linear flow line, (*b*) 0.36-slope biradial flow line, (*c*) infinity-acting radial

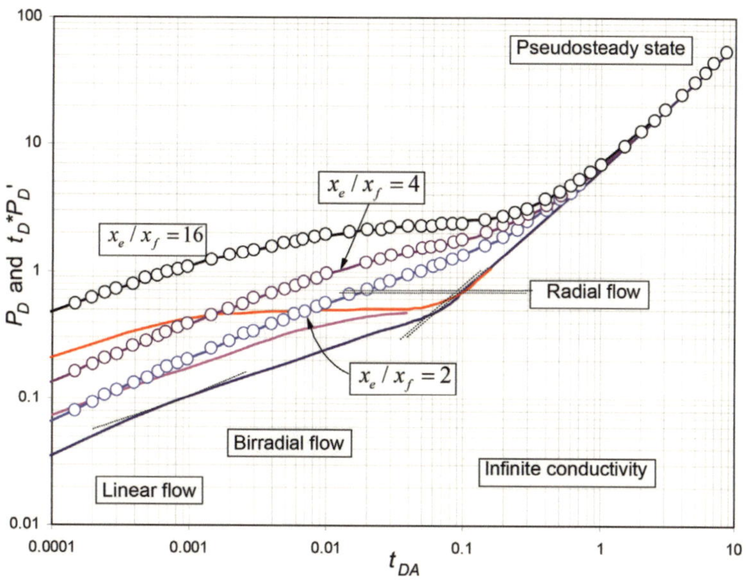

Figure 7.11
Pressure derivative behavior for an infinite-conductivity fracture inside a square reservoir, after [35].

flow line (horizontal line), and (*d*) unit-slope pseudosteady-state flow line. For $x_e/x_f > 8$, the linear flow regime is almost nonexistent, and the biradial flow line is observed first. For $x_e/x_f < 8$, the biradial flow line disappears [34]. Only the characteristics of the biradial flow regime will be discussed here. The characteristics and interpretation of the other three flow regimes (linear, radial, and pseudosteady state) are the same as discussed above for uniform-flow fracture.

(1) The equation of the biradial flow regime line introduced by [35] and modified by Bettam et al. [2] is:

$$t_{DA}{}^{*}P_D{}' = 0.769\left(\frac{x_e}{x_f}\right)^{0.72}\left(\frac{t_{DA}}{\xi}\right)^{0.36} \tag{7.74}$$

Taking logarithm to both members of the above equation leads to:

$$\log(t_{DA}{}^{*}P_D{}') = 0.36\log\left(\frac{t_{DA}}{\xi}\right) + \log\left[0.769\left(\frac{x_e}{x_f}\right)^{0.72}\right] \tag{7.75}$$

In dimensional form, Eq. (7.74) becomes:

$$t^{*}\Delta P' = 0.769C_{BR}(x_e/x_f)^2 t^{0.36} \tag{7.76}$$

where

$$C_{BR} = 7.268\frac{q\mu B}{kh}\left(\frac{k}{\xi\phi\mu c_t A}\right)^{9/25} \tag{7.77}$$

Taking logarithm to both sides of Eq. (7.76) yields:

$$\log(t^{*}\Delta P') = 0.36\log t + \log\left(0.7699C_{BR}\left(\frac{x_e}{x_f}\right)^{0.72}\right) \tag{7.78}$$

Thus, the biradial flow line can be identified by its slope of 0.36. Let $(t^{*}\Delta P')_{BR1}$ be the value of pressure derivative at a time $t = 1\text{hr}$ in the straight line (extrapolated if necessary). An expression to find the half-fracture length is found from Eq. (7.76) when linear flow regime is absent:

$$x_f = 0.694x_e\left[\frac{C_{BR}}{(t^{*}\Delta P')_{BR1}}\right]^{1.388} \tag{7.79}$$

C_{BR} is found from Eq. (7.77).

(2) The time of intersection between the linear flow and biradial flow regimes is given by Eqs. (7.56) and (7.74):

$$t_{DALBRi} = 0.00257\left(\frac{x_f}{x_e}\right)^2 \tag{7.80}$$

Substituting the dimensionless time and solving for x_f^2/k yields:

$$\frac{x_f^2}{k} = \frac{t_{LBRi}}{39\xi\phi\mu c_t} \tag{7.81}$$

If the radial flow is too short, permeability can be found from an expression obtained by combining the above expression and Eq. (7.61):

$$k = \left(\frac{12.67q\mu B}{h(t^*\Delta P')_{L1}}\right)\frac{1}{\sqrt{t_{LBRi}}} \tag{7.82}$$

(3) The time of intersection between the biradial and radial flow regime lines can be used to verify k and x_f:

$$t_{DArBRi} = 0.3023\left(\frac{x_f}{x_e}\right)^2 \tag{7.83}$$

Substituting the dimensionless time in the above expression and solve for x_f^2/k:

$$\frac{x_f^2}{k} = \frac{t_{BRri}}{4587\xi\phi\mu c_t} \tag{7.84}$$

(4) The time of intersection between the biradial flow regime line and the pseudosteady-state line (Eqs. 2.96 and 7.74) provides:

$$t_{DABRpssi} = 0.03755\left(\frac{x_e}{x_f}\right)^{1.125} \tag{7.85}$$

After substituting the dimensionless time based on area, Eq. (1.100) in Eq. (7.85) leads to:

$$k = \frac{142.3\xi\phi\mu c_t A}{t_{BRpssi}}\left(\frac{x_e}{x_f}\right)^{1.125} \tag{7.86}$$

(5) Combination of Eqs. (7.81) with (7.84) and (7.83) with (7.85) will, respectively, yield:

$$t_{BRri} = 117.6t_{LBRi} \tag{7.87}$$

$$t_{BRri} = 8\left(\frac{x_f}{x_e}\right)^{2.125} t_{BRpssi} \tag{7.88}$$

which can be used for either verification or test design purposes.

7.7.4 Rectangular systems

For both types of fractures in rectangular systems, the transition between the infinite-action radial flow and the pseudosteady-state period is much longer than for a square system since, in the first one, formation linear flow exists as described in Section 2.7. The equation of this straight line proposed by Tiab [35] and modified by Escobar et al. [16] to include naturally fractured formations is:

$$t_{DA}{}^*P_D{}' = 3.545\sqrt{\frac{t_{DA}}{\xi}} \tag{7.89}$$

Substituting the dimensionless terms:

$$t^*\Delta P' = m_{CB}\sqrt{t} \tag{7.90}$$

where

$$m_{CB} = 8.128\frac{qB}{h}\left(\frac{\mu}{\xi\phi c_t A}\right)^{0.5} \tag{7.91}$$

where $(t^*\Delta P')_{CB1}$ is the value of $(t^*\Delta P')$ at a time $t = 1$hr on the formation linear flow straight line (extrapolated if necessary). Permeability can be solved from the above equation, so that:

$$k = 66.0712\left(\frac{qB}{h(t^*\Delta P')_{CB1}}\right)^2 \frac{\mu}{\xi\phi c_t A} \tag{7.92}$$

Tiab [35] presents step-by-step procedures for the interpretation of pressure tests in fractured wells. These procedures are not included here.

Example 7.2

Tiab [35] presented an example of a pressure test in a highly productive fractured well. Pressure and pressure derivative [13] data versus time are reported in **Figure 7.12** and **Table 7.2**. Other relevant data are given below:

$q = 2000$ STB/D, $\phi = 0.24$, $m = 0.3$ cp, $c_t = 14.8 \times 10^{-6}$ psia^{-1}, $B = 1.5$ bbl/STB, $h = 50$ ft, $r_w = 0.4$ ft, $P_i = 5200$ psia

Find permeability, skin factor, and half-fracture length. Verify the value of the half-fracture length.

Solution

Since it is a homogenous reservoir, then $\xi = 1$. The following characteristic points were read from **Figure 7.12** and **Table 7.3**:

$t_r = 48$ hr, $\Delta P_r = 507$ psia, $(t^*\Delta P')_r = 105.5$ psia, $t_{LRi} = 1.2$ hr, $t_{LBRi} = 0.047$ hr, $t_{rBRi} =$ 4.5 hr

Estimate permeability using Eq. (2.71) and skin factor with Eq. (2.92):

$$k = \frac{70.6q\mu B}{h(t^*\Delta P')_r} = \frac{70.6(2000)(0.3)(1.5)}{50(105.5)} = 12 \text{ md}$$

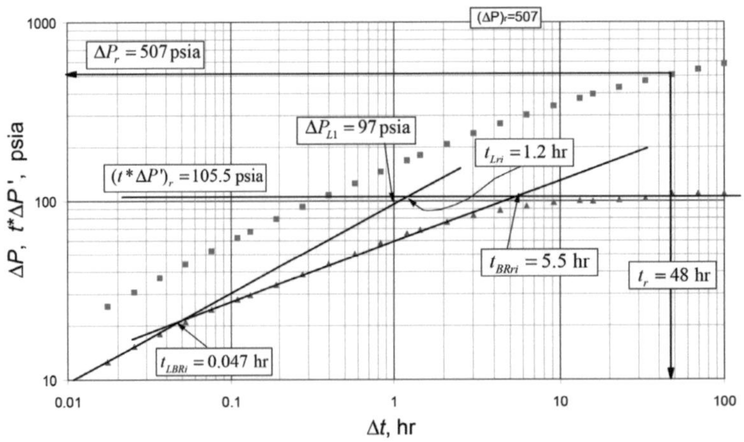

Figure 7.12
Pressure and pressure derivative against time log-log plot of Example 7.3.

t, hr	P, psia	t*DP', psia	t, hr	P, psia	t*DP', psia
0.017378	5174.322	12.605	1.44544	5019.68	68.901
0.025119	5169.16	15.287	2.089296	4992.606	76.039
0.036308	5163.033	18.027	3.019952	4962.879	83.251
0.052481	5155.836	21.039	4.365158	4930.874	88.885
0.075858	5147.48	24.541	6.309573	4897.025	93.405
0.109648	5137.868	28.044	9.120109	4861.749	98.014
0.131826	5132.556	29.705	13.18257	4825.404	100.147
0.190546	5120.828	33.792	15.84893	4806.922	99.820
0.275423	5107.467	38.787	22.90868	4769.489	100.962
0.398107	5092.232	44.292	33.11311	4731.58	103.280
0.57544	5074.815	50.627	47.86301	4693.335	109.456
0.831764	5054.879	58.306	69.1831	4654.853	109.456
1.202264	5032.142	65.746	100	4616.205	109.456

Table 7.3.
Pressure and pressure derivative against time data of Example 7.3.

$$s = 0.5\left[\frac{507}{105.5} - \ln\left(\frac{12 \cdot 48}{(0.24)(0.3)(14.8\times 10^{-6})(0.4)^2}\right) + 7.43\right] = -4.85$$

Find half-fracture length with Eq. (7.63):

$$x_f = \frac{2.032qB}{h(t^*\Delta P')_{L1}}\sqrt{\frac{\mu}{\phi\, c_t k}} = \frac{2.032(2000)(1.5)}{50(97)}\sqrt{\frac{0.3}{0.24(14.8\times 10^{-6})(12)}} = 105.4 \text{ ft}$$

Recalculate the half-fracture length with Eqs. (7.70), (7.81), and (7.84), respectively.

$$x_f = \sqrt{\frac{t_{Lri}k}{1207\xi\phi\mu c_t}} = \sqrt{\frac{(1.2)(12)}{1207(0.24)(0.3)(14.8\times 10^{-6})}} = 105.81 \text{ ft}$$

$$x_f = \sqrt{\frac{t_{BRLi}k}{39\xi\phi\mu c_t}} = \sqrt{\frac{(0.047)(12)}{39(0.24)(0.3)(14.8\times 10^{-6})}} = 116.5 \text{ ft}$$

$$x_f = \sqrt{\frac{t_{BRri}k}{4587\xi\phi\mu c_t}} = \sqrt{\frac{(5.5)(12)}{4587(0.24)(0.3)(14.8\times 10^{-6})}} = 116.2 \text{ ft}$$

7.7.5 Finite-conductivity fractured vertical wells

A log-log plot of pressure and pressure derivative versus test time for a fractured well in a closed system may reveal the presence of several straight lines corresponding to different flow regimes, excluding wellbore storage, such as (a) bilinear flow characterized by a slope of ¼ in the pressure and pressure derivative curve, (b) linear flow, (c) infinite-action radial flow, and (d) pseudosteady-state period. The slopes and points of intersection of these straight lines are unique and were used by Tiab et al. [36, 38] to find expressions for well test interpretation.

The characteristics of bilinear flow were first discussed by Cinco et al. [6]. It is called bilinear flow because it is the result of two linear flow regimes. A flow regime is the incompressible linear flow of the fracture and the other flow regime is the compressible linear flow in the formation, as shown in **Figure 7.9**. They showed mathematically that bilinear flow exists whenever (a) most of the fluid entering the well face comes from the formation and (b) the effects of the fracture do not affect well behavior.

During the bilinear flow regime, the behavior of the dimensionless pressure of the well given by Tiab et al. [36] and modified by Bettam et al. [2] is:

$$P_D = \frac{2.45}{C_{fD}^{1/2}} \sqrt[4]{\frac{t_{Dxf}}{\xi}} \tag{7.192}$$

Replacing the dimensionless parameters given by Eqs. (1.89), (7.3), and (7.15), Eq. (7.92) becomes:

$$\Delta P = m_{BL} \sqrt[4]{t} \tag{7.93}$$

$$m_{BL} = \frac{44.13}{(\xi\phi\mu c_t k)^{0.25}} \frac{q\mu B}{h_f \sqrt{k_f w_f}} \tag{7.94}$$

The fracture conductivity is solved from Eq. (7.93):

$$k_f w_f = 1947.46 \frac{1}{\sqrt{\xi\phi\mu c_t k}} \left(\frac{q\mu B}{h(\Delta P)_{BL1}}\right)^2 \tag{7.95}$$

The derivative of Eq. (7.92) is:

$$t_D{}^*P_D{}' = \frac{0.6125}{C_{fD}^{1/2}} \sqrt[4]{\frac{t_{Dxf}}{\xi}} \tag{7.96}$$

Replacing the dimensionless quantities, given by Eqs. (2.57), (7.3), and (7.15), in Eq. (7.96) and solving for the fracture conductivity:

$$k_f w_f = \frac{121.74}{\sqrt{\xi\phi\mu c_t k}} \left(\frac{q\mu B}{h(t^*\Delta P')_{BL1}}\right)^2 \tag{7.97}$$

Since this is linear flow, Eqs. (7.61) and (7.63) also apply for finite-conductivity fractures.

The intercept between linear flow and bilinear flow lines given by the governing pressure derivative solutions (Eqs. 7.4 and 7.92) leads to:

$$t_{BLLi} = 13910\xi\phi\mu c_t \left(\frac{x_f^2\sqrt{k}}{k_f w_f}\right) \tag{7.98}$$

Solving for *k*,

$$k = \left(\frac{k_f w_f}{x_f^2}\right)^2 \frac{t_{BLLi}}{13910\xi\phi\mu c_t} \tag{7.99}$$

The pressure derivative of Eq. (7.4) is:

$$t_D{}^*P_D = \frac{1}{2}\sqrt{\frac{\pi t_{Dxf}}{\xi}} \tag{7.100}$$

The intercept between linear flow and bilinear flow lines given by the governing pressure derivative solutions (Eqs. 7.100 and 7.96) leads to:

$$k = \left(\frac{k_f w_f}{x_f^2}\right)^2 \frac{16t'_{BLLi}}{13910\xi\phi\mu c_t} \tag{7.101}$$

Eqs. (7.99) and (7.101) can be used for verification purposes, if all three flow regimes are observed. If the test is too short to observe the radial flow line, or a prefracture test is not possible as in the low permeability formation, then Eqs. (7.99) and (7.101) can be used to calculate the permeability of the formation. Also, needless to say that Eqs. (2.71) and (2.93), along with many other relationships in Chapters 2 and 3, will apply to fractured wells.

The intersection of Eq. (2.70), neglect C_D, and Eq. (7.96) leads after rearranging:

$$0.25m_{BL}t_{BLri}^{0.25} = \frac{70.6q\mu B}{kh} \tag{7.102}$$

Solving for the intersection time,

$$t_{BLri} = 1677\frac{\xi\phi\mu c_t}{k^3}(k_f w_f)^2 \tag{7.103}$$

which Tiab et al. [36] recommend to be used for verification purpose.

The intersection of the biradial flow regime pressure derivative (Eq. 7.74) with the bilinear flow regime pressure derivative (Eq. 7.96), lines will result in:

$$t_{BLBRi}^{0.11} = \frac{0.25m_{BL}}{0.7699C_{BR}\left(\dfrac{x_e}{x_f}\right)^{0.72}} = \frac{1.197k^{0.39}(\xi\phi\mu c_t k)^{0.11}x_f^{0.72}}{\sqrt{k_f w_f}\left(\dfrac{x_e}{x_f}\right)^{0.72}} \tag{7.104}$$

being m_{BL} and C_{BR} defined by Eqs. (7.94) and (7.77), respectively. Either half-fracture length or conductivity can be solved from Eq. (7.104).

The intersect of the pressure derivative bilinear governing expression (Eq. 7.96) with the pressure derivative pseudosteady-state period Eq. (2.96) will lead to:

$$k_f w_f = 2220.603A^2k\left(\frac{\xi\phi\mu c_t}{kt_{BLpssi}}\right)^{3/2} \tag{7.105}$$

7.7.6 Special Cases

As also mentioned by Tiab et al. [36, 38], the above assumption assumes that all three flow regimes (bilinear, linear formation, and radial) are observed during the pressure test and that these are well defined in the pressure derivative curve. In many instances, at least one of the flow regimes is not observed or defined. For example, when the fracture has low conductivity, let us say, $C_{fD} < 5$, probably linear flow regime is not seen. In the contrary case, when $C_{fD} > 50$, probably bilinear flow is absent or maybe masked by wellbore storage effects. In such cases, the below correlations [37], which are excellent, can be used to find one parameter (Eq. 7.107) as a function of the other one or vice versa (Eq. (7.108)).

$$x_f = \frac{1.92173}{1/r_{wa} - 3.31739k/k_f w_f} \tag{7.106}$$

where r_{wa} is the effective wellbore radius given by Eq. (7.54), so Eq. (7.106) can be rewritten as:

$$x_f = \frac{1.92173}{e^s/r_w - 3.31739k/k_f w_f} \tag{7.107}$$

$$k_f w_f = \frac{3.31739k}{e^s/r_w - 1.92173/x_f} \tag{7.108}$$

When radial flow is absent, this can be artificially from Eq. (2.71) by solving for the pressure derivative during the radial flow regime. Once estimated, a horizontal line can be drawn through this value, which corresponds to the place where radial flow really exists. Intersection of this line with others can be used without any problem. However, to find skin factor, the below correlation developed by Economides et al. [9] is recommended to be used:

$$s = \ln\left[r_w\left(\frac{1.92173}{x_f}\right) - \frac{3.31739}{k_f w_f}\right] \tag{7.109}$$

The internal result between parentheses may be considered in absolute value. Skin factor can also be estimated by a graphical procedure formulated by Cinco-Ley and Samaniego [7], type-curve matching, or the following correlation [9]:

$$s = \ln\frac{r_w}{x_f} + \frac{1.65 - 0.32u + 0.11u^2}{1 + 0.18u + 0.064u^2 + 0.005u^3} \tag{7.110}$$

where

$$u = \ln\left(\frac{k_f w_f}{k \cdot x_f}\right) = \ln C_{fD} \tag{7.111}$$

Finally, Tiab et al. [36, 38] also provided more relationships, which are not reported because of their relevance. Neither the step-by-step procedures are reported.

Alternatively, Eqs. (7.40), (7.41), and **Figure 7.8** can be used.

Fracture conductivity can be found by a graphical correlation (**Figure 7.13**), given by Economides et al. [9], which polynomial fitting is given here:

$$x = s + \ln\left(\frac{x_f}{r_w}\right);\ 0.67 \le x \le 2.8 \tag{7.112}$$

$$C_{fD} = 10^{\left(\frac{0.59222806 - 1.77955x + 0.86571983x^2}{1 - 1.5944514x + 0.010112x^2}\right)} \tag{7.113}$$

Example 7.4

Tiab et al. [36, 38] presented pressure data for a buildup test run in a fractured well. Pressure and pressure derivative data are reported in **Table 7.4** and **Figure 7.14**. Other important information concerning this test is given below. Find the fracture and reservoir parameters for this well.

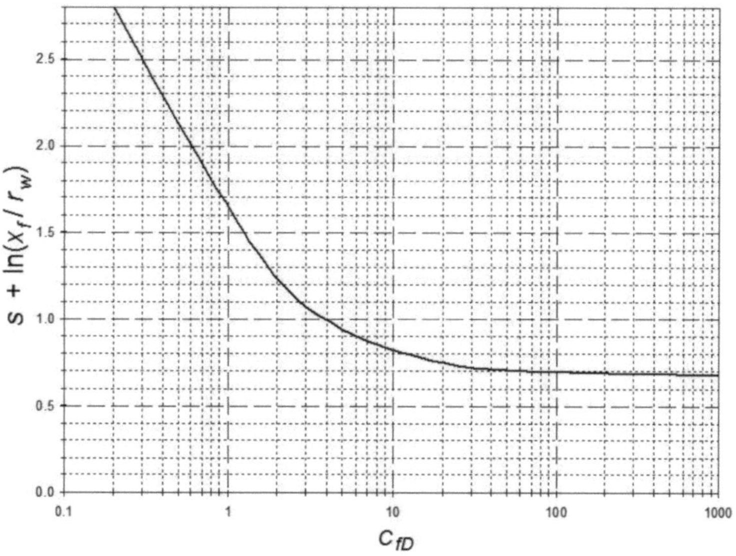

Figure 7.13
Effect of skin factor on fracture conductivity, after [9].

t, hr	DP, psia	*t**DP', psia	*t*, hr	DP, psia	*t**DP', psia
0.23	102	26.3	15	390	117
0.39	115	30	20	423	112
0.6	130	35.8	25	446	120
1	145	40.8	30	471	141
1.8	183	57.2	35	493	136.5
2.4	195	67	40	510	132
3.8	260	83.3	45	526	135
4.1	265	69.2	50	540	150
4.96	280	96.9	55	556	137.5
6.2	308	102.3	60	565	144
8.5	320	103.3	65	580	121.1
10	345	149	71	583	

Table 7.4.
Pressure data for Example 7.4. Derivative digitized from [38].

$q = 101$ STB/D, $\phi = 0.08$, $\mu = 0.45$ cp, $c_t = 17.7 \times 10^{-6}$ psia^{-1}, $B = 1.507$ bbl/STB, h = 42 ft, $r_w = 0.28$ ft, $t_p = 2000$ hr, $P_i = 2200$ psia, $\xi = 1$

Solution

The following characteristic features were read from **Figure 7.14**:

$t_r = 30$ hr, $\Delta P_r = 471$ psia, $(t^*\Delta P')_r = 150$ psia, $(t^*\Delta P')_{BL1} = 160$ psia, $\Delta P_{BL1} = 40$ psia, $\Delta P_{L1} = 120$ psia, $t_{Lri} = 8.2$ hr, $t_{BLri} = 195$ hr

Estimate permeability and skin factor from Eqs. (2.71) and (2.92):

$$k = \frac{70.6 q\mu B}{h(t^*\Delta P')_r} = \frac{(70.6)(101)(0.45)(1.507)}{(42)(150)} = 0.76 \text{ md}$$

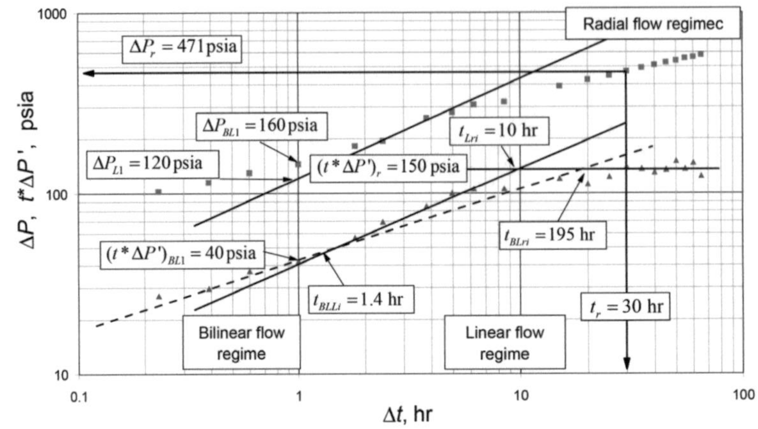

Figure 7.14
Pressure and pressure derivative against time log-log plot for Example 7.4.

$$s = \frac{1}{2}\left[\frac{471}{150} - \ln\left(\frac{(0.76)(30)}{(0.08)(0.45)(17.7\times10^{-6})(0.28^2)}\right) + 7.43\right] = -4.68$$

Estimate fracture conductivity using Eqs. (7.95) and (7.97):

$$k_f w_f = \frac{1947.46}{\sqrt{(0.08)(0.45)(17.7\times10^{-6})(0.76)}}\left(\frac{(101)(0.45)(1.507)}{(42)(160)}\right)^2 = 290.7 \text{ md-ft}$$

$$k_f w_f = \frac{121.74}{\sqrt{(0.08)(0.45)(17.7\times10^{-6})(0.76)}}\left(\frac{(101)(0.45)(1.507)}{(42)(40)}\right)^2 = 290.77 \text{ md-ft}$$

Find the intercept between bilinear and biradial flow regimes with Eq. (7.103):

$$t_{BLri} = 1677\frac{(0.08)(0.45)(17.7\times10^{-6})}{(0.76)^3}(310.8)^2 = 235 \text{ hr}$$

This is in the range of 195 hr read from **Figure 7.13**. Use Eqs. (7.63) and (7.70) to find half-fracture length:

$$x_f = 4.064\left(\frac{(101)(1.507)}{(42)(120)}\right)\sqrt{\frac{0.45}{(0.08)(17.7\times10^{-6})(0.76)}} = 79 \text{ ft}$$

$$x_f = \sqrt{\frac{k t_{Lri}}{1207\xi\phi\mu c_t}} = \sqrt{\frac{(0.76)(10)}{1207(0.08)(0.76)(17.7\times10^{-6})}} = 76.5 \text{ ft}$$

From Eq. (7.98), the time intercept of bilinear and linear flow regimes is found to be:

$$t_{BLLi} = 13910(0.08)(0.45)(17.7\times10^{-6})\left(\frac{79^2\sqrt{0.76}}{290.7}\right)^2 = 1.48 \text{ hr}$$

This is very close to the value of 1.4 hr found from the derivative plot. Use Eq. (7.107) to estimate the half-fracture length:

t, hr	DP, psia	t*DP′, psia	t, hr	DP, psia	t*DP′, psia
0.017	27.45	26.62	3.78	93.59	24.85
0.019	42.39	24.98	4.78	99.56	26.61
0.082	48.5	5.19	5.78	104.26	28.49
0.28	56.18	10.50	7.78	113.36	30.37
0.33	61.87	9.67	9.78	121.04	33.32
0.78	63.72	10.91	11.78	126.87	35.17
1.08	72.11	13.18	13.78	131.85	35.67
1.78	76.38	15.36	17.78	142.66	38.51
2.78	86.34	22.58	19.78	146.07	40.70

Table 7.5.
Pressure data for Example 7.5. Derivative digitized from [38].

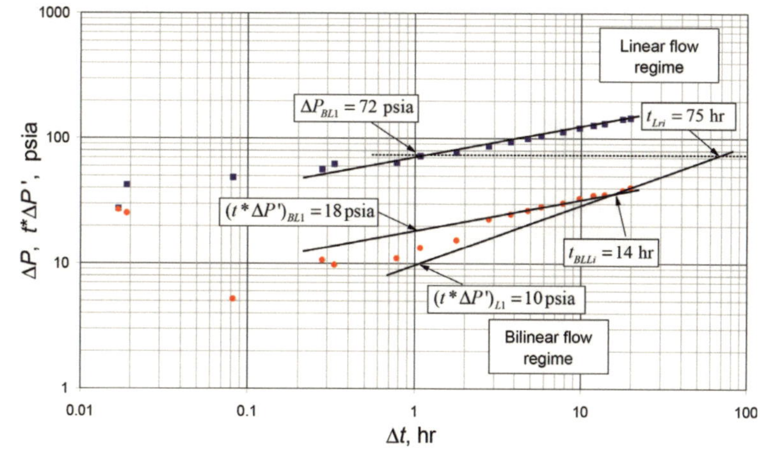

Figure 7.15
Pressure and pressure derivative against time log-log plot for Example 7.5.

$$x_f = \frac{1.92173}{\frac{e^s}{r_w} - \frac{3.31739k}{w_f k_f}} = \frac{1.92173}{\frac{e^{-4.6844}}{0.28} - \frac{3.31739(0.76)}{290.7}} = 79 \text{ ft}$$

And the dimensionless fracture conductivity is found from Eq. (7.15), so that:

$$C_{fD} = \frac{w_f k_f}{x_f k} = \frac{290.7}{79(0.76)} = 4.8$$

Example 7.5

Tiab et al. [36, 38] presented a short buildup test run in a fractured well. Radial flow was not developed, but the reservoir permeability was measured from another to be 12.4 md. The pressure and pressure derivative [13] data are reported in **Table 7.5** and **Figure 7.15**. Additional data:

$q = 411.98$ STB/D, $f = 0.2$, $m = 0.53$ cp, $c_t = 101 \times 10^{-6}$ psia^{-1}, $B = 1.258$ bbl/STB, $h = 21$ ft, $r_w = 0.689$ ft, $t_p = 3000$ hr, $P_i = 479.61$ psia, $\xi = 1$

Solution

Since permeability is known, the pressure derivative during infinite-acting radial flow is found from Eq. (2.71):

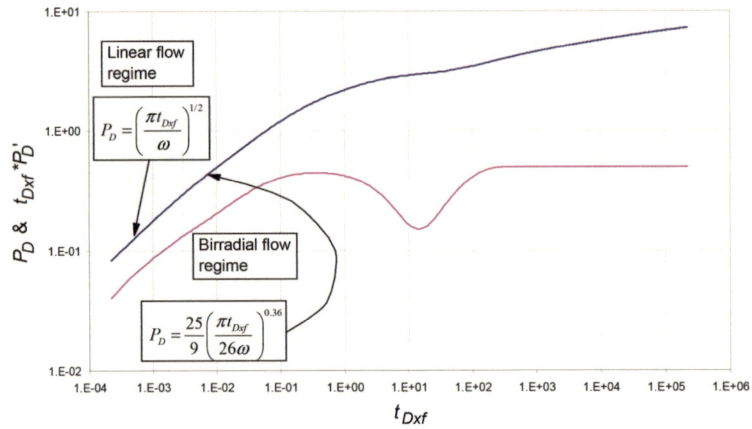

Figure 7.16
Pressure and pressure derivative behavior against dimensionless time for a vertical well with infinite-conductivity fracture in a heterogeneous reservoir with $\lambda = 1 \times 10^{-8}$ *and* $\omega = 0.1$*, after [23].*

$$(t^*\Delta P')_r = \frac{70.6q\mu B}{hk} = \frac{(70.6)(411.98)(0.53)(1.258)}{(21)(12.4)} = 74.5 \text{ psia}$$

A horizontal line is drawn throughout $(t^*\Delta P')_r$ of 74.5 psia. This corresponds to an arterially created radial flow regime line. The following data were then read from **Figure 7.15**:

t_{Lri}=75 hr, t_{BLLi}=14 hr, $(t^*\Delta P')_{BL1}$=18 psia, ΔP_{BL1}=72 psia, ΔP_{L1}=10 psia,
Use Eqs. (7.95) and (7.97) to determine fracture conductivity:

$$k_f w_f = \frac{1947.46}{\sqrt{(0.2)(0.53)(101\times10^{-6})(12.4)}}\left(\frac{(411.98)(0.53)(1.258)}{(21)(72)}\right)^2 = 5578.35 \text{ md-ft}$$

$$k_f w_f = \frac{121.74}{\sqrt{(0.2)(0.53)(101\times10^{-6})(12.4)}}\left(\frac{(411.98)(0.53)(1.258)}{(21)(18)}\right)^2 = 5579.44 \text{ md-ft}$$

Find the half-fracture length with Eqs. (7.61) and (7.71):

$$x_f = 4.064\left(\frac{411.98(1.258)}{(21)(10)}\right)\sqrt{\frac{0.53}{(0.2)(101\times10^{-6})(12.4)}} = 260.5 \text{ ft}$$

$$x_f = \sqrt{\frac{kt_{Lri}}{1207\xi\phi\mu c_t}} = \sqrt{\frac{(12.4)(75)}{1207(0.2)(0.53)(101\times10^{-6})}} = 268.3 \text{ ft}$$

Use Eq. (7.109) to find skin factor:

$$s = \ln\left[\left|0.689\left(\frac{1.92173}{264} - \frac{3.31739(12.4)}{5578.9}\right)\right|\right] = -9.6$$

Estimate the dimensionless fracture conductivity by means of Eq. (7.15):

$$C_{fD} = \frac{k_f w_f}{k x_f} = \frac{5578.9}{12.4(264)} = 1.7$$

7.8. New elliptical or biradial flow model

It was not possible to use Eq. (7.79) to find half-fracture length in Example 7.2. This is because Eq. (7.74) depends on area, which should not be the case since this causes the test to be very long and therefore costly, which is not well accepted by most operators who in many circumstances do not allow fractured wells to develop radial flow during a well test. This implies the impossibility of determining the mean fracture length by means of Eq. (7.79) when at early times only biradial or elliptical flow is observed. In cases where the radial flow is observed or the permeability is known, it is possible to determine the mean fracture length using Eq. (7.84).

To overcome the above issue, Escobar et al. [23] presented a new model (see **Figure 7.17**) for biradial/elliptical flow, which excludes the drainage area and is presented below for homogeneous reservoirs ($\xi = 1$) or heterogeneous reservoirs ($\xi = \omega$):

$$P_D = \frac{25}{9}\left(\frac{\pi t_{Dxf}}{26\xi}\right)^{0.36} \tag{7.114}$$

which pressure derivative is given by:

$$t_D{}^*P_D{}' = \left(\frac{\pi t_{Dxf}}{26\xi}\right)^{0.36} \tag{7.115}$$

7.8.1 *TDS* technique for the new biradial flow model

Once the dimensionless parameters given by Eqs. (1.89), (7.55) and (2.57) in the above expressions solve for the half-fracture length, it yields:

$$x_f = 22.5632\left(\frac{qB}{h(\Delta P)_{BR}}\right)^{1.3889}\sqrt{\frac{t_{BR}}{\xi\phi c_t}\left(\frac{\mu}{k}\right)^{1.778}} \tag{7.116}$$

$$x_f = 5.4595\left(\frac{qB}{h(t^*\Delta P')_{BR}}\right)^{1.3889}\sqrt{\frac{t_{BR}}{\xi\phi c_t}\left(\frac{\mu}{k}\right)^{1.778}} \tag{7.117}$$

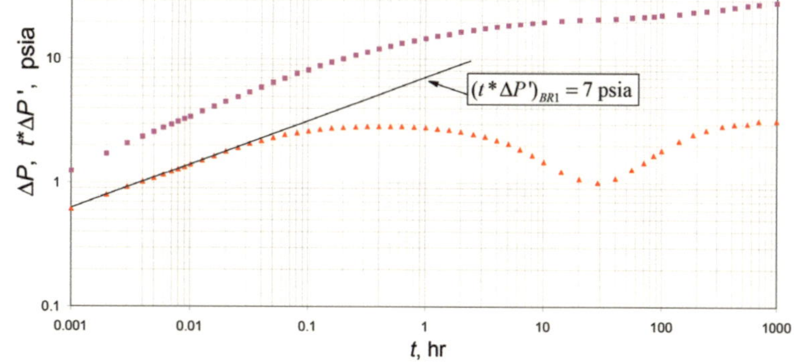

Figure 7.17
Pressure and pressure derivative against time log-log plot for Example 7.6, after [18].

Normally, the well test data are affected by noise, so it is recommended to draw the best line on the dealt flow (in this case, biradial) and read the value on that straight line at a time $t = 1$ hr (extrapolated if required), which leads to Eqs. (7.116) and (7.117) being rewritten as:

$$x_f = 22.5632\left(\frac{qB}{h(\Delta P)_{BR1}}\right)^{1.3889}\sqrt{\frac{1}{\xi\phi c_t}\left(\frac{\mu}{k}\right)^{1.778}} \tag{7.118}$$

$$x_f = 5.4595\left(\frac{qB}{h(t^*\Delta P')_{BR1}}\right)^{1.3889}\sqrt{\frac{1}{\xi\phi c_t}\left(\frac{\mu}{k}\right)^{1.778}} \tag{7.119}$$

The intercept between the straight lines of the derivatives of bilinear and biradial flows t_{BLBRi} (Eqs. 7.96 and 7.115) allows obtaining an expression to determine the half-fracture length, x_f,

$$k_f w_f = 10.5422\left(\frac{\xi\phi\mu c_t k^{3.5454} x_f^{6.5454}}{t_{BLBRi}}\right)^{0.22} \tag{7.120}$$

The intercept between the straight lines of the derivatives of linear and biradial flows t_{LBRi} (Eqs. 7.89 and 7.115) also allows obtaining an expression to determine the half-fracture length, x_f,

$$x_f = \sqrt{\frac{k t_{BRLi}}{39.044\xi\phi\mu c_t}} \tag{7.121}$$

Another way to obtain the half-fracture length is the intercept of the straight lines of the radial flow derivatives (Eq. 2.70) and biradial flow regime (Eq. 7.115) t_{BRri},

$$x_f = \frac{1}{4584.16}\sqrt{\frac{k t_{BRri}}{\xi\phi\mu c_t}} \tag{7.122}$$

The intersection formed by the line of the derivative of the biradial flow with the line of the derivative of pseudosteady state (Eq. 2.96), called t_{BRpssi}, leads to:

$$x_f = 41.0554 A^{1.3889}\left(\frac{\xi\phi\mu c_t}{k t_{BRpssi}}\right)^{0.8889} \tag{7.123}$$

For circular/square constant pressure systems which pressure derivative is governed by Eq. (2.349), when intercepts with Eq. (7.115), called t_{BRSSi}, also leads to:

$$x_f = \frac{1}{4247.92 A^{25/18}\sqrt{\omega}}\left(\frac{k t_{BRSS_i}}{\phi\mu c_t}\right)^{\frac{17}{9}} \tag{7.124}$$

Other recent publications dealing with elliptical/biradial flow regime in horizontal and vertical wells can be found in Refs. [15–18, 28].

7.8.2 Conventional analysis for the new biradial flow model

After replacing Eqs. (1.89) and (7.55), kin Eq. (7.114) leads to:

$$\Delta P = 9.4286 \frac{q\mu B}{kh} \left(\frac{k}{\xi \phi \mu c_t x_f^2} \right)^{0.36} t^{0.36} \tag{7.125}$$

or

$$\Delta P = m_{bir} t^{0.36} \tag{7.126}$$

Eq. (7.126) implies that a Cartesian graph of ΔP versus $t^{0.36}$ (for drawdown tests) or ΔP versus ΔP versus $[(t_p+\Delta t)^{0.36} - \Delta t^{0.36}]$ (for buildup tests) provides a line which slope, m_{ell}, allows obtaining the half-fracture length,

$$x_f = \sqrt{9.4286 \frac{q\mu B}{khm_{bir}} \left(\frac{k}{\xi \phi \mu c_t} \right)^{0.36}} \tag{7.127}$$

Example 7.6

Determine the half-fracture length for a pressure test which data are reported in **Table 7.6** and plotted in **Figure 7.17** for a hydraulically fractured well in a heterogeneous reservoir. Other relevant information for this test is given below:

$B = 1.25$ bbl/STB, $q = 350$ STB/D, $h = 100$ ft, $\mu = 3$ cp, $r_w = 0.4$ ft, $c_t = 1 \times 10^{-5}$ psia^{-1}, $P_i = 5000$ psia, $\varphi = 20\%$, $k = 300$ md, $\omega = 0.1$, $\lambda = 1 \times 10^{-7}$, $x_f = 100$ ft

t, hr	Δ*P*, psia	*t**Δ*P*, psia	*t*, hr	Δ*P*, psia	*t**Δ*P*, psia	*t*, hr	Δ*P*, psia	*t**Δ*P*, psia
0.001	1.246	0.608	0.101	8.159	2.624	8.021	19.566	1.680
0.002	1.722	0.797	0.127	8.775	2.696	10.098	19.932	1.484
0.003	2.063	0.920	0.160	9.405	2.753	14.264	20.397	1.234
0.004	2.336	1.015	0.201	10.048	2.798	20.148	20.777	1.074
0.005	2.569	1.088	0.254	10.697	2.832	28.460	21.108	1.012
0.006	2.773	1.166	0.319	11.354	2.852	40.202	21.442	1.093
0.007	2.956	1.224	0.402	12.015	2.863	55.107	21.797	1.283
0.008	3.122	1.286	0.506	12.677	2.863	70.107	22.121	1.474
0.009	3.277	1.341	0.637	13.336	2.849	85.107	22.423	1.678
0.010	3.433	1.398	0.802	13.992	2.825	100.107	22.707	1.838
0.013	3.768	1.521	1.010	14.641	2.788	140.107	23.390	2.200
0.016	4.132	1.650	1.271	15.280	2.737	185.107	24.051	2.472
0.020	4.528	1.790	1.600	15.906	2.671	240.107	24.734	2.691
0.025	4.956	1.928	2.015	16.514	2.590	330.107	25.637	2.903
0.032	5.418	2.066	2.537	17.102	2.488	420.107	26.355	2.999
0.040	5.911	2.200	3.193	17.667	2.367	545.107	27.147	3.009
0.051	6.434	2.324	4.020	18.198	2.226	685.107	27.852	3.194
0.064	6.985	2.437	5.061	18.695	2.063	985.107	28.973	3.194
0.080	7.562	2.538	6.372	19.153	1.879			

Table 7.6.
Pressure and pressure derivative versus time data for Example 7.6, after [16].

Solution

The pressure derivative value during biradial flow regime at a time of 1 hr, $(t^*\Delta P')_{BR1}$=7 psia. Use Eq. (7.119) to find the half-fracture length:

$$x_f = 5.4595\left(\frac{350(1.25)}{100(7)}\right)^{1.3889}\sqrt{\frac{1}{(0.1)(0.2)(1\times 10^{-5})}\left(\frac{3}{300}\right)^{1.778}} = 105.96\text{ ft}$$

The estimation of the naturally fractured parameters can be found in [18].

7.9. Horizontal wells

This last topic was left last since horizontal wells and fractured wells behave similarly. Actually, Escobar et al. [19] presented an approach for estimating the average reservoir pressure for horizontal wells under multirate testing using the mathematical solution of a vertical fractured well. The use of the *TDS* technique for horizontal wells is so extensive and deserves more than a chapter, for that reason it is only mentioned here. The pioneer papers on *TDS* technique for horizontal wells where presented by Engler and Tiab for naturally fractured deposit [10] and for anisotropic homogeneous formations [11]. The reader may not understand later publications without going them first. *TDS* technique for horizontal wells is so rich. Just to name so few cases, let us refer to treatment of zonal isolations by Al Rbeawi and Tiab [1] that even has conventional analysis by Escobar et al. [21]. Lu et al. [27] dealt with double permeability systems, and Escobar et al. [22] presented *TDS* technique for heterogeneous and homogeneous formations when the threshold gradient plays an important role for the flow to start flowing. Some applications on shale formations are summarized by Escobar [24].

Nomenclature

A	area, ft^2
B	oil volume factor, bbl/STB
b	fraction of penetration/completion
c	compressibility, 1/psia
C	wellbore storage coefficient, bbl/psia
C_A	reservoir shape factor
C_{fD}	dimensionless fracture conductivity
c_t	total or system compressibility, 1/psia
F_{cor}	correction factor
h	formation thickness, ft
k	permeability, md
kh	reservoir flow capacity, md-ft
h_f	fracture height, ft
k_c	uncorrected reservoir permeability, md
k_f	fracture permeability, md
$k_f w_f$	fracture conductivity, md-ft
m	slope of P versus log t plot, psia/hr/cycle
m_{bir}	slope of P versus $t^{0.36}$ plot during elliptical/biradial flow, psia$^{0.36}$/hr
m_{lf}	slope of P versus $t^{0.5}$ plot during linear flow, psia$^{0.5}$/hr
m_{vf}	slope of P versus $t^{0.5}$ plot during linear flow, psia$^{0.5}$/hr

m_{bf}	slope of P versus $t^{0.25}$ plot during bilinear flow, psia$^{0.25}$/hr
P	pressure, psia
$P_D{}'$	dimensionless pressure derivative
P_D	dimensionless pressure
P_i	initial reservoir pressure, psia
P_{wf}	well flowing pressure, psia
P_{ws}	well shut-in or static pressure, psia
P^*	false pressure, psia
ΔP_s	pressure drop due to skin conditions, psia
q	liquid flow rate, bbl/D
q_{fD}	dimensionless flow rate
r_D	dimensionless radius
r	radius, ft
r_e	drainage radius, ft
r_w	well radius, ft
r_{wa}	apparent wellbore radius, ft
s	skin factor
t	time, hr
t_p	production (Horner) time before shutting-in a well, hr
t_D	dimensionless time based on well radius
t_{DA}	dimensionless time based on reservoir area
t_{Dxf}	dimensionless time based on half-fracture length
t_{pDA}	dimensionless Horner time based on area
X	distance along the x direction
x_e	half-reservoir side, ft (square system)
x_f	half-fracture length, ft
$t^*\Delta P'$	pressure derivative, psia
w_f	fracture width, ft

Greek

Δ	change, drop
Δt	shut-in time, hr
η	diffusivity constant, $\varphi\mu c_t/k$
ξ	indicator of either heterogeneous, $\xi = \omega$, or homogeneous, $\xi = 1$, reservoir
φ	porosity, fraction
ρ	fluid density, lbm/ft^3
μ	viscosity, cp
ω	dimensionless storativity coefficient

Suffices

$1h$	read at a time of 1 hr
D	dimensionless
DA	dimensionless with respect to area
Dxf	dimensionless with respect to area
BL	bilinear flow
$BL1$	bilinear flow at 1 hr
$BLLi$	intercept of bilinear and linear lines in pressure curve
$BLLi$	intercept of bilinear and linear lines in pressure derivative curve
$BLBRi$	intercept of bilinear and biradial lines in pressure curve

BLri	intercept of bilinear and radial lines in pressure curve
blf	beginning of linear flow
brf	beginning of radial flow
bsl	beginning of semilog line
BRLi	intercept of biradial and linear lines
BRri	intercept of biradial and radial lines
BRSSi	intercept of biradial and steady-state lines
BRpssi	intercept of biradial and pseudosteady-state lines
CB	formation linear flow regime
ebf	end of bilinear flow
el	end of linear flow
f	fracture
i	intersection or initial conditions
L	linear flow
*L*1	linear flow at 1 hr
Lpssi	intercept of linear and pseudosteady state lines
Lri	intercept of linear and radial lines
p	production
pss	pseudosteady state
*pss*1	pseudosteady state at 1 hr
r	radial flow
Lri	intercept of linear and radial lines
rSSi	intersection between the radial line and the −1-slope line
s	skin
SS	steady
vf	vertical fracture
w	well, water
wa	apparent wellbore
wf	well flowing
ws	well shut-in

References

[1] Al Rbeawi SJH, Tiab D. Effect of the number and length of zonal isolations on pressure behavior of horizontal wells. 2011. DOI: 10.2118/142177-MS

[2] Bettam Y, Zerzar A, Tiab D. Interpretation of multi-hydraulically fractured horizontal wells in naturally fractured reservoir. 2005. DOI: 10.2118/95535-MS

[3] Cherifi M, Tiab D, Escobar FH. Determination of fracture orientation by multi-well interference testing. Society of Petroleum Engineers. 2002. DOI: 10.2118/77949-MS

[4] Chukwu IF. Pressure derivative analysis of vertically fractured wells in bounded reservoirs [M.Sc. thesis]. Norman, OK: The University of Oklahoma; 1989. p. 132

[5] Cinco-Ley H, Samaniego F. Evaluation of hydraulic transient pressure analysis for fractured wells. Society of Petroleum Engineers. 1982. DOI: 10.2118/7490-PA

[6] Cinco LH, Samaniego VF, Dominguez AN. Transient pressure behavior for a well with a finite-conductivity vertical fracture. SPE Journal. 1978;**18**(4):253-264. DOI: 10.2118/6014-PA

[7] Cinco-Ley H, Samaniego F. Transient pressure analysis for fractured wells. Society of Petroleum Engineers. 1981. DOI: 10.2118/7490-PA

[8] Earlougher RC, Jr. Advances in Well Test Analysis, Monograph Series. Vol. 5. Dallas, TX: SPE

[9] Economides MJ, Watters LT, Dunn-Norman S. Petroleum Well Construction. New York: John Wiley & Sons. 622 p

[10] Engler T, Tiab D. Analysis of pressure and pressure derivative without type-curve matching, 5. Horizontal well tests in naturally fractured reservoirs. Journal of Petroleum Science and Engineering. 1996;**15**:139-151

[11] Engler T, Tiab D. Analysis of pressure and pressure derivative without type-curve matching, 6. Horizontal well tests in anisotropic media. Journal of Petroleum Science and Engineering. 1996;**15**:153-168

[12] Escobar FH, Tiab D, Berumen-Campos S. Well pressure behavior of a finite-conductivity fracture intersecting a finite sealing-fault. Society of Petroleum Engineers. 2003. DOI: 10.2118/80547-MS

[13] Escobar FH, Navarrete JM, Losada HD. Evaluation of pressure derivative algorithms for well-test analysis. Society of Petroleum Engineers. 2004. DOI: 10.2118/86936-MS

[14] Escobar FH, Tiab D, Jokhio SA. Pressure analysis for a well intersected by a hydraulic fracture with multiple segments. Society of Petroleum Engineers. 2001. DOI: 10.2118/71035-MS

[15] Escobar FH, Muñoz OF, Sepúlveda JA. Horizontal permeability determination from the elliptical flow regime for horizontal wells. CT&F—Ciencia, Tecnología y Futuro. 2004; **2**(5):83-95. ISSN: 0122-5383

[16] Escobar FH, Montealegre M, Cantillo JH. Conventional analysis for characterization of bi-radial (elliptical) flow in infinite-conductivity vertical fractured wells. CT&F—Ciencia, Tecnología y Futuro. 2006;**3**(2):141-147. ISSN: 0122-5383

[17] Escobar FH. Montealegre M. Determination of horizontal permeability from the elliptical flow of horizontal wells using conventional analysis. Journal of Petroleum Science and Engineering. 2008;**61**:15-20. ISSN: 0920-4105

[18] Escobar FH, Montealegre M. Conventional analysis for the determination of the horizontal permeability from the elliptical flow of horizontal wells. Society of Petroleum Engineers. 2007. DOI: 10.2118/105928-MS

[19] Escobar FH, Cantillo JH, Santos N. A practical approach for the estimation of the average reservoir pressure from multi-rate tests in long horizontal wells. Fuentes: El Reventón Energético. 2011; **9**(1):13-20. First Semester

[20] Escobar FH, Zhao YL, Zhang LH. Interpretation of pressure tests in horizontal wells in homogeneous and heterogeneous reservoirs with threshold pressure gradient. Journal of Engineering and Applied Sciences. 2014;**9**(11):2220-2228

[21] Escobar FH, Meneses AR, Losada LM. Straight-line conventional transient pressure analysis for horizontal wells with isolated zones. Dyna. 2014;**81**(185): 78-85

[22] Escobar FH, Zhao YL, Zhang LH. Interpretation of pressure tests in horizontal wells in homogeneous and heterogeneous reservoirs with threshold pressure gradient. Journal of Engineering and Applied Sciences. 2014;**9**(11):2220-2228

[23] Escobar FH, Ghisays-Ruiz A, Bonilla LF. New model for elliptical flow regime in hydraulically-fractured vertical wells in homogeneous and naturally-fractured systems. Journal of Engineering and Applied Sciences. 2014;**9**(9):1629-1636

[24] Escobar FH. Recent Advances in Practical Applied Well Test Analysis. New York: Nova publishers; 2015. 422 p

[25] Gringarten AC, Ramey HJ, Raghavan R. Applied pressure analysis for fractured wells. Society of Petroleum Engineers. 1975. DOI: 10.2118/5496-P

[26] Kamal MM. Transient Well testing. SPE Monograph. Vol. 23. Henry L. Doherty Series. Richardson, TX; 850 p. Chapter 11

[27] Lu J, Zhu T, Tiab D, Escobar FH. Pressure behavior of horizontal wells in dual-porosity, dual-permeability naturally-fractured reservoirs. Journal of Engineering and Applied Sciences. 2015;**10**(8):3405-3417. ISSN: 1819-6608

[28] Martinez JA, Escobar FH, Bonilla LF. Reformulation of the elliptical flow governing equation for horizontal wells. Journal of Engineering and Applied Sciences. 2012;7(3):304-313. ISSN: 1819-6608

[29] Tiab D, Puthigai SK. Pressure-derivative type curves, for vertically fractured wells. Society of Petroleum Engineers. 1988. DOI: 10.2118/11028-PA

[30] Noureddine R, Tiab D, Azzoughen A, Mazouzi A, Escobar FH, Zhu T. Evaluation of hydraulic fracturing and re-entries in the Hassi-Messaoud field, Algeria. 2002. DOI: 10.2118/76775-MS

[31] Raghavan R, Cady GV, Ramey HJ. Well-Test Analysis for Vertically Fractured Wells. Society of Petroleum Engineers. 1972. DOI:10.2118/3013-PA

[32] Russell DG, Truitt NE. Transient pressure behavior in vertically fractured reservoirs. Journal of Petroleum Technology. 1964:1159-1170. Trans. AIME, 231. DOI: 10.2118/967-PA

[33] Stewart G. Well Test Design & Analysis. Tulsa, Oklahoma, USA: PennWell Corporation; 2011. 1940 p

[34] Tiab D. PE-5553: Well Test Analysis. Lecture Notes. Norman, OK: The University of Oklahoma; 1993

[35] Tiab D. Analysis of pressure derivative without type curve matching: vertically fractured wells in closed systems. Journal of Petroleum Science and Engineering. 1994;**11**:323-333. DOI: 10.2118/26138-MS

[36] Tiab D, Azzougen A, Escobar FH, Berumen_Campos S. Analysis of pressure derivative data of finite-conductivity fractures by the "Direct Synthesis" technique. Society of Petroleum Engineers. 1999. DOI: 10.2118/52201-MS

[37] Tiab D. Advances in Pressure Transient Analysis—TDS technique. Lecture Notes Manual. Norman, Oklahoma, USA: The University of Oklahoma; 2003. 577 p

[38] Tiab D. Analysis of pressure derivative data of hydraulically fractured wells by the Tiab's direct synthesis technique. Journal of Petroleum Science and Engineering. 2005;**49**:1-21. DOI: 10.1016/j.petrol.2005.07.001

[39] Wong DW, Harrington AG, Cinco Ley H. Application of the pressure derivative function in the pressure transient testing of fractured wells. 1986. DOI: 10.2118/13056 PA